The Random House Guide to Technical and Scientific Communication

The Random House Guide to Technical and Scientific Communication

DONALD E. ZIMMERMAN
Colorado State University

DAVID G. CLARK
Colorado State University

Random House
New York

First Edition
987654321
Copyright © 1987 by Random House, Inc.

All rights reserved under International and Pan-American Copyright Conventions. No part of this book may be reproduced in any form or by any means, electronic or mechanical, including photocopying, without permission in writing from the publisher. All inquiries should be addressed to Random House, Inc., 201 East 50th Street, New York, N.Y. 10022. Published in the United States by Random House, Inc., and simultaneously in Canada by Random House of Canada Limited, Toronto.

Library of Congress Cataloging-in-Publication Data

Zimmerman, Donald E.
    The Random House guide to technical and scientific communication.

    Bibliography: p.
    Includes index.
    1. Technical writing.   2. Communication of technical information.   3. Communication in science. I. Clark, David G.   II. Title.   III. Title: Guide to technical and scientific communication.
T11.Z56   1987   808'.0666021   86-26291
ISBN 0-394-33260-1

Cover Illustration: Oil painting, "Have a Chew on me," by Manny Farber, courtesy Quint Gallery, San Diego, CA. Cover photograph by George Legrady.

Cover Design and Text Design: Lorraine Hohman

Manufactured in the United States of America

To Marietta, Rachel, and Jeramy
and
to Alice, Sarah, and Matthew

Preface

The Random House Guide to Technical and Scientific Communication provides the most in-depth application of the process approach to technical or scientific communication available today. The book explains in detail all the steps a technical writer must go through. We show students how to define a problem, how to collect and evaluate evidence, and how to prepare a report—or present a speech—so that the audience will readily understand it. Our approach is relevant to students in virtually any technical or scientific field.

The Random House Guide evolved from a decade of teaching technical communication, writing, and editing. We tested parts of the text in our classes for students in 30 different technical, scientific, engineering, and communication fields. Elements of the text were also tested in technical writing classes at the University of Wisconsin-Madison by Domenic Fuccillo. The results, confirmed by the successes of our students, form the heart of the book.

We have found that students need to learn that clear writing requires more than mastering basic language skills. Effective communication emerges from clear thinking and sharpening questions and testing evidence. To help students develop the knowledge and skills they will need on the job, the book combines discussions of communication *theory* and *practice*. First, we provide a comprehensive treatment of basic communication principles and theory, based on current research. Then we illustrate each concept with detailed and varied accounts of the techniques professionals use in practice.

As a tested, practical guide to technical and scientific communication, the text includes several special features:

- ■ A rich variety of **real-world examples.** Panels and in-text examples are drawn from engineering, computer science, nursing, agronomy, horticulture, animal science, natural resources, biological sciences, and many other technical and scientific fields.

- Up-to-the-minute discussions of **computer applications** in literature reviewing, illustration preparation, word processing, printing, and presentations.
- Full coverage of **special topics** important to professionals:

  Interviewing and observation techniques (Chapter 4)
  Word processing and printing (Chapter 14)
  Preparing slide presentations (Chapter 15)
  Legal and ethical issues (Chapter 18)
  Public understanding of science (Chapter 19)
  Evaluating technical and scientific communication (Chapter 20)

The book begins by addressing your students' likely question, "Why must I take a technical communication course?" by showing the importance of communication skills in a variety of technical and scientific careers (Part 1). Key communication concepts are then introduced. Part 2 discusses the first steps in the communication process: problem solving, using the library, interviewing, observing, and conducting surveys.

Part 3 describes audience analysis techniques and explains a variety of ways to draft a manuscript; a number of professionals discuss how they write. A separate chapter (8) explains how to prepare illustrations. Part 4 is devoted to strategies for revision, and it gives detailed advice on how to edit manuscripts and check mechanics, spelling, and style. Part 5 first addresses the particular problems of writing definitions, instructions, and technical reports. Then it discusses word processing and printing, professional presentations, correspondence, and job searching.

The Random House Guide is the first technical communication text to cover the material in Part 6: legal and ethical issues, public understanding of technical and scientific fields, and evaluating communications.

To help students understand key points and principles, each chapter begins with an **overview,** uses **lists** for key concepts, and closes with a **summary.** Each chapter also contains a number of **projects** so students can apply what they have learned. An extensive and diverse selection of **panels, figures,** and **tables** illustrate specific points with real-world examples. For students seeking additional information, a **bibliography** ("For More Help") at the end of each chapter lists useful references.

The instructor's manual includes suggested syllabuses for 10- and 15-week courses, resources for teaching technical communication, teaching tips, suggested semester assignments, and chapter aids. Chapter aids include chapter overviews, additional writing assignments, teaching hints for the chapter, answers to text assignments, and test questions. For those adopting the text, Random House will provide a packet of overhead transparencies to supplement class lectures.

As you use the text, we would welcome your reactions, suggestions, ideas, and questions.

Donald E. Zimmerman
David G. Clark
Fort Collins, Colorado

Acknowledgments

Any work of this size owes debts to many people. We would like to thank some of the people whose time, suggestions, generosity, and support have made this book possible.

Members of the technical communication profession in northern California opened their companies to David Clark, and welcomed him into their professional lives. Joe Williams and Jim Detrick of Hewlett-Packard's Santa Rosa plant gave detailed descriptions of their technical communication procedures. Sylvia Onalfo-Wybrant of ASK Computer Systems described how her company evaluates its instruction manuals. Harriett Duzet, Jim Overholt, Fidel Salinas, Jim Vreeland, and Terry Allard of IBM's Santa Teresa plant discussed at length their writing strategies. Lyle Settle, Dix McGuire, and Armand Varteressian of Tandem Computers described the practice of technical communication at that company. Ron Teunis, Jack Saunders, Bonnie Schofield, Sue Stephenson, and Arnie Heller of Lawrence Livermore National Laboratory's Public Affairs Office provided insight into the difficult task of conducting public relations for a nuclear weapons research center. Walter Alvarez of the University of California-Berkeley reflected on how a scientist's life changes when he becomes a media celebrity.

Many other people shared their experiences and helped shape the ideas presented in the following chapters. Members of the technical and scientific communication profession, professional public relations practitioners, companies, and agencies, and scientists and technical specialists in Colorado's Front Range provided innumerable ideas, suggestions, and support. Fred McGehan explained the roles of an information officer in a federal agency and arranged interviews with scientists at the National Bureau of Standards in Boulder, Colorado. Russell B. Stoner shared his technical communication expertise and arranged interviews with staff members of the National

ACKNOWLEDGMENTS

Telecommunications and Information Administration in Boulder. George Hoerter and his staff outlined their approaches to technical communication at Martin Marietta Denver Aerospace. Carol Darr of Hewlett-Packard discussed her technical communication experiences and suggested references for writing manuals. Mary Ann Moore shared her technical communication experiences with university extension services, the Agency for International Development, the U.S. Forest Service, and Hewlett-Packard. At Colorado State University, Frank Moore provided data for selected illustrations and Carol Marander prepared selected figures.

Domenic Fuccillo, friend, visiting lecturer and a former managing editor of the American Society of Agronomy, offered advice, encouragement, and ideas for examples.

The following professors, scientists, and journalists played key roles in influencing Don Zimmerman's contributions: John Knowles, Robert J. Robel, Lowell Brandner, Thomas Haig, V. E. Suomi, Bill Shaner, and D. Stanley Eitzen.

A number of technical communication professionals reviewed the manuscript, provided useful suggestions, and answered our questions. They include Jim Bolick, Gene Decker, Derry Eynon, Oguz Nayman, Ursula Lord, Diane Monaghan, and Musser Moore, all of Colorado State University; Joy Yunker, Metropolitan State College, Denver; JoAnn Hackos, University of Colorado at Denver; and Bill Mattingly, Colorado School of Mines.

We'd also like to thank the many teachers, technical and scientific professionals, technical communicators, public relations practitioners, communication researchers, and journalists cited in the text who provided ideas, guidelines, suggestions, and examples. Without their help, the book could not have shown the practice of technical communication in the real world.

Colorado State University's librarians solved many problems in locating and retrieving useful literature. We'd like to thank librarians Antoinette Lueck, Judith Berndt, Emily Taylor, Barbara Burke, Fred Schmidt, Marianna Wagers, the interlibrary loan staff, and Director of the Libraries Joan Chambers.

Students in our classes, who have always taught us at least as much as we have offered them, informed our approach throughout. Several students were especially helpful—Mickie Calkins, Ann Carr, Lisa Curtis, Kristen Ellis, Miriam Flood, Sherri Heckel, Carol Kellett, Deb Klemsz, Pat Larson, Marilee Long, Nancy Nydegger, Marty Petzel, and Charles Sorensson.

During work on the manuscript, we switched from writing on electric typewriters to writing on personal computers. Ann Molison, Lois Hunziker, Linda McNamara, Clifford Scherer, Nancy Wilson, and Steve Burrell helped us make that transition and answered our many questions. Margaret Neff Withey, Lois Hunziker, and Marcella Fuentes provided word processing support on selected chapters.

ACKNOWLEDGMENTS

We are deeply indebted to the Random House staff. Jerry Arni, field representative, encouraged us to submit the prospectus to Random House. Steve Pensinger, executive editor, encouraged us to develop the process approach to technical communication. With her close attention to detail, project editor Jennifer Sutherland vastly improved our manuscript. Cynthia Ward and Ed Maluf answered our many questions. Copyeditor Susan Friedman made sure we followed our own guidelines on clarity and conciseness. Lorraine Hohman is responsible for the attractive design of the book, and Karen Lumley ably supervised its production.

We are grateful to the following reviewers for their extremely constructive criticism, suggestions, and recommendations: Douglas Catron, Iowa State University; Bertie Fearing, East Carolina University; Dean Hall, Kansas State University; John Harris, Brigham Young University; Debra Journet, Louisiana State University; Mark Larson, Humboldt State University; John Patterson, East Carolina University; and Richard Profozich, Prince George's Community College.

And to six who waited, Marietta, Rachel and Jeramy, and Alice, Sarah and Matthew, we owe thanks beyond our ability to express.

Contents

Preface — vii
To the Student — xxv

PART 1
TECHNICAL COMMUNICATION THEORY — 1

CHAPTER 1
COMMUNICATION PRINCIPLES AND THEORY — 3

A Definition of Technical and Scientific Communication — 3
   Why Are Communication Skills Important to Your Career? 4

A Closer Look at Technical and Scientific Communication — 5
   Terminology 5
   Communication Forms 6
   Characteristics 6
   Competition between Messages—the Communicator's Burden 7

Major Functions of Technical and Scientific Communication — 8
   Function: To Inform 8
   Function: To Instruct 8
   Function: To Persuade 9
   Function: To Document 9

How Rhetorical and Communication Theory Can Help — 9

CONTENTS

Basic Communciation Theory   10
   Key Ingredients in the Communication Process 10
   Additional Communication Concepts 14
   Theory Applied to an Example 15
Major Steps in the Technical Communication Process   16
Highlights   18
Projects   19
For More Help   20

PART 2
GATHERING INFORMATION AND IMPROVING CONTENT   21

CHAPTER 2
PROBLEM SOLVING: PRECISE PHRASING LEADS TO GOOD SOLUTIONS   23

What Is a Problem?   23
   A Seven-Step Problem Solving Method 25
Minimizing Content Errors   33
   Descriptive? Correlational? Causal? 33
   Numbers and Their Misuses 34
   Errors in Generalization 37
Highlights   38
Projects   38
For More Help   39

CHAPTER 3
USING THE LIBRARY AND REVIEWING LITERATURE   40

The Library's Resources   41
   Libraries and Reading Rooms 41
   Librarians 41
   Card Catalog 41
   General References 46
   Abstracting Journals and Indexes 47
   Computer Searches and Electronic Data Bases 49
   Serials—Journals, Magazines, and Periodicals 52
   Books and Publications 55
   Government Documents 55

  Special Collections 57
  Electronic and Photographic Forms 57
  Interlibrary Loans 57

 Developing a Literature Search Plan 58
  Write Your Question 58
  Asking Librarians for Help 58

 Executing Your Search 59
  Taking Notes 59
  Work from General to Specific Literature 61

 Library Use beyond Graduation 62
 Highlights 63
 Projects 63
 For More Help 64

CHAPTER 4
COLLECTING INFORMATION IN THE FIELD 65

 Learning by Observing 65
 What Is Observation? 66
  Distortions 66
  Developing Good Observation Techniques 68
  Overcoming Observational Barriers 68

 Interviewing 69
  Ten Steps to Productive Interviews 69
  Note Taking 71

 Surveys 72
  Should You Survey? 73
  Conducting the Survey 74

 Highlights 81
 Projects 82
 For More Help 82

PART 3
CREATING DRAFTS 83

CHAPTER 5
AUDIENCE ANALYSIS: IMPROVING THE ODDS FOR EFFECTIVE
COMMUNICATION 85

 To Understand Audiences, Apply Communication Theory 85

Determine Your Audience 87
Analyze Audience Characteristics 90
Level I Audience Analysis 91
Level II Audience Analysis 94
Level III Audience Analysis 97
Using Audience Information 100
Highlights 104
Projects 105
For More Help 106

CHAPTER 6
HOW PROFESSIONALS WRITE: A LOOK AT WRITING PROCESSES 107

How Technical and Scientific Professionals Write 107
Surveys of Technical Professionals 109
How Professional Writers Practice Their Craft 110
How Professors and Instructors Write 112
Commonalities across Fields 114
Highlights 115
Projects 116
For More Help 117

CHAPTER 7
A CLOSER LOOK AT DRAFTING MANUSCRIPTS 118

Writing Processes 119
Free Writing 119
The Document Design Model 120
The Problem Solving Model 126
Planning, Selecting Content, and Organizing: Preparations for Writing 128
Planning: Content, Writer, and Purpose 128
Selecting Content 129
Organizing Content 131
Drafting Paragraphs 136
Paragraph Length 138

Developing Paragraphs   138
  Writing Paragraphs   140

 Committing Ideas to Paper   140
  Paper, Pencil, or Pen   140
  Electric or Electronic Typewriter   140
  Word Processing Units   140
  Dictating   141

 Highlights   142
 Projects   142
 For More Help   143

CHAPTER 8
ILLUSTRATIONS: TECHNOLOGY'S UNIVERSAL LANGUAGE   144

 Illustration Functions   145
 Selecting, Planning, and Producing Illustrations   146
  Selecting Illustrations   146
  Planning Illustrations   148
  Obtaining Illustrations   149
  Combining Illustrations and Narrative   150
  Handling Illustrations   152

 Guidelines for Illustrations   152
  General Principles   153
  Specific Guidelines: Tables   153
  Specific Guidelines: Line Graphs   155
  Specific Guidelines: Bar Graphs   156
  Specific Guidelines: Circle Graphs   158
  Specific Guidelines: Isotypes   159
  Specific Guidelines: Notations   160
  Specific Guidelines: Flow Charts   160
  Specific Guidelines: Organizational Charts   160
  Specific Guidelines: Algorithms   160
  Specific Guidelines: Time Charts   163
  Specific Guidelines: Maps   165
  Guidelines for Line Art   168
  Guidelines for Photographs   170

 Highlights   176
 Projects   176
 For More Help   178

PART 4
REVISING AND REWRITING: STRATEGIES FOR IMPROVING TECHNICAL AND SCIENTIFIC WRITING 179

CHAPTER 9
REVISIONS: MAKING GOOD INTO BETTER 181

- General Guidelines for Revising and Rewriting — 181
  - *Allow Time between Successive Drafts* 181
  - *Assume the Editor's Role* 182
  - *Revise at Different Levels* 182
  - *Make Several Trips through Your Manuscript* 182
  - *Use Copyediting Symbols* 182
- Strategies for Editing Content — 182
  - *What Are the Goals and Questions behind the Content?* 183
  - *How Did You Collect the Information?* 183
  - *What Information or Evidence Did You Provide?* 184
  - *Which Data Analysis Techniques Did You Use?* 184
  - *How Did You Verify Your Results?* 185
  - *What Conclusions Did You Draw?* 185
- Highlights — 186
- Projects — 186
- For More Help — 187

CHAPTER 10
EFFECTIVE EDITING 189

- Reviewing and Editing Overall Organization — 189
- Reviewing and Editing Paragraphs — 192
- Guidelines for Clear, Concise Writing — 193
  - *Strategies for Strengthening Verbs* 193
  - *Strategies for Simplifying Constructions* 196
  - *Strategies for Speaking Plainly* 200
- Highlights — 205
- Projects — 206
- For More Help — 209

CHAPTER 11
REVISING TO IMPROVE APPEARANCE 210

    Appearances: How Does Your Draft Look? 210
    Editorial (Stylebook) Style 212
        Resolving Editorial Style Issues   *212*
        Developing a Reference Library   *215*
    Typographic Style 215
    Idiomatic Style 215
    Personal Style 218
    Peer Reviews 218
    Highlights 219
    Projects 219
    For More Help 220

PART 5
PREPARING PROFESSIONAL COMMUNICATION PRODUCTS 221

CHAPTER 12
DEFINITIONS, DESCRIPTIONS, INSTRUCTIONS 223

    The Importance of Definitions 223
    Toward Precise Definitions 224
        Defining by Examples   *226*
        Defining with Words   *226*
        Defining by Explication   *226*
        Defining by Comparison   *227*
        Defining by Visuals   *227*
        Defining by Symbols   *227*
        Defining by Word History   *229*
    Placing Definitions 229
        Placing Definition in the Narrative   *229*
        Placing Definitions in Footnotes   *230*
        Placing Definitions in a Glossary   *230*
        When to Define?   *230*
    Descriptions 230

Preparing Descriptions ... 231
    Consider Your Audience   231
    Consider Your Purpose   232
    Organize Your Descriptions   232
    Consider Your Accuracy   233
    Consider Words and Illustrations   235
    Consider Your Format   236
    Consider Your Narrative   237

Common Descriptions ... 237
    Describing Hardware and Conditions   237
    Describing Spatial Relationships   239
    Describing Events   239
    Describing Classifications   241
    Descriptions that Compare and Contrast   242
    Preparing Process Descriptions   245

Instructions ... 246
    Preparing Instructions   246
    Considering Your Audience   248
    Planning and Producing Instructions   248

Highlights ... 254

Projects ... 255

For More Help ... 256

CHAPTER 13
DOCUMENT FORMAT ... 258

Document Format: A Key to Effective Communication ... 258
Shifting from "In School" to "On the Job" Writing ... 259
Good Format Anticipates Readers' Needs ... 259
Key Parts of Technical Documents ... 261
    Title   262
    Proposal   263
    Problem Statement   266
    Justification/Literature Review   267
    Methods   267
    Timetable and Budget   267
    Bibliography/Reference List   268
    Results   268
    Discussion/Conclusion   268
    Recommendations   269
    Abstract   269
    Citations   270
    Progress Reports   271

    Principles for Format Development   273
    Revising a Business Proposal Format   276
    Highlights   279
    Projects   279
    For More Help   280

CHAPTER 14
WORD PROCESSING AND PRINTING   281

    Word Processing   281
        Working with a Word Processing Operator   282
        Doing Your Own Word Processing   284
        Using Computer Graphics   287
    The Printing Process   288
        The Basic Steps in the Printing Process   289
    The Continuing Communication Revolution   295
    Highlights   295
    Projects   296
    For More Help   296

CHAPTER 15
PROFESSIONAL TALKS AND SLIDE PRESENTATIONS   298

    The Professional Talk   299
    Producing Slide Presentations   312
        Planning the Presentation   312
        Preparing the Slide Set   314
        Preparing for and Practicing the Slide Presentation   324
        Presenting the Slide Set   325
        Evaluating the Slide Presentation   325
    Highlights   325
    Projects   326
    For More Help   326

CHAPTER 16
CORRESPONDENCE: PUTTING YOUR BEST SELF FORWARD   328

    Letters Can Show You at Your Best   328
    Major Categories of Correspondence   329
        Letter of Transmittal   329

    Claim or Complaint Letters 329
    Letters to Colleagues 333
    Inquiries or Requests for Information 334
    Memoranda 334

Parts of a Letter   335

Helpful Hint   336

Letter and Memoranda Formats   336
    Letters Outside an Organization 336

Page Arrangement and Spacing   339

Letter Punctuation   342

Addressing the Envelope   342

Avoid Sexist Language   342
    When You Don't Know the Addressee 343

Making Your Letters "Sing"   344

Form Letters   345

Guide Letters   345

Electronic Mail   346

Highlights   346

Projects   347

For More Help   348

CHAPTER 17
JOB SEARCHING FROM A COMMUNICATOR'S PERSPECTIVE   349

Job Searching Strategies   349

Resources for Job Searching   351
    Career Services 351
    Department Support 351

How Employers Find and Evaluate Candidates   352

Job Searching Techniques   354
    Assessing Yourself 354
    Analyzing Employers 355
    Preparing Your Resume 355
    Correspondence 359
    Handling Interview Offers and Interviews 361
    Accepting a Job and Handling Rejections 362

Highlights   363

Projects   363

For More Help   364

PART 6
ETHICS, PUBLIC PERCEPTION, AND EVALUATING COMMUNICATION 365

CHAPTER 18
LEGAL AND ETHICAL ASPECTS OF TECHNICAL COMMUNICATION 367

Copyright Law and Technical Communication 367
Copyright Law of 1976 368
Limitations on a Copyright Owner's Rights 369
Plagiarism: Literary Theft 373
Plagiarists Sometimes Get Away with It 375
Two Well-Known Plagiarists 375
How to Avoid Even the Suspicion of Plagiarism 377
Ethics and Competitive Pressure 377
Other Ethical Problems 378
Concocting Data 378
"Cooking" Results 378
Withholding Adverse Information 379
Sharing Credit for Authorship 380
Products Liability 380
Development of Products Liability 381
Communication's Role in Products Liability 382
Highlights 385
Projects 385
For More Help 386

CHAPTER 19
CREATING BETTER PUBLIC UNDERSTANDING OF TECHNICAL FIELDS 387

Science, Media, and the Public 387
Public Support Is Essential 388
People Love Science . . . 389
. . . And Resent It, Too 389
Science and Technical Professionals Are Suspicious 390
Media Are Skeptical 391

What Is Public Relations, and Why Is It Important? 394
    Components of Effective Public Relations   395
    An Effective Public Relations Campaign in
        Applied Science   395
Actively Improving Public Understanding 398
    Helping Media   398
    "No Comment"   400
    Relying on the Public Relations Department   400
Popularizing Your Work Yourself 401
How to Complain 403
Highlights 404
Projects 404
For More Help 405

CHAPTER 20
EVALUATING YOUR COMMUNICATION 406

Evaluation Techniques—Seeking Feedback 407
An Information Evaluation Strategy 407
    Gathering Information   407
Formal Evaluations 411
    A Precautionary Note   417
Highlights 418
Projects 418
For More Help 419

Appendix 421
References 423
Index 435

To the Student

As they begin their careers, engineers, scientists, and other professionals in technical fields sometimes assume that command of their disciplines is all they need for success. They tell themselves that mastery of mathematics, laboratory procedures, and techniques for testing and analysis make the difference in this competitive world.

Important as such technical skills are, mid-career professionals know that other factors are almost equally important. Especially as one's expertise in a discipline matures and promotions come, people skills emerge as crucial. The most important? Good communication.

As you make the transition from college to your professional career, communicating about your discipline to others in your field and outside it will become increasingly important. The following chapters will help you develop a professional perspective on communicating and enhance your skills in that critical area.

We assume you have completed a basic composition course and may have had a basic speech course. And we assume you're well along in developing your professional skills, knowledge, and background.

Furthermore, we know that you and your classmates come from many technical and scientific fields—engineering, medicine, biological sciences, physical sciences, computer science, agriculture, forestry, physical education, nutrition, and dozens of related fields.

Up to now, your writing may have used only one approach or followed a closely prescribed model with little leeway. If so, you'll find the approach we suggest quite different. As we explain, before beginning any report or speech, you must first carefully consider the audience it is intended for. Only then can you choose the kind of communication that will work best. We do not prescribe a single way of writing or speaking; we present a number of alternative strategies for you to try. We think you can learn and adapt a variety of communication techniques to use throughout your career. After all, no two professionals go about their communication tasks in exactly the same way.

You'll find the organization of this book is somewhat linear: step one, followed by step two, and so on, roughly following the technical communication process outlined in Chapter 1. We realize, and so will you if you don't already, that in practice people may perform several steps nearly simultaneously. But a textbook's linear nature requires either a beginning, middle, and end or a reference manual–dictionary approach. We opted for the former and assume that you'll start with Chapter 1 and proceed through the book more or less in order. Most likely your instructor will assign readings sequentially.

Moreover, developing communication skills seems to require a linear approach. Often, you need to develop one skill before beginning to try another. For example, it does little good to polish your drafts if your content and overall organization are weak. So we talk about broader issues before getting down to specifics.

We have divided the book into six parts. Part 1 provides an overview of communication principles with an emphasis on technical communication. Part 2 explains how you can improve your content through a better understanding of problem solving, recognizing weaknesses in your thinking, using the library more effectively, and observing, interviewing, and surveying people. Part 3 concentrates on writing strategies and processes. First we discuss ways to learn more about your readers and audiences and take a look at how several professionals go about writing. Then we discuss various writing processes and how to obtain good illustrations. Part 4 describes revising and editing strategies, and how to improve your writing by becoming your own editor.

Part 5 focuses on specific communication products: definitions, descriptions, and instructions; effective formats; correspondence; and oral presentations and slide sets. We also devote chapters to word processing and printing and job searching.

Part 6 rounds out the book by considering legal and ethical issues and how to improve public understanding of technology and science. The last chapter discusses techniques for evaluating technical and scientific communication.

The Appendix provides a bibliography of selected dictionaries and style manuals.

Throughout the book, we use examples from diverse fields. If an example doesn't fit your field, consider how the principle could apply.

And from time to time we use examples of how specialists in technical and scientific communication work. We do so to show you how successful professionals approach and solve communication problems. Even though you probably will not become a career technical or scientific communicator, you can use the same processes, techniques, and approaches to enhance the quality and effectiveness of your professional communications.

It's much like learning a sport or musical instrument: study the pros and you'll spot ways of improving.

PART ONE
Technical Communication Theory

What is the role of communication in the practice of scientific and technical professions? Do people in technical fields read, write, speak, and otherwise exchange information in ways that differ from those the rest of us use? Does an engineer or scientist, when sending and receiving messages on the job, make assumptions that differ from those made in reading a newspaper, watching television, or conversing with friends?

Almost always. Yet communication is perhaps the most human of traits. Can something so basic as communication also be highly specialized? We argue that it can be and is. Furthermore, communication is crucial to the everyday practice of science and technology. In his extremely influential *The Structure of Scientific Revolutions,* historian of science Thomas S. Kuhn spoke of the role communication has played in making previously invisible scientific revolutions visible to society. Textbooks and popularizations of scientific and technical work make developments known to scientists and to the public, Kuhn pointed out. The communication becomes as important as the science itself.

Any technical or scientific development owes its adoption by public or industry to effective communication that makes its value clear. The great instrument of visualization in medicine, the X-ray, was disclosed on December 28, 1895. Its discoverer, the German physicist Wilhelm Roentgen, announced his find in a paper he published in the Proceedings of the Physico-Medical Society of Würzburg. Four days later, he sent reprints of the article, with X-ray prints, to a number of well known physicists. Word of the discovery immediately spread to newspapers around the world. By February 1896, Thomas Edison was attempting to X-ray the human brain. Within weeks X-ray machines were being manufactured for use in hospitals and doctors' offices. Science produced a great invention; communication sped its worldwide adoption.

1
Communication Principles and Theory

OVERVIEW

In Chapter 1, we

- ☐ Define technical and scientific communication
- ☐ Show the importance of effective communication to professionals of all fields
- ☐ Look closely at the terminology, forms, and characteristics of technical communication
- ☐ Describe the functions and goals of technical communication
- ☐ Describe the components of the communication process
- ☐ Identify the eight major steps in professional communication

A DEFINITION OF TECHNICAL AND SCIENTIFIC COMMUNICATION

A working definition of technical and scientific communication should recognize the technical and scientific nature of the subject, be precise, and acknowledge differences in audiences and in their needs for information. Furthermore, the definition should reflect the usefulness of all communication forms—written, visual, and oral—in conveying technical information.

In this book, technical and scientific communication means *specialized communication about technical and scientific topics that uses precise written, visual, or oral methods to reach various audiences seeking specific information on those topics.*

Why Are Communication Skills Important to Your Career?

Professionals from diverse fields attest to the importance of communication. Just after being appointed president of Apple Computer, John Scully told the *Wall Street Journal* that he had a two-point program for improving still further that remarkably successful company. One point dealt with the company's technical expertness: avoid wasting energies on too many products and too many areas. Scully's second point stressed concise communication: "I will not accept any memorandum over one page. I'll throw it away without reading it," he said (*Wall Street Journal* 1983, 38).

A former dean of the Yale University School of Forestry, George Garratt, underscored the importance to foresters of good communication when he observed: "Professional foresters seldom fail on the job for lack of technical competence. Rather, the inadequacies are much more often related to lack of ability to communicate" (Garratt 1968, 553).

Richard M. Davis (1978), of the Air Force Institute of Technology at Ohio's Wright-Patterson Air Force Base, surveyed 350 engineers listed in *Engineers of Distinction* with an average of more than 30 years of experience. The engineers told Davis they spent nearly one fourth of their time writing and nearly one third of their time reading others' writing. They rated their ability to write effectively from very important to critically important. When considering job candidates, 87 percent said that the candidate's writing ability affected their selection.

At the University of California at Berkeley, technical writing instructor Charlene Spretnak (1982) surveyed 1,000 engineering alumni. They spent 25 percent of their job time writing, 23 percent reading technical material, 11 percent supervising others' writing, and 7 percent giving oral presentations. They had just 34 percent—fewer than three hours in an eight hour day—for engineering.

Professor James Paradis of the Massachusetts Institute of Technology surveyed 265 people from more than 20 industrial research and development firms (Weaver and Lampe 1984). Some 60 percent of the respondents were engineers; the rest were scientists. He found that managers spent more than 50 percent of their time, supervisors 41 percent, and staff people 35 percent of their time in technical communication.

The ability to communicate clearly is a highly valued skill, as these instances also show:

- North of Denver, on Colorado's Front Range, a control supervisor for Kodak Colorado prepares a report to her boss, a business school graduate, on the emulsion density of the latest batch of X-ray film. Lately he has complained that her reports themselves are too dense. She is careful to use nontechnical language.
- Farther east, at the Rockford, Illinois, plant of the Woodward Governor company, workers prepare maintenance instructions for the speed

governor of a locomotive. Crews must maintain the locomotive engine as required, or the governor will malfunction. If instructions are not perfectly clear, the company may be held liable for any resulting damages.
- In Oregon, the president of Tektronix writes a memo commending his staff. A large airline, having tried various computer systems, has chosen Tektronix and closed a multimillion-dollar sale. The airline did so because of the "user friendliness" of Tektronix's manuals. Good communication, writes the president, brought high dividends.

On and on goes the testimony, from engineers, agronomists, building contractors, and from virtually every professional field.

The principles of clear and effective technical communication apply across the board. Therefore, we have chosen the examples in the following chapters from as many fields as possible. These examples show how professionals from many different fields handle communication problems. Whatever your field may be, you need to become a professional in a second field, communication.

A CLOSER LOOK AT TECHNICAL AND SCIENTIFIC COMMUNICATION

A close look at technical and scientific communication will show you its similarities to and differences from other forms of communication. Consider specifically

- Terminology
- Communication forms
- Characteristics
- Competition

Terminology

Throughout this book, we use the terms *technical* and *scientific* as though they were synonyms, although they are not. Science involves asking basic questions; technology involves applications. Many subjects can be addressed either scientifically or technically. This book discusses purposive communication—communication that conveys information about and within a given field. *The fundamental communication principles and techniques are essentially the same for technical and scientific fields.*

The evolution of "technical writing" into "technical communication" may not be complete, but it has come far. University courses such as "Basic Technical Writing," "Writing for Engineers," and "Writing for the Professions" have given way to "Basic Technical Communication." The professional association of technical communicators changed its name from the "Society of Technical Writers and Publishers" to the more inclusive "Society for Technical Communication." The change is significant. It suggests the need to consider communication as being simultaneously written, spoken, and visual. Unlike some other forms of communication, technical and scientific communication has long relied on visual information to convey its message. Pythagoras, Leonardo da Vinci, and others relied on visual forms of communication. Today's effective communicators not only must know what makes a good visual and how to obtain or create one, but also how to talk effectively about their work.

In this book, we therefore use the words *communication* and *message* over the more specific *writing*.

Communication Forms

Scientific and technical professionals use several forms of communication to reach their audiences, depending in part upon their position and in part on the audiences they seek to reach. The three basic forms of communication are

> *Print*—memos, letters, proposals, progress and final reports, brochures, instructions, documentation, research articles, trade journal articles, and books;
> *Oral*—speeches, oral reports, meetings, and narration for slide shows, films, and videotapes;
> *Visuals*—slides, transparencies, photographic prints, videotapes, and films.

Of course, these categories are not mutually exclusive. Technical reports often contain photographs, line art, or other visuals, and presentations mix oral and visual forms. Today's well-equipped professional understands all forms.

Characteristics

Consider three points about effective professional communication:

1. The content must withstand the critical review of your profession.
2. Effective communication requires a clear grasp of the process and factors influencing it.
3. Appearance must be polished.

Admittedly, the three points are more easily enumerated than applied. All are necessary to effective communication. By itself, no single point ensures effective communication.

Solid Content. Having something to say is mostly your job and that of the people who have helped you gain professional competence. Courses in your major field should ensure that you produce well thought out and properly executed content for your reports.

Understanding Communication Principles. Assuming you have something worthwhile to say, all the attributes of good communication should come into play. The basics of communication include planning, structuring, executing, and evaluating your message. Much of this chapter covers these basics.

Polished Appearance. In today's sophisticated world, advertising, television, and all kinds of print media are extremely well produced. Audiences expect polished, professional looking reports, memos, slide presentations, and other communication products.

"Don't judge a book by its cover" may be good advice, but people do judge appearances first. If your work does not conform to accepted standards—correct grammar, spelling, style, form—your audience may, subtly but to your disadvantage, question your professional competence.

In a recent Denver, Colorado, election, the Republican candidate mailed a campaign brochure to 32,000 voters. His Democratic opponent obtained a copy and spotted spelling errors, bad grammar, and poor syntax. The Democrat sent a corrected version, along with his own campaign literature, to the district's voters. The Associated Press (1982) quoted the opponent as saying, "The real issue is not spelling or grammar. Instead, this sloppiness reflects poorly on [the Republican candidate's] overall competence and his ability to represent the district." In the end, the sloppy appeal for voter support detracted from the Republican's message, and he lost the election.

Competition between Messages— the Communicator's Burden

With two million published communications and uncounted millions of memoranda, minutes, letters, audio-visual presentations, speeches, films, tapes, drawings, slides, transparencies, and other message forms produced each year, two main facts stand out:

1. Someone out there sends a lot of messages.
2. Lots of people get a lot of these messages.

So bear the following in mind:

- Your communication faces competition from that of your colleagues.
- To compete well, your messages must be excellent. If not, you and the ideas they present will be inadequately represented.
- You, your organization, and your profession deserve the most articulate and effective communication possible.

For these reasons, cultivating communication skills makes as much sense as working to master a chosen profession.

MAJOR FUNCTIONS OF TECHNICAL AND SCIENTIFIC COMMUNICATION

When you prepare a memo, letter, report, or presentation, your purpose will be one or more of the following:

- to inform
- to instruct
- to persuade
- to document or record

Any message may serve several or all functions, but its dominant function heavily influences its form. Form should follow function.

Function: To Inform

A model of informative communication, the *Random House College Dictionary* (1982) informs readers of the meaning and spelling of 170,000 words. Its function dictates its form: an alphabetical listing, with each entry beginning a new line and set in boldface type. We take the alphabetized format entirely for granted. But the fellow who originated the format, the compiler of one of the world's first encyclopedias, felt divinely inspired when he put together the thirteenth century *Catholicon:* " 'Amo' comes before 'bibo' because 'a' is the first letter of the former and 'b' is the first letter of the latter and 'a' comes before 'b' . . . by the Grace of God working in me I have devised this order!" (Eisenstein 1981, 89). He was justly proud of himself, for the form he created made sense.

At its best, communication has several components, as Wilbur Schramm (1972), director of Stanford University's Institute for Communication Research, once pointed out. You want to draw attention to the information, have it accepted, interpreted correctly, and stored for later use. Accomplishing those objectives can be difficult.

Function: To Instruct

If you want the user to do something with the information you provide, your function is to instruct, so let that purpose influence your message's form. For example, an IBM instruction sheet (1982) for unpacking and setting up a new typewriter contains 16 photographs and no words (because IBM sells worldwide). The pictures provide step-by-step procedures for opening the carton and operating the typewriter.

When you provide instructions, encourage your audience to practice and use them.

Function: To Persuade

Most communicators want to persuade, although some seek only to persuade an audience that they are competent. When communicators try to change attitudes or behavior, they may face great difficulties. Their message's form becomes exceedingly important. If communicators indicate that they have searched through the relevant literature and know what others have done on the topic, they increase their persuasiveness. If they also estimate the time and cost that will be required and provide the tools and skills necessary for a new approach, they grow even more persuasive. As in the law, evidence persuades. The more evidence you have, the better its quality, and the better it is organized and presented, the more likely you are to persuade your audience. Evidence usually persuades better than rhetoric.

Function: To Document

Much technical and scientific communication consists of documenting how something was built, developed, or repaired. A laboratory journal recording how experiments were conducted provides an excellent example of documentation. A caution: in the professional field of technical communication, the words *documentation* and *to document* often refer to all the messages that accompany a product, including manuals, specifications, and promotional material.

HOW RHETORICAL AND COMMUNICATION THEORY CAN HELP

Some 23 centuries ago, the Greek philosopher Aristotle described a system of communication based on the way effective orators persuade their audiences. This "rhetorical" method stressed learning as much as possible about the audience and adjusting the message to the particular characteristics of each audience. The rhetorical method has thus heavily influenced writers as well as speakers ever since. Even today it remains a key part of much instruction in writing and speaking. Increasingly, though, scholars and researchers have pointed out that for all its usefulness, the rhetorical method takes a too simple view of the relationship between communicator and audience. For one thing, it assumes that all communication is persuasive. Proceeding on that assumption means putting what has been called an "argumentative edge" on the message. Sometimes argumentative messages are clearly inappropriate. For another, as social science, in particular psychology, has demonstrated over the past half century, simply being persuasive does not guarantee that you will be understood. Nor does being persuasive guarantee your audiences will do what you want them to do. Modern communication theory, building upon the insights of Aristotle, has investigated many barriers that hinder perfect understanding. Some of these we discuss below.

BASIC COMMUNICATION THEORY

If technical communication involves the transfer of specialized information among specialists, what are the elements that make that exchange possible? The answer to that complex question requires an examination of the communication process. In the nearly half century that researchers have been examining communication scientifically, they have learned much about what can go wrong, and why, when people try to communicate. They have learned that understanding the communication process and its components has strong practical value.

Some of the earliest and most significant scientific thinking about what communication *is* occurred in the late 1940s at Bell Telephone Laboratories. There, engineer Claude Shannon and a colleague, Warren Weaver, developed a mathematical model of communication (Shannon and Weaver 1949).

Key Ingredients in the Communication Process

Shannon and Weaver were interested in what happens physically when a message is transmitted, as through a telephone wire. The model they came up with for this process describes human communication as well (Figure 1-1). Even though communication is a dynamic process, the model depicts communication at one point in time. Modified slightly, the communication elements they saw consist of

- *Sender,* who originates the message
- *Message* itself
- *Channel,* or means chosen to transmit the message
- *Noise,* which interferes with the message
- *Receiver,* who accepts the message
- *Encoding/decoding,* or the processes of phrasing the message for transmission and receiving it in understandable form

Other researchers later added two other elements:

- *Interpretation* of the message by the receiver
- *Feedback,* or reaction to the message, itself a new message

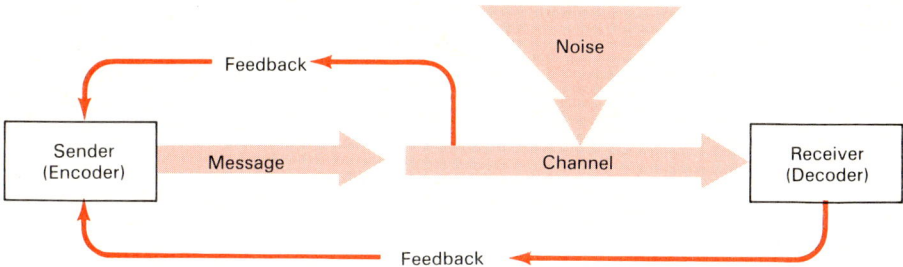

FIGURE 1-1. Sender–receiver communication model.

By breaking the communication act into its various parts, we can examine the factors that help or hinder understanding.

Sender. As you prepare a message, consider your role as the message sender. Don't you automatically take certain of your own characteristics into account in writing a paper or preparing a speech? If you were asked to speak about your chemistry lab experiment, perhaps the one that shut down the lab and brought the fire department, you would probably consider some important sender characteristics. You might realize that your appearance could help or hurt your cause. Probably you would dress to convey seriousness of purpose. If you know that you are not good at thinking on your feet, you might prepare notes on major points. You might make a note to yourself to smile once in a while. If you know that you are nervous in front of senior people, such as a panel of chemistry professors, you might rehearse your talk several times.

Knowing things about yourself helps you to plan your communication. Knowing what you do and don't know about your topic can help you to design your message. The way a sender prepares and executes a message establishes his or her credibility. A sender who seems credible goes a long way toward ensuring the success of any communication. Appearing authoritative, *being* authoritative, often can win the day.

Message. Phrasing the message in just the correct way occupies most people's attention, to the exclusion of other considerations. How should the message be organized? How long should it be? What level of language should it use? What kinds of symbols will best convey meaning? Errors in formulating the message can easily lead the audience to draw unkind conclusions about the intellectual quality of the work. Unsuitable vocabulary, grammar, and style impede understanding and send messages of their own.

Channel. The *channel* is the means by which a message is transmitted, such as print, videotape, speech, and others. Some channels convey certain types of information better than others. Thus mathematicians and chemists use chalk and board or other written channels. Veterinarians usually describe surgical procedures with slides and other visual channels.

Noise. In communication theory, *noise* includes both mechanical and semantic barriers to a message. Noise often reflects the environmental setting in which the communication act occurs. When you hear it said, "It was an idea ahead of its time," or "Times weren't right," you are hearing someone's judgment that noise prevented a communication from accomplishing its purpose. Mechanical noise includes static on a radio or telephone line, music and voices in a crowded restaurant, a messily reproduced written report, and the missing paragraphs that make a memo impossible to follow.

But far more difficult to recognize and deal with than mechanical noise is semantic noise. Communication scholars John C. Merrill and Ralph L. Lowenstein (1979) have suggested a few examples: divergent backgrounds of the participants; differences in education, in interest in the message, in intelligence, in language, in age, sex, race, and class; lack of skill by sender or receiver; mental or physical stress at the time of communication; and lack of mutual respect. Many more kinds of semantic noise exist, and all hamper or prevent understanding.

Receiver. Ernest Hemingway once observed, "Easy writing makes hard reading" (Bohle 1976, 460). A message may be perfectly clear to you, but unless you have studied and acted on receiver characteristics, the message may be meaningless to your reader.

A California psychiatrist forgot that point when he devised a computer program to take the medical histories of his patients. His intended message was: I want to save time taking histories so that I can better treat people. But many of his patients felt alienated by the impersonal computer. He had misjudged the receivers of his message. Understanding your audience is, as Aristotle knew, the most important step in the communication process. All the careful polishing of messages in the world is a waste of time, if you don't know your audience. Consider your audience above all else, design your communication for *that* audience, and you greatly increase your chances for success.

Encoding/decoding. We all use symbols—letters, numbers, words, pictures, and others—in formulating and translating messages. If communicators do not use symbols that have meaning for the receiver, communication cannot occur. Communicators must precisely define terms and possibly even include a glossary. As a communicator, be greatly concerned about reaching your audience in understandable terms.

Interpretation. When people communicate technical information, they usually want to send one particular meaning. Warning signs, such as those

COMMUNICATION PRINCIPLES AND THEORY 13

PANEL 1-1. International Traffic Signs

(Source: Colorado Drivers' Manual and Supplemental Motorcycle Drivers' Manual (Denver, Colo.: Colorado Motor Vehicle Division, 1984), p. 22.)

in Panel 1-1, must have a single interpretation, for example. Most technical and scientific communication strives for one interpretation. Some messages have more complex interpretations and multiple meanings. Great art, whether fiction, poetry, painting, music, drama, or sculpture, has many different levels of meaning. Indeed, our sustained interest in art comes largely from our seeing something new each time we regard it.

Professional technical communicators rarely paint paintings, however. They strive to design their messages with a single interpretation, so that all who receive them agree on their meaning. For instance, when technical writer Joe Williams (1982) composes a user manual for a network analyzer at Hewlett-Packard's Santa Rosa, California, plant, he knows that only a few customers a year use that extremely sophisticated scientific instrument. Some of them are electrical engineers who are his fellow workers. He knows them well, down to what they like for breakfast. Often he's tempted to address his readers in the electrical engineers' idiomatic language. But he cannot risk using an idiom, because it might be misinterpreted beyond that small circle. What if it were translated into French, Japanese, or another language? Williams cannot risk losing that one interpretation his readers must make. No communicator can take that risk.

Feedback. Although it is really an entirely new message, because it is a response from the receiver, feedback provides clues to whether your first message got through and was correctly interpreted. Your feedback may come in the form of a puzzled look, vigorous agreement or disagreement, modified behavior, radio signals, or other responses. Feedback tells you if communication has occurred.

The principles of communication theory can help you to formulate, transmit, and diagnose where and why your communication has gone wrong—or hit the mark.

Additional Communication Concepts

Behavioral scientists have added to basic communication theory. The following concepts prove especially important to technical and scientific communication.

The Gatekeeper. If you visualize a road with a series of gates that are opened to some travelers and closed to others, you have a good grasp of the *gatekeeper* concept. In communication, some gates are closed and others opened because of the assumptions and attitudes held by people—"gatekeepers"—along the way. Suppose that you are asked to write a five-page paper on a 500-page environmental impact statement. In deciding what to put in and what to leave out, you function as a gatekeeper for your readers. Of course, the authors of the original report acted as gatekeepers themselves in deciding what to put in and what to leave out. Readers also function as gatekeepers. They may not have the background or interest to read all sections of your report.

Frame of Reference. Communication can occur only when sender and receiver have common frames of reference, when they share language, culture, experience, education, and other attributes. True understanding occurs when their frames of reference overlap (Figure 1-2). By studying your audience carefully, you will learn where your frames of reference overlap. Then, by using that information, you can recognize the boundaries between you and the receivers of your message. In doing so, you have taken an extremely important step toward effective communication. You have begun to establish the commonalities necessary for understanding.

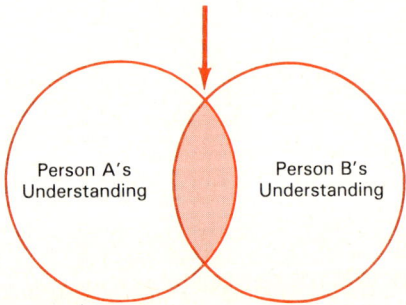

FIGURE 1-2. Frame of reference.

Theory Applied to an Example

Randy is hired for a summer job as tour guide at an oil shale demonstration project in western Colorado. The project shows how oil-bearing shale, which lies on or close to the surface, is crushed and oil extracted. Tours will consist of groups from major oil companies, members of the general public (especially on weekends), government officials from the U.S. Department of Energy, visitors from oil-producing countries, and various other groups. Randy will guide tours through the plant, answer questions, and try to make clear the feasibility of oil shale extraction.

Applying the communication theory points covered above, Randy creates a set of questions the answers to which will help assure successful tours.

Theory Point	*Questions Raised*
Sender characteristics	Does young college student appearance raise questions about credibility? How to overcome that? Dress? Adjusting speech? Need to build background on topic?
Message	Give the same message to each tour group, or tailor language to each group's technical knowledge? How to handle foreign visitors, whose technical knowledge may be large but whose command of English small? Rely on their interpreter? Combine talk with visuals, to enhance understanding? How long should message be for maximum effectiveness? Provide printed or other take-home materials for later study?
Channel	Will guide rely strictly on oral presentation or combine with photos, slides, or information from experts at various locations? If other people will speak, who chooses and trains them?
Noise	How to deal with activity and plant noise that might distract tour members? Will tour group size prevent guide from speaking to group while moving from one point to another? How to deal with psychological noise, such as tiredness of touring visitors or their need to use restrooms?
Receiver/audience	How to learn interests of various audiences so tour can cover those topics? Chemical engineers—refining aspects? Mechanical engineers—machinery? Geologists—formations? Conservationists—effects on environment? How to verify *common frame of reference?*

(Continued)

Theory Point	Questions Raised
Encoding/decoding	How to frame message to cover necessary points? How to test for *single interpretation?* How to test for word accuracy to ensure *semantics* present no problem to audience?
Selectivity	How to maximize attention paid during tour, and how to reinforce learning to maximize retention?
Dysfunctional aspects	Do language or visuals contain sexist, racial, or cultural biases? Do parts of tour detract from its central purpose?
Feedback	How to get feedback on strengths and weaknesses of tour?

By asking and trying to answer these questions, most of which bear on audience analysis, the tour guide has begun to think like an effective technical communicator. This example shows only a few of the many questions relevant to communication.

This book's remaining chapters explain how to raise and answer questions like these. Before proceeding to those details, we will consider the technical communication process.

MAJOR STEPS IN THE TECHNICAL COMMUNICATION PROCESS

Effective technical communication involves eight steps. We will consider them briefly here. In later chapters, we will discuss these steps at length.

1. *Define and solve the problem.* Your competence in your field gives you a starting point in the technical and scientific communication process. No matter your field, if you do not have a clear question, your search for answers remains fuzzy, you gather unnecessary information, and you waste time. With a clear, precise, and concise statement of the problem you are addressing, you give yourself a focused guide for searching the literature, collecting and analyzing data, and communicating the results.

2. *Select the audience.* Do this at the start of your project, and you will save time and effort. If, at the outset, you know to whom your finished work will be addressed, you also know a great deal about the form, language, level of knowledge, and other factors your project will use.

3. *Analyze the audience.* The time you spend describing on paper the characteristics of your audience will pay good dividends. Audience analysis, similar to market analysis, seeks knowledge about people. Whereas market analysis focuses on economic factors, audience analysis focuses more on attitudes and educational levels.

4. *Determine what to communicate.* Your selection of your audience and your analysis of its characteristics will guide you on what the audience

needs to know. Remember, you are dealing with people who have a specific problem that you are going to help solve or who have the answer to your problem. In either case, your message should be as concise as possible. Audiences do not wish to be told everything about a given topic. Spend time early on figuring out what to leave out as well as what to put in.

5. *Select the proper format*. Consider the available alternatives in message formats and choose those that will best help you reach your audience. If you consider only one alternative, likely to be the one that's as familiar as the proverbial old shoe, you needlessly limit your chances of being understood. When Joe Williams of Hewlett-Packard, whom we mentioned previously, thought through how people actually used H-P's manual on network analyzers, he realized that new users consulted the manual often, but users familiar with the equipment consulted it less often. So Williams developed a three-part manual with: (1) a dictionary of definitions and descriptions, (2) a tutorial for the new user, and (3) a pocket-sized, quick-reference card for the experienced user. That three-format approach soon became standard throughout the computer industry.

6. *Produce the product*. After you have attended to steps 1 through 5, you need to produce your product. The production entails going through the processes, such as writing and editing, necessary to create a report, an article, a slide presentation, or other communication. The production entails considering the overall appearance of your communication and the impression it creates. You need to consider good organization, spelling, grammar, and style, the proper paper, typeface, margins, methods of reproduction, and type of binding. If you need a photograph, you need to consider whether you will use a black and white print, a color print, or a slide. When you need visuals for an oral presentation, will you use slides or overhead transparencies? You will need to take many different factors into account as you prepare the actual communication "package."

7. *Distribute the product cost-effectively*. In many cases, others determine how you will distribute your message, and you cannot change the system. A master's thesis may go into a computerized abstracting/indexing service. If so, the thesis writer should carefully choose the key words by which the study may be retrieved. An interoffice memo may have to be written on a standard company form and follow a required distribution list. The form may have been designed to keep memoranda short. As a wise communicator, you'll need to pay careful attention to how your message will be distributed. Remember historian Henry Adams's observation (1918): it makes little difference whether you are read by five hundred readers or by 500,000, *if* you can pick the five hundred. Pick the right five hundred, and you will reach 500,000 and more.

8. *Evaluate the communication*. Lyle Settle, a senior technical writer for Tandem Computers, stresses three priorities that cover evaluation. He says the project must be done on time. It must be effective (that is, complete, well organized, well written and edited, well produced). Third, and well worth special mention, the product must be "ascendant," that

is, better than the competition's. Always remember to check the competition's product as well as yours.

Evaluation takes place at two levels. At one you assess the production of the message—its content, phrasing, appearance, time required for production, and distribution. Naturally, try to identify points at which you encountered difficulties. Perhaps you failed to allow enough time for production of the graphics you wished to include. People often get so wrapped up in the early stages of a communication project that they neglect to plan for its evaluation.

At the second level, you evaluate how your message influenced your audience. You assess changes in information levels, attitudes, and even the behavior of your audience. Your assessment can be an informal reflection on audience reactions, after which you add or drop a slide from your show, or it can be as formal and systematic as the evaluations discussed in Chapter 20.

As you apply the principles covered in this and subsequent chapters, try to keep two points in mind. First, although you may be doing the work for a class and therefore learning and practicing, soon you will be doing similar activities as part of your professional career. Second, if you approach studying communication seriously, you will soon reap professional benefits.

HIGHLIGHTS

1. Technical and scientific communication is specialized communication about technical and scientific topics that uses precise written, visual, or oral methods to reach audiences seeking specific information on those topics.
2. Professionals from diverse fields attest to the value of communication skills. Communication takes up a considerable part of their time.
3. *Communication* is a broader term than *writing;* professional communication entails not just writing but also oral and visual forms.
4. Professional communication must withstand critical review of its appearance, content, and effectiveness.
5. The application of communication principles is of practical everyday value.
6. No matter the field, technical communication functions to inform, instruct, persuade, and document.
7. Technical communication is aimed toward a controlled interpretation of its content.
8. Communication may be broken down into its components: sender, message, channel for transmission, noise that deters clear reception, receiver, and feedback, which itself is another message.

9. Professional communication entails eight steps: (1) defining and solving the problem, (2) selecting audiences, (3) analyzing those audiences, (4) deciding what to communicate, (5) selecting the proper format, (6) producing the product, (7) distributing it, and (8) evaluating the communication.

PROJECTS

For each project write a short report (of not more than three typed, double spaced pages), or share what you learn in a classroom discussion.

1. Interview a professor in your major department on the kinds of technical communication, besides teaching, the professor does. You might ask how the professor presents information—by reading a paper, talking about it, submitting it to publications? What can the professor tell you about his or her audiences?

2. Ask a professional in your major field what kinds of communication products—reports, articles, proposals, presentations, visuals, slide sets—that person most often has to prepare. How does the professional communicate with audiences at different levels of understanding about the topic?

3. To see how audiences are targeted, watch several children's educational television programs. Note how much time programs devote to various topics, whether they use repetition to teach. Are real objects or other devices used as visuals, photographs, models, and animation? Is the dominant function to inform, persuade, instruct, or document?

4. Apply the questions in Project 3 to several commercial television programs for children. How do they inform, persuade, and instruct their audiences?

5. To see how technical communication varies with audiences, compare the content, treatment, and approach of a children's with an adults' scientific television program (such as *Nova*). Use the questions in Project 3 as a guide. How do the functions of the programs differ? How have the functions been worked into the entertainment aspect of television programming?

6. Bring to class an effective or ineffective technical communication. Select something from your experience, such as instructions for repairing a bicycle wheel, sewing a down sleeping bag, or operating a machine. Analyze the communication according to the principles discussed in Chapter 1. In what ways do the instructions succeed? Fail? Does the presence or absence of illustrations aid understanding?

7. A Japanese cabinetmaker visits a New England cabinet shop. Despite a language barrier, visitor and host have a fine time demonstrating

woodworking techniques. The shop owner asks his new friend home for dinner. The evening is a flop, punctuated by long silences. Use communication theory to explain what happened.

8. After several weeks of class, a teacher gives a brief test. From the standpoint of communication theory, why must the teacher do that? What different method(s) might the teacher use?

9. Select an effective or ineffective communication in your intended career, and analyze its components using communication theory. You might choose something that you observed during a summer job or something that happened in your major department.

FOR MORE HELP

ANDERSON, P. V., R. J. BROCKMAN, AND C. R. MILLER, eds. 1983. *New essays in technical and scientific communication: Research, theory, practice.* Farmingdale, N.Y.: Baywood Publishing.
The 12 articles discuss such areas as the sociology of science and technical communication.

ANDERSON, W., AND D. COX. 1980. *The technical reader: Readings in technical, business, and scientific communication.* New York: Holt, Rinehart and Winston.
Anderson and Cox provide short, readable articles that describe professional communication with clear models.

HARTY, K. 1980. *Strategies for business and technical writing.* New York: Harcourt Brace Jovanovich.
The strategies cover the areas common to technical and business communication.

MORAN, M. G., AND D. JOURNET, eds. 1985. *Research in technical communication.* Westport, Conn: Greenwood Press.
The editors have compiled an up-to-date collection of articles that synthesize research and theory behind technical communication.

POOL, I. DE S., W. SCHRAMM, F. FREY, N. MACCOBY, AND E. PARKER. 1973. *Handbook of communication.* Chicago: Rand McNally.
The 31 in-depth articles will help readers develop a fuller understanding of communication's complexities.

ROGERS, E. M. 1983. *Diffusion of innovations,* 3rd ed. New York: Free Press.
Rogers explains why and how innovations are (or are not) adopted throughout the world.

SPARROW, W. K., AND D. H. CUNNINGHAM, eds. 1978. *The practical craft.* Boston: Houghton Mifflin.
Sparrow and Cunningham have collected articles on communication by professionals in business and technical fields.

PART TWO
Gathering Information and Improving Content

The eight-step communication process described in the opening chapter included defining the content and solving the problem. We observed that the content must withstand critical review by members of the communicator's profession. In the three chapters that follow, we provide an overall problem solving strategy, techniques for effectively reviewing literature, and guidelines for gathering information from people—observing, interviewing, and conducting surveys. Most scientific and technical specialists find themselves observing events, interviewing people, and on occasion conducting surveys. If they are untrained in such information gathering techniques, they may collect erroneous data and draw faulty conclusions.

The methods we describe will allow you to deepen and broaden your content as you gather information from many sources. When you use many sources, you have the distinct advantage of cross-checking your content. You also will be able to evaluate reports, studies, and other communication more accurately.

2
Problem Solving: Precise Phrasing Leads to Good Solutions

OVERVIEW

In Chapter 2, we

- ☐ Treat problem solving's role in technical and scientific communication
- ☐ Introduce a seven-step problem solving method
- ☐ Describe common thinking errors to guard against in beginning a project or writing up results

Reading, writing, and *thinking*. In today's world, they go by different names: data collection, communication, and problem solving. No matter which terms you refer to them by, they occupy most of the professional's working time. Of the three, thinking, or problem solving, holds primary importance, because it determines how a person goes about the other two. Clear thinking leads to clear writing, your writing teacher may argue. But clear thinking does more. It helps people to phrase the problem at hand so that they can address it.

WHAT IS A PROBLEM?

But what is a problem? It's the unanswered question. At a simple conceptual level, it's the distance between where you are and where you want to be. In *Concepts in Problem Solving* (1980), Moshe F. Rubinstein and Kenneth R. Pfeiffer use a simple model (Figure 2-1). Most problems have more than one correct solution, though possibly only one *best* solution, as the authors illustrate:

"Now give me another problem."

Alex thought for another moment, and said, "What about what I am going to do for a career?"

"Here the present state is your uncertainty about what you are going to do with your life. The goal state is for you to have a satisfying career, and the solution paths are for you to pursue any of a number of satisfying careers. In this problem, you want to pick the best available path." (p. 1)*

The engineer at a nuclear power plant must solve a problem when the flexible shaft connecting a motor and a coolant pump breaks down every three months. The physician must solve a problem when her patient complains of being light-headed before meals. The county agricultural extension agent must solve a problem when a farmer reports that soybean leaves have turned yellow and are wrinkled, oddly shaped, and stunted.

Engineer, physician, and county agent go through a similar process when confronted with those problems. Training and experience help each person define the problem and propose a set of responses. They will consider and eliminate certain explanations at once, perhaps so quickly that the proposed answer does not even consciously register. Then three or four probable causes may be singled out for more detailed examination. Tests to confirm or rule out explanations may be needed. When the answers appear close, trial and error may be employed. Usually, success rewards their effort. Like these professionals, you'll spend a lot of time solving problems in college and afterward. Most of these problems will be obvious. They will slow you or prevent you from proceeding until you solve them. Occasionally, instead of problems finding you, you may have the freedom to choose the problem you will address.

Suppose you are facing the problem of what to do on an important paper for a technical communication course. Getting from that state to your goal state—a successful paper—requires solving a series of problems along the way: How to select a topic. How to phrase the problem. Where and how to gather materials. How to test for the correct answers. How to

* Reprinted by permission of Prentice-Hall, Inc., Englewood Cliffs, N.J.

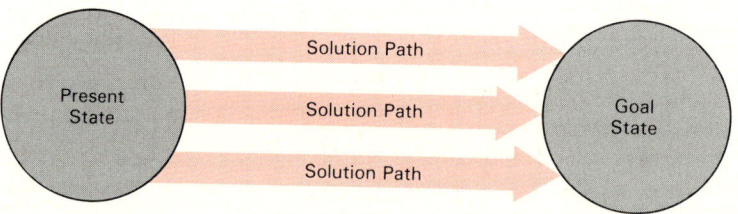

FIGURE 2-1. Rubinstein and Pfeiffer's model of a general problem. The model shows that any problem is likely to have more than one acceptable solution. (Source: M. F. Rubinstein and K. R. Pfeiffer, Concepts in Problem Solving, © 1980, p. 2. Reprinted by permission of Prentice-Hall, Inc., Englewood Cliffs, N.J.)

PROBLEM SOLVING: PRECISE PHRASING LEADS TO GOOD SOLUTIONS

present results. This chapter and the two that follow offer many suggestions on improving the content of your communication, both as you generate it and as you present it.

A Seven-Step Problem Solving Method

The seven-step problem solving method outlined below will help you organize your communication. It will make your data and information available for substantiating your interpretations and recommendations. The method (Figure 2-2) comes from the military and was adapted by scientists Verner Suomi and Thomas Haig (1974) for use at the University of Wisconsin's Space Science and Engineering Research Center.

The model, useful for individual and team problem solving, does not preclude the sudden insights or flashes of inspiration that sometimes solve a problem. It provides the structure within which illuminating insights or flashes of inspiration develop.

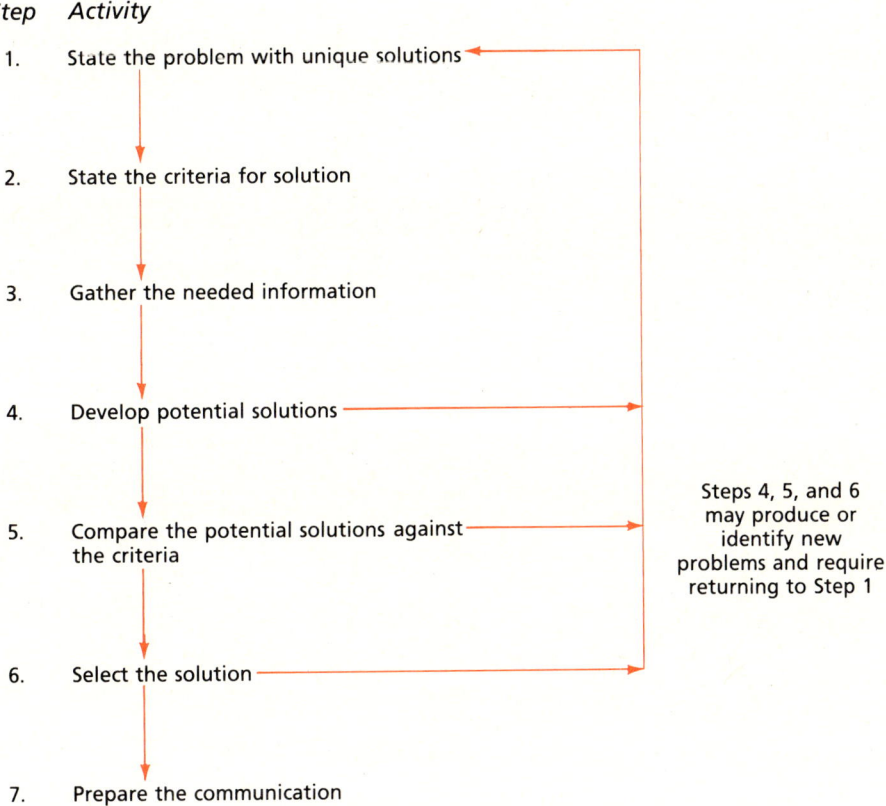

FIGURE 2-2. A problem solving strategy. (Source: Adapted from a presentation by Thomas O. Haig, former Associate Director, Space Science and Engineering Center, University of Wisconsin at Madison, October 1974.)

Step 1. State the Problem. State your problem as carefully as you can so that you understand it. If you are working on a team, state the problem so that every team member understands and agrees on it. People in different professions call problem statements "a statement of purpose," "objectives," "the research question," or "research hypotheses," as Panel 2-1 illustrates.

Developing a problem statement requires familiarity with the problem. So start broad, and then narrow your focus. The work of poultry scientist Roland Leach (1983) offers a good example of going from a broad, general topic to a more narrow (and more researchable) one. When Leach joined

---

PANEL 2-1. Problem Statements

Professional fields vary in the styles they use to phrase problem statements, as the examples below illustrate. However, all fields seek precision and tight focus.

Statement of Purpose

This report brings together information on relative proportions of energy and labor used by all manufacturing industries, rural and urban. The study identifies industries that use the most energy in relation to their labor inputs, where those industries are located, and which are concentrated in rural areas.

(Source: E. J. Smith, Energy and labor use by rural manufacturing industries, *Rural Development Research Report No. 26* (Washington, D.C.: Economics and Statistics Service, United States Department of Agriculture, 1981), p. 1.)

Objectives

Objectives of the hazard tree survey in Region 3 are to:

1. Assist landowners in providing an acceptable level of public safety within developed recreation areas.
2. Provide on-site training in detection and assessment of hazard trees for individuals responsible for maintenance and safety in recreation areas.
3. Establish base information on current disease situation in developed recreation areas.

(Source: E. M. Sharon, Biological Evaluation Recreation Site Examinations, *Santa Fe National Forest, New Mexico* (Albuquerque, N.M.: Southwestern Region, U.S. Forest Service, U.S. Department of Agriculture, 1977), p. 1.)

Research Questions

... the study includes the results for the following research questions: (1) Are there any differences between the importance attached to certain managerial tasks by male and female managers? (2) Do women managers feel sexual discrimination in the newspaper business is a serious problem? (3) How aggressively do women managers seek out opportunities for a better job? (4) How well trained are women managers? (5) What kind(s) of personal sacrifices have been made by women to achieve success on the job?

(Source: A. B. Sohn, Women in newspaper management: An update, *Newspaper Research Journal* 3 (October 1981): 95. Reprinted by permission of A. B. Sohn.)

Research Hypotheses

(1) Smokers are less likely than nonsmokers to watch the anti-smoking film ...
(2) Smokers are less likely than nonsmokers to choose the "threatening message" on the topic ...

(Source: J. Bertrand, Selective avoidance on health topics: A field test, *Communication Research* 6 (July 1979): 276. Reprinted with permission of *Communication Research*.)

> ### PANEL 2-2. Shell-Shocked
>
> How many cartons, you wonder angrily, do you have to open before you find one with 12 honest-to-goodness unbroken eggs?
>
> Five to 10 percent of U.S. eggs break even before reaching the market, according to Roland Leach, professor of poultry science at Penn State. And in 1981 broken eggshells cost the nation's poultry industry more than $200 million at the wholesale level (no doubt reflected in higher retail prices).
>
> Now Leach and his colleagues are trying to build a tougher egg and have begun to study the makeup of the eggshell. Initial experiments have shown that a lack of the mineral manganese in chicken feed dramatically alters the shell structure. Manganese plays an important role in the synthesis of connective tissues in both humans and animals, Dr. Leach explains.
>
> After a 12- to 15-week period on a manganese-deficient diet, "the shells were 10 percent thinner and weaker, and there were dramatic alterations in the shell structure," he reports. After 40 weeks, the manganese-deficient hens were laying eggs without shells. All that remained was a thin membrane coating.
>
> Why not give chickens more manganese in their feed? It isn't that simple. Leach wants to rule out other factors first in his research to develop a mightier egg. Do keep us posted.
>
> (Source: Family Weekly, *January 9, 1983, p. 34. Reprinted by permission from* Family Weekly/USA Weekend.)

the Pennsylvania State University poultry science department, his colleagues urged him to investigate eggshell breakage (Panel 2-2). Leach says it took him five years to understand the problem as he finally conceived it. He variously investigated the egg's structure, the calcification of eggs, copper deficiencies, the biochemistry of eggshells, and finally manganese's role. Leach had been the first scientist to observe manganese's role in bone calcification, and so his background in bone research led him to research manganese's role in the structure of eggshells. As Professor Leach's poultry research showed him, developing a problem statement can take considerable time. Often the scope of the problem does not become clear until the research is well under way.

As a student, you may feel that such sorting out of research topics is beyond your capability. In fact, though, you follow much the same process in choosing a term paper project. Your curiosity as a scientist leads you to ask lots of questions: Why does this happen? How does this mechanism work? How can we improve the quality of life? What can we do to minimize harm and maximize benefits? Questions such as these lead you to general topics.

In a basic technical communication class at Colorado State University, a student once compared initial costs, operating costs, and efficiencies of air-cooled and water-cooled air conditioning units. The student developed a proposal, outline, rough draft, and final copy. Only when reading the student's final paper did the professor realize that the student had conceptualized the wrong research question. The student had improperly matched

air conditioning units. One unit performed two functions—heating and cooling—while the other only cooled. For his study, the student should have compared units that only cooled.

Such insights are not unique to professors reading student papers. Thomas Kuhn (1970) argues that problems may not really be clear until you've tried to solve them. Only when they have struggled with a problem for an extended time do most people understand their problem well enough to solve it. Alexander Graham Bell worked for years under the impression that he was investigating deafness. Finally it became clear to him that he was in fact inventing the telephone, which had great commercial potential. He shifted his emphasis, and within a few months he and his backers were on the way to developing American Telephone and Telegraph.

Hardly anyone states the problem absolutely right the first time. For example, a water quality class was asked to test for sewage pollution of the Cache la Poudre River near Fort Collins, Colorado. Here's the initial problem statement that one student drafted:

> *Version A:* The purpose of this study is to determine the possible changes in four parameters (fecal coliforms, phosphates, ammonium, and nitrates) resulting from the discharge of sewage effluents from the Fort Collins Sewage Treatment Plant Number One.

Although the statement appears precise, in that it limits the study to four parameters, it raises a question. What does "possible changes" mean? Changes that may or may not occur? Can a change that is possible but that does not occur be measured? After further thought, the student presented this tightened statement:

> *Version B:* My study will examine changes in fecal coliforms, phosphates, ammonium, and nitrates resulting from the discharge of effluents from the Fort Collins Sewage Treatment Plant Number One into the Cache la Poudre River.

This version is more precise, because it declares that changes that actually occur (and thus can be measured) will be examined. But notice that Version B assumes that changes do result. A more scientifically objective phrasing would be to cast the problem as a question:

> *Version C:* Does sewage effluent from the Fort Collins Sewage Treatment Plant Number One change the levels of fecal coliforms, phosphates, ammonium, and nitrates in the Cache la Poudre River?

Perhaps the student anticipated pollution and saw no reason to hide that conviction. If the student knew that a pollution problem already existed, from a literature review or talking with students from previous years' classes and the instructor, a more pointed research question might have been:

PROBLEM SOLVING: PRECISE PHRASING LEADS TO GOOD SOLUTIONS

PANEL 2-3. Guidelines for Refining Your Question

Once you have your general idea, develop a problem statement to direct your problem solving. To narrow your problem statement, write it down and define the terms. Then ask:

- Will my statement effectively direct my problem solving?
- Will my project be descriptive, associative, or causal?
- Have I posed a narrowly defined question?
- Is my question clear and concise?
- Which factors (variables) will I need to consider?
- Are my definitions of terms adequate?
- Will my peers agree with my terms and their definitions?
- Will my peers understand what I'm about to do?
- Can I design an investigation based on my problem statement?
- Will my problem statement suggest other problem statements that must be answered before I can answer the first?

Version D: Does sewage effluent from the Fort Collins Sewage Treatment Plant Number One increase levels of fecal coliforms, phosphates, ammonium, and nitrates in the Cache la Poudre River?

Panel 2-3 presents guidelines for narrowing a question or problem statement. Once you have the problem statement, define how you'll use each word. Then explain how you will measure the terms you're using. Remember that your problem statement directs

- reviewing literature
- defining concepts and variables
- gathering data and information
- analyzing data and information
- interpreting data and information
- presenting data and information
- illustrating data and information

Step 2. State the Criteria for Solutions. Those factors that restrain, limit, or restrict will determine the potential solutions to your problem. At times the criteria will require redefining the problem and ruling out some solutions. University of Wisconsin's Thomas Haig (1983) identifies two types of criteria, those that (1) limit the possible solutions and (2) specify the desired performance. Limiting criteria keep you out of trouble, and desired criteria help you recognize the solution. To illustrate, Haig uses a simple problem: "Which automobile should I buy?"

Let's say you're married and have two children. The limiting criteria for purchasing an automobile might include

- Cost: not more than $7,000
- Passenger seating: self, spouse, two children, including one child's safety car seat
- Financing: over 24 months
- Heating: good, or we'll be uncomfortable
- Additional carrying capacity: 36 cubic feet
- Towing capacity: can pull 3,000 pound trailer

The desired performance criteria would include

- Gas mileage: 40 miles per gallon, more for highway driving
- Long-term guarantee against body rust
- Better than average maintenance record as reported by independent consumer groups
- Washable upholstery, preferably textured
- Neutral color that won't show road dust

By developing a detailed criteria list, you improve the chances of solving the problem.

For many technical and scientific problems, the criteria restraining solutions include limited funding, staff size and competence, available time, equipment, and facilities, and ethical, moral, and legal restrictions. By stating your criteria before you collect information, you preclude letting the information influence your thinking about potential solutions. Furthermore, your criteria keep you focused on your problem.

Important limiting criteria for a term project or thesis include such items as

- Do you have the time to carry out the project?
- Do you have the skills and knowledge required? If statistics, a foreign language, or other skills are needed, are you competent?
- Are the resources available? Does the library have the books you need, or must you use interlibrary loan? Is the local expert you must consult available to you or out of the country for several weeks?
- Do you have the funds for any out-of-pocket expenses or travel needed?

An outdoor recreation major at Colorado State University once proposed to do as a class project a tour-guide pamphlet for the Badlands

National Monument in South Dakota. The student had the whole winter quarter to work on the project, good skills, and even a travel budget, because the United States Park Service offered to pay the cost. Everything looked fine, until halfway through the project, when the student realized that the quarter would end before the trees leafed out. That meant no chance for pictures to illustrate the pamphlet. Had the student listed desired criteria at the project's outset, the quarter would not have been wasted.

Step 3. Gather the Needed Information. You may find brainstorming a useful starting point to identify the information needed to develop potential solutions. Brainstorming entails quickly listing as many ideas as possible without evaluating them.

Gather the information you need to solve your problem from many sources. Chapters 3 and 4 explain techniques—literature reviewing, interviewing, observing—that will supplement the skills you're acquiring in your major. When do you know that you've gathered enough information? When you begin turning up the same information from different sources, you may have reached a point where further information gathering proves inefficient. Your understanding of the topic will help you to determine when you have enough information.

As you gather material, keep accurate and detailed records. For literature reviewing, keep detailed notes, as Chapter 3 recommends. In many professional problem solving settings, you can purchase or request free copies of major reports, articles, books, and other documents.

If you interview experts or make observations, follow the note-taking guidelines in Chapter 4. Be sure to include names, correct spellings, dates, locations, and other pertinent information. For surveys, prepare summary tables of the needed information.

If you conduct field or laboratory work, keep detailed and accurate records. Many professional fields have specific guidelines on preparing field notes, laboratory reports, and records. Collect and record data carefully.

Develop an organized way of filing the information so you can easily retrieve it when you prepare communications. Detailed, accurate, and well organized records provide easy access should you later need to substantiate your decisions.

Step 4. Develop Potential Solutions. From your information, you can develop the more likely solutions in light of the criteria you established under Step 2. You can eliminate considerations that do not meet your criteria. For example, you would waste your time collecting information on a $10,000 automobile when you specified a $6,000 limit. If you have only two months to solve the problem, you can eliminate a solution requiring 18 months.

Thomas Haig recommends that you develop *at least* three solutions for careful, in-depth consideration. You need several potential solutions because a careful analysis will eliminate some. Then carefully write down your proposed solutions so that you understand them. If you're working on a team, be sure that all team members understand the solutions.

Step 5. Compare the Potential Solutions against the Criteria. Next, develop a detailed comparison of your potential solutions against the criteria you establish under Step 2. For many applied, day-to-day problems, a table may help you consider the alternatives and the criteria. You can list the potential solutions across the top and the criteria along the left-hand side. Let's say you've finished college and you're seeking a job. You're offered jobs in Florida, California, and Minnesota. Which job will you take? Your criteria include: working environment, potential for professional development, climate, salary, fringe benefits, outdoor recreational opportunities, and travel time from relatives (Table 2-1). You define each criterion. For example, would you consider an hour's travel time from your parents positive, negative, or neutral? Would you consider an hour's travel time from brothers and sisters positive, negative, or neutral?

TABLE 2-1. Comparison of Florida, California, and Minnesota Job Offers

Criteria	*Florida*	*California*	*Minnesota*
1. Working environment			
2. Potential professional development			
3. Climate			
4. Salary			
5. Fringe benefits			
6. Outdoor recreational opportunities			
7. Travel time from relatives			

Step 6. Select the Solution. Out of a comparison of solutions with criteria, a clearly defined solution should emerge. If so, fine. If not, you may need to return to Step 1 and begin the process over again. For the job searching example, you may find your outdoor recreation criterion inadequate. You would have to redefine what you mean by outdoor recreation opportunities, specify the kinds of recreation you want, collect further information, and repeat the problem solving process.

Step 7. Prepare Communication and Act on the Solution. Most projects require that you prepare a final report, article, or other document presenting your findings. Step 7 entails implementing the communication process detailed in Chapter 1 and expanded upon in other chapters. If you've kept accurate records, the seven-step problem solving method will help you prepare your final product.

MINIMIZING CONTENT ERRORS

In the *CHEMTECH* of December 1974, Thomas Kilgore Sherwood, a chemical engineer and professor at the University of California at Berkeley, pointed out a dilemma facing scientists and technical professionals:

> The experimentalist's job is only half done when the laboratory or pilot-tests have been completed and the data recorded. He [or she] must then attempt to derive . . . meaning from the results, or at least summarize the study and suggest correlations of results . . . that others can use. The art of doing this is a tricky business. (Sherwood 1974, 736)

After you prepare a problem statement, collect and analyze data and information, you must interpret those data and information. Part of interpreting your data includes checking for content errors.

Errors of content do slip by professionals. On July 22, 1962, a Mariner I rocket rose from its launch pad and headed for Venus. It soon veered wildly off course, and the ground controllers had to destroy it. What had gone wrong? In *Thinking Better* (1982), authors David Lewis and James Green point out that a minus sign had been missing from Mariner's computer program. That content error cost American taxpayers $18.5 million.

Effective technical communication begins with clear thinking, and clear thinking begins as you carefully apply your professional communication skills. By developing a questioning attitude toward your communication's content, you'll minimize content errors.

Descriptive? Correlational? Causal?

Avoid attributing more to the data than they actually support. Data may be descriptive, correlational, or causal. It is important that you know which of these your data are. Your problem solving may be descriptive—you describe an event, equipment, procedure, or observation. Your problem solving may document correlational or associative information—the occurrence of two variables together, such as thunderstorms and hail. Your problem solving may be causal—you document cause and effect relationships.

A descriptive study is Swedish botanist Carolus Linnaeus's classification system of all flowers and plants, or the work of Russian chemist Dmitri Ivanovich Mendeleev, who described elements using atomic numbers, weights, electronegativity, and symbols, or a study of the number of people in a nation, their ages, religion, sex distribution, and other demographics.

Correlational or associative studies relate one factor to another. For example, there is a positive correlation between the height and weight of high school students. As height increases, weight usually increases.

In a cause and effect study, a microbiologist and an entomologist might want to determine if a new biological control agent, perhaps a protozoan, reduces grasshopper populations. As a test, the scientists would need to conduct a controlled field experiment in which they treated one population of grasshoppers with the protozoan and maintained an untreated population under similar conditions. The scientists would then measure the number of grasshopper deaths in each population during the study. By subtracting the number of grasshopper deaths in the control group from the number of grasshopper deaths in the treatment group, the scientists could infer whether the protozoans reduced grasshopper numbers.

Professionals sometimes collect descriptive or correlational information and make erroneous inferences about cause and effect. An internationally known sociologist once surveyed farmers in a developing country. The survey's purpose was to learn which characteristics were *associated* with farmers' adoptions of new farming practices. The sociologist found that farmers who adopted new practices more frequently had traveled to larger cities, knew more about credit and market conditions, and were younger than farmers who did not adopt the new practices. The sociologist then proposed training to increase farmers' knowledge of credit and market conditions, and he proposed giving them opportunities to travel to larger cities.

In making such recommendations, the sociologist made causal inferences based on correlational data—an error. His data gave him no basis for inferring that increasing farmers' knowledge about credit and market conditions or having them travel to larger cities would make them more likely to adopt new farming practices. Luckily, the sociologist did not recommend making farmers younger—that's neither possible nor logical.

Know the limitations of your project, and draw conclusions carefully. If you want to speculate beyond your data, clearly label your speculations.

Numbers and Their Misuses

A scathing comment on statistical evidence has been variously attributed to Mark Twain and the British political leader Benjamin Disraeli: "There are three kinds of lies: lies, damn lies, and statistics." You certainly don't want others to categorize your statistics in such a way.

Collected, handled, analyzed, and interpreted carefully, numbers provide strong, compelling arguments that support your conclusions. But collected carelessly, handled sloppily, and misinterpreted, numbers can lead you astray. Proceed carefully.

In *The Nature of Statistics* (1962), W. Allen Wallis and Harry V. Roberts identify 6 common errors:

Shifting Definitions. Define your terms clearly for a project, and don't change your terms or their definitions later on. For example, a professional doing a study comparing range conditions, grasslands, and vegetation might encounter problems because various government agencies have different ways of defining how to sample vegetation. If one agency clips vegetation samples at ground level, another agency clips them one inch above ground, and a third clips them two inches above ground, the three data sets are not comparable.

Inaccurate Measurements or Classification of Cases. Measurements in technical and scientific fields are never exact. The more precision needed, the more the imprecision becomes apparent. A teaspoon measures sugar well enough for coffee, but in chemistry class you learn to read the level of fluids in a graduate cylinder by holding the cylinder at eye level and reading the bottom of the meniscus. Similarly, national unemployment reports, when examined carefully, seem little more than estimates. In *Social Research for Consumers* (1982), Earl Babbie points out that unemployed people, within the U.S. Department of Labor's definition, are those *not working, available for work, or looking for work.* "Looking for work" means an individual has registered with an unemployment agency within the past four weeks and has not found work. Thus, individuals who have given up looking for a job and have not registered with an unemployment agency are not counted as unemployed.

Inappropriate Comparisons. The cliché "comparing apples and oranges" cautions against inappropriate comparisons. For example, comparing unemployment statistics of the early 1980s with those of the 1930s is inappropriate. People who were unemployed during the Depression did not have social nets—unemployment benefits and welfare—to fall back on as the unemployed of the 1980s had.

Disregard of Dispersion. Reporting averages or statistical means can be misleading. For instance, electrical wind generation mills, or Wind Energy Conversion Systems (WECS), need winds between 9 and 35 miles per hour, or they will not produce electricity. If you were considering buying one of these $30,000 systems for your home, you should first have checked with the nearest weather station for wind speeds. Let's say you do and find an average wind speed in your area of 18.3 m.p.h. for the year. Sound good? Think again. You'd better check dispersion over the year. Let's say a check shows monthly averages of 40, 40, 40, 40, 10, 10, 10, 10, 5, 5, 5, and 5 m.p.h. The monthly averages suggest that your system would only operate four months out of the year. A closer look at the monthly averages might even produce a similar problem. Besides the mean (average), you'd need to review the standard deviation, minimum, and maximum wind speeds to estimate electricity production.

Technical Errors. Errors in calculations, mathematics, manipulation, computers, and data analysis can easily alter your interpretation of data. In

1982, for example, David S. Walonick, a computer programmer and consultant in Minneapolis, was testing his new IBM personal computer. It showed that 0.1 divided by 10 equaled 0.001; the correct answer was, of course, 0.01. Walonick called IBM, and IBM later released a corrected version of the operating system, according to *Science News* (Peterson 1982). Computers and calculators can produce technical errors on occasion, so check yours (Panel 2-4). When you use computers, calculators, and other equipment, double check, and check again the values you obtain. Do those values fall within the expected and logical range of similar studies?

Misuses in Selecting Cases. Biases known or unknown may cause professionals to select cases that are not representative of the population they are studying. Some people knowingly select cases improperly. Some err out of ignorance. For example, without having had adequate instructions on sampling, a graduate student in horticulture once erroneously selected

---

PANEL 2-4. Calculators Can Do Funny Things

BY I. PETERSON

You can check your own calculator or personal computer to see what kinds of arithmetic anomalies arise. Some calculators are better than others; older calculators, even in the same product line, tend to make more mistakes.

Try this. Start with 1, divide by 3 and then multiply your answer by 3. Subtract 1. What answer do you get? On a TI-55 (made by Texas Instruments), for example, the answer is zero. On a TI-25, it's -1×10^{-8}. Repeat the process, but this time in the last step, instead of subtracting 1, subtract 0.5 twice. On a TI-25, the new answer is -1×10^{-9}, while the TI-55 gets -1×10^{-11}.

Although the differences in the answers appear trivial, the difficulty is that if a programmable calculator or computer performs arithmetic in the same way (and many do), then it may come across division by zero depending on how the machine has rounded off earlier results. The problem is the lack of a proper guard digit.

Is 2^3 exactly equal to 8? Calculate 2^3, then subtract 8. On a TI-55, the answer is 2×10^{-10}. On your calculator, does $\pi \times e$ give the same result as $e \times \pi$? Is 1 divided by 3 the same as 9 divided by 27?

If your calculator has trigonometric functions like sin, cos and tan, check to see whether trigonometric identities (mathematical relationships among the functions) are preserved. For example, $\tan 20° = \tan 200° = \tan(2 \times 10^n)°$ (for any power of ten). Compare your answer for $\tan 20°$ to, say, $\tan(2 \times 10^6)°$. Does your calculator give the correct values for $\sin \pi = 0$, $\cos \pi = 1$ and $\tan \pi = 0$, although the calculator cannot represent the number π exactly?

In more complicated calculations, other peculiar behaviors can show up. Although today's calculators are very good and in most applications will give accurate, reliable results, users must be wary of exceeding the sometimes arbitrary limits of any machine.

(Source: Science News, *July 31, 1982, p. 75. Reprinted with permission from* Science News, *the weekly newsmagazine of science, copyright 1982 by Science Service, Inc.)*

PROBLEM SOLVING: PRECISE PHRASING LEADS TO GOOD SOLUTIONS

plants for an experiment. She selected the largest plants for the controls and the smallest plants for the treatment. The researchers noticed the size difference and questioned the graduate student. All of the data collected in the experiment had to be abandoned because of the selection bias—the plants had not been randomly selected.

Errors in Generalization

Besides these six kinds of errors in using statistics, professionals sometimes make errors in generalization. But our professional and personal lives would be impossible without generalizations. They help us to operate without having to question everything about us. We generalize when we look at specific facts and then make a statement covering all similar events. When you do generalize, generalize with care.

David Fischer, in *Historians' Fallacies* (1970), and Dwight Ingles, in *Is It Really So?* (1976), discuss errors in generalization. They consider overgeneralizations and false extrapolations.

Overgeneralization. In an overgeneralization, a person exaggerates facts or goes beyond the available information. When a student went home for a visit from college his freshman year, he discussed his courses with a neighbor. When he mentioned a chemistry course, the neighbor responded, "But aren't all chemicals bad?" The neighbor had read newspaper accounts of chemical pollutants and had concluded that all chemicals were bad. Not having taken any chemistry courses, the neighbor had overgeneralized from limited information.

False Extrapolation. Fischer defines false extrapolation as a statistical series stretched beyond the breaking point. He shows Mark Twain having fun with the concept:

> The Mississippi between Cairo and New Orleans was 1,200 miles long 170 years ago. It was 1,180 after the cut-off of 1722. It was 1,040 after the American Bend cut-off. It has lost 67 miles since. Consequently, its length is only 973 miles at present. . . .
>
> Therefore, any calm person, who is not blind or idiotic, can see that in the Old Oolitic Silurian Period, just 1 million years ago next November, the lower Mississippi River was upward of 1,300,000 miles long, and stuck out over the Gulf of Mexico like a fishing rod. And by the same token any person can see that 742 years from now the Lower Mississippi will be only a mile and three-quarters long. . . . (p. 121)

Chapter 9 will give you strategies for evaluating the content of your writing. For now, carefully consider the points made in this chapter to develop and narrow your problems.

HIGHLIGHTS

1. Problem solving plays a crucial role in technical communication.
2. You have a problem when you're trying to find a way to move from your present state to a goal state.
3. The seven-step problem solving method includes (1) stating the problem, (2) stating the criteria for solution, (3) gathering the needed information, (4) developing potential solutions, (5) comparing the potential solutions against the criteria, (6) selecting the solution, and (7) carrying out the solution and communicating your results.
4. Avoiding content errors begins with assessing whether your data or information are descriptive, associative, or causal.
5. Common misuses of numbers include (1) shifting definitions, (2) inaccurate measurements or classification of cases, (3) inappropriate comparisons, (4) disregard for dispersion, (5) technical errors, and (6) misuses in selecting cases.
6. Avoid overgeneralizations and false extrapolation of data and information.

PROJECTS

Review items 1 through 6 below, analyze the content, and then write a brief analysis describing the errors in content or logic.

1. In the spring of 1983, an Associated Press article's headline read, "Scientists Expect Huge Crop of Mosquitoes in Nation." The story recounted problems in the Northeast, Arizona, Florida, and eastern South Dakota (Associated Press 1983). What's wrong with the headline?
2. Maps of the Great Lakes prepared in 1755 for King Louis XV showed four islands in Lake Superior—Pontchartrain, Maurepas, Philippeaus, and Ste. Annex. The Treaty of 1783 gave two islands to the United States and two to Canada. But explorers searched in vain for the islands. They were never found (Carlin 1982). What explanations might be offered for the failure to find the islands?
3. In a study for a technical communication class, Pam Fishlein (1981), a natural resources major at Colorado State University, asked 10 women, "What were the problems, obstacles, or barriers you encountered when you entered your first job with a natural resources agency?" Of the 45 problems identified, 70 percent related to communication difficulties, 50 percent to men's expectations of women. How would a different question have changed the results?
4. In trying to convince you to ride with him, a pilot of a small airplane argues that flying is the safest way to travel. He says that

flying is safer than driving a car. Would you ride with him? Why? Why not?

5. The January 14, 1979, issue of *Family Weekly* reported, "The heftier the hips, the smarter the student, according to a somewhat scientific study of DePauw University coeds. According to the study, which related students' measurements to their grade point averages, hip size turned out to be the vital statistic in predicting good grades." Should a student who wants better grades eat more to increase hip size?

6. In checking the quality of circuit boards being manufactured, an engineer reported using a somewhat random sampling procedure and found a mean rate of 2.5 faulty boards per 1,000 manufactured. He reported to his supervisor that the shipment of 1,000 units to XYZ computer manufacturer was all right. What error(s) did the engineer make?

7. Select a research publication or journal applicable to your profession. Review a current issue, and select an article with a clear, clean research question. Photocopy the article. Review the article, and then write a succinct report discussing in what ways the research question influenced the subsequent problem.

FOR MORE HELP

ADAMS, J. L. 1976. *Conceptual blockbusting*. San Francisco: San Francisco Book Company.
  Adams provides useful insights and guidelines for rethinking problems.

CAMPBELL, D. T., AND J. C. STANLEY. 1973. *Experiments and quasi-experimental designs for research*. Chicago: Rand McNally.
  A classic outlining basic research designs and threats to the validity of causal inferences.

KUHN, T. S. 1970. *The structure of scientific revolutions*. 2nd ed. Chicago: University of Chicago Press.
  Kuhn describes how paradigms or disciplinary matrices develop and explains why texts need to be developed as they do.

LEWIS, D., AND J. GREENE. 1982. *Thinking better*. New York: Rawson, Wade.
  The authors provide different kinds of learners alternative approaches to learning and solving problems.

MCCAIN, G., AND E. M. SEGAL. 1982. *The game of science*. Monterey, Calif.: Brooks/Cole.
  McCain and Segal provide readers with a succinct overview of the complexities directing science.

RUBINSTEIN, M. F., AND K. P. PFEIFFER. 1980. *Concepts in problem solving*. Englewood Cliffs, N.J.: Prentice-Hall.
  The book provides techniques for boosting problem solving skills.

3
Using the Library and Reviewing Literature

OVERVIEW

In Chapter 3, we
- ☐ Provide an overview of the library's resources
- ☐ Explain how to develop a literature search plan
- ☐ Discuss how to execute your search

The information age is upon us, and information proliferates exponentially. The National Science Foundation calculates that the number of scientific articles published each year increased from under 200,000 in 1960 to more than 400,000 in 1980. In such an information storm, the challenge becomes how to find what you need before you grow too old to use it.

You choose a problem to investigate, or one chooses you. Following the seven-step problem solving method, you state it as precisely as you can. You list the criteria for the problem's solution. By this time you are already involved in Step 3, gathering the needed information. That means, in part, consulting the library, or as industry calls it, "the technical information center."

If you're like many college students, your first experiences with the college library may have been bewildering. Part of your difficulties may arise from the number of publications in your library. It probably contains hundreds of thousands, if not millions, of publications. Harvard's libraries have more than 10 million books, Stanford's more than 4 million. Even an average university library has more than 1 million.

To use a library efficiently and effectively, you need to develop an understanding of the library's basic resources, plan a searching strategy, and then break your search into small tasks. Invest a few hours developing your library skills, and your investment will pay for itself over and over throughout your career.

USING THE LIBRARY AND REVIEWING LITERATURE 41

THE LIBRARY'S RESOURCES

Many university and college libraries have several collections spread across a campus. Some house less frequently used materials off campus. Most college libraries have all of the following.

Libraries and Reading Rooms

Because many campuses have several libraries, reading rooms, or branches, check the main library for its directory of your school's libraries. Such directories identify the libraries, branches, and reading rooms, and list the open hours, loan periods, and other vital information.

Librarians

Most libraries have subject librarians who have training in physical sciences, biological sciences, education, government documents, and other specialties. They're experts in retrieving literature in their specialized areas. They can help you learn how to use the technical and scientific literature quickly and efficiently.

But don't assume that everyone working in the library is a librarian. Part-timers and clerks have diverse assignments. Some will be helpful; others won't be. If you ask for help and the person can't answer your question, ask who *can* help you.

Card Catalog

Learn about your library's card catalogs. Your library may have combined or separate card catalogs for books, serials, and government documents. The card catalog may have several formats—printed or typed cards, a computerized system, or both. Libraries organize the printed card catalog by authors, titles, and subjects. Each section has its own style of card, as illustrated in Panel 3-1.

Serials—magazines, newspapers, journals, and other periodicals—have a title card and a serials record card, as shown in Panel 3-2. You'll find the serial title cards in the title section of the card catalog; the serial record cards may be housed in a separate card catalog.

Many libraries are switching to computerized card catalogs. You search the computer data base by author, title, or subject. Because of the high costs, many libraries will not transfer existing card catalog information to computerized systems. They are entering data for new acquisitions into the computerized system. Thus you may find yourself learning both manual and electronic literature searching.

Each catalog card contains the publication's call numbers. These numbers correspond to a pattern for arranging documents in the library's stacks. Most libraries use either the Library of Congress Classification

PANEL 3-1. Author, Title, and Subject Cards

how to find books by using the card catalog

Libraries telephone number: 491-5911

The card catalog is a bibliographic tool and a location device for both **books** and **periodicals.** It is located in the Reference Area on the First Floor. Through its use it can be determined if the book or periodical needed is owned by the Libraries. The catalog is divided into three major sections: **Author, Title,** and **Subject.** Informational signs designate the sections of the card catalog. It is important to pay particular attention to which part of the card catalog is consulted to avoid needless waste of time in the search for material.

Each catalog card gives many pieces of information. The following sample card indicates some of these.

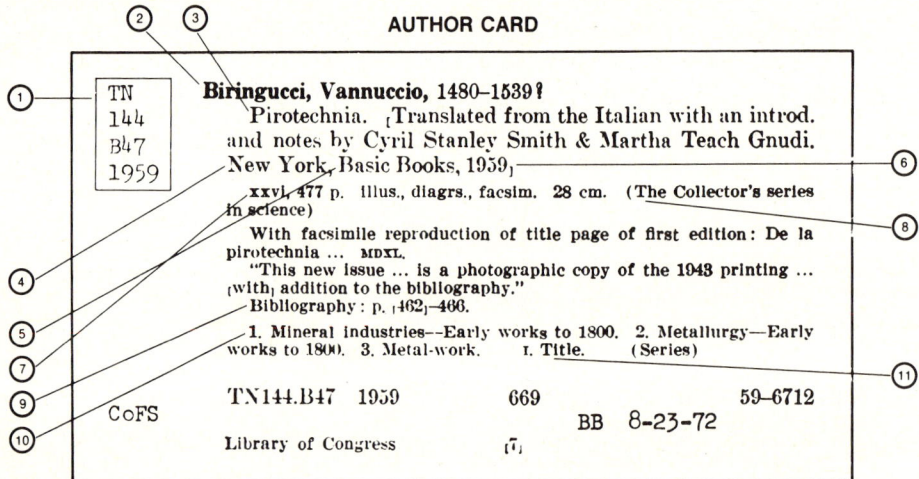

1. Call number
2. Author's name
3. Title
4. Place of publication
5. Publisher
6. Date of publication
7. Physical description of book
8. Series of which this book is a part
9. A note: it contains important additional information
10. Subject headings
11. Added entries. There will be a similar card for this book under each of these entries in the author and/or title sections

The call number in the upper left hand corner of the card is the key to physically locating the desired material. Most of the collection is classified according to the Library of Congress (LC) classification system, but some books are still in the Dewey Decimal system. It is easy to distinguish between the two systems. The LC call number always starts with one or two letters followed by numbers. *The top line of the call number should be read as a whole number and the "author" or bottom line should be read as if it were a decimal.* The Dewey call numbers always start with numerals.

The word FOLIO (which means oversize) may appear above the call number; words such as PERIOD or SPECIAL may appear under the call number. These words are important location aids and must be copied down as part of the total call number.

(Source: Used with permission of Morgan Library, Colorado State University.)

PANEL 3-1. (continued)

The call number can be considered as a code to establish the physical location of desired material. The various places where materials are shelved according to call number are listed in the schedule that follows. This schedule is also on the wall to the left of the Loan Desk, and on the boxes of scratch paper on the consultation tables at the card catalog. Leaflets showing this schedule are available at the Loan Desk, the Reserve Desk, the General Reference Desk, the Science Reference Desk, and the Documents Reference Desk.

AUTHOR

The author section of the card catalog contains a card for each of the Libraries' holdings arranged alphabetically by the author's name. The author can be a person, a corporation, institution, organization, government agency, or other nonpersonal author. Cross reference cards will lead from a form of a name which is not used to the correct entry for that name. Books by the same author are arranged alphabetically by title.

TITLE

The title section of the card catalog contains a card for each of the Libraries' holdings arranged alphabetically by title. Most of the titles are **underlined in green ink,** but older cards, cards for additional titles of a book, and the newer computer-produced cards have the title typed at the top of the card. The articles **A, An,** and **The** and their foreign equivalents are disregarded if they are the first word of the title. If the Libraries have several editions of the same title the most recent edition will be filed first. Cards in the title section also show the "holdings" CSU has for any particular work, such as: **volumes 1 and 2** of a four volume set.

In addition to the title cards, colored "order slip copies" are also filed in the title section of the catalog. These slips indicate titles which are on order or in process. More information about these titles can be secured at the General Reference Desk.

TITLE CARD

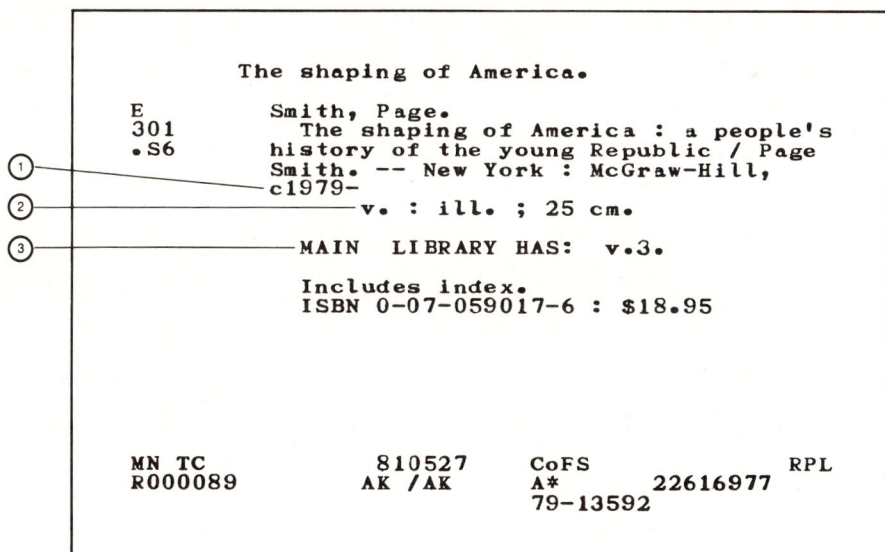

1. Open publication date.
2. Undetermined volumes in set.
3. Holdings note: the library owns only volume 3 in the set.

PANEL 3-1. (continued)

SUBJECT

The subject section of the card catalog arranges by subject the books and periodicals in the Libraries. It groups together material on the same subject. Each subject heading is typed in **red** at the **top** of the first card (a guide card) for that subject; the various titles on that subject are then filed behind it with the **most recently published title first,** followed by the others in reverse order of their publishing dates. There are cross reference cards which lead from a subject heading which is not used to one that is. For example:

Literature, American

See

American literature

The Library of Congress *List of Subject Headings* may help in determining which subject heading should be examined. The *List of Subject Headings* is located on a stand at the east end of the card catalog. It also contains "see also" references, indicating related subject headings where additional material may be found.

Many subjects are divided by dates, geographical or political divisions, or by forms of materials. A few of the most common form subdivisions are **bibliographies, collected works,** and **periodicals.** If assistance is required in locating a particular subject, ask at the Reference Desk.

SUBJECT CARD

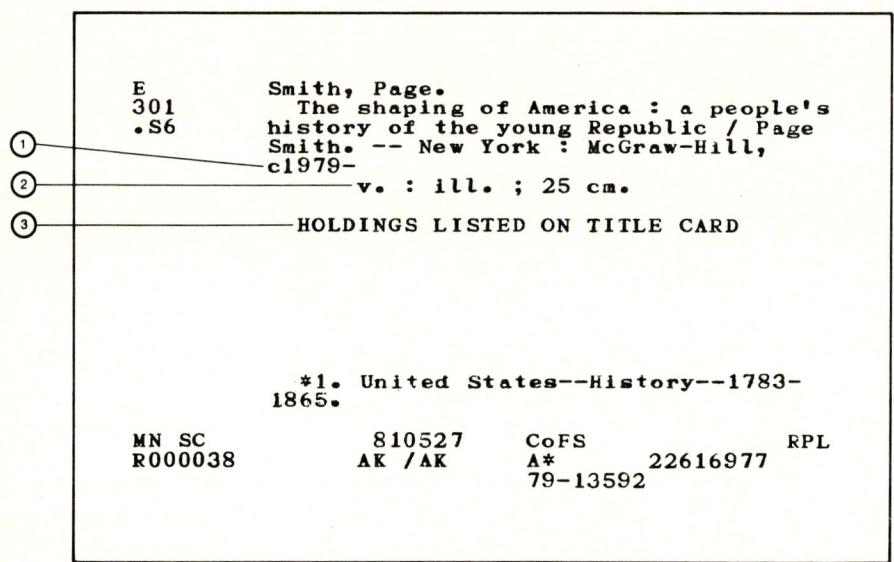

1. Open publication date.
2. Undetermined volumes in set.
3. Note explaining that card filed in title section of card catalog shows the specific volumes owned by the library.

This subject card is filed behind a **guide** card with the following subject heading typed in red at the top of the card: **"United States — History — 1783-1865."** Note that this is the **identical** heading with the one shown behind the numeral "1" at the bottom of the card.

PANEL 3-2. Serials Cards

how to identify and locate periodicals

Libraries Telephone Number 491-5911

Periodicals, like books, are listed in the card catalog. Most periodicals do not have a specific author so they are entered (filed) under their title in the **title** section of the card catalog. Since periodicals sometimes change their titles, information regarding these changes is noted on the cards. **Cards for earlier titles are also found in the title section of the catalog.** Below are two examples of cards for periodicals and the information that can be gathered from them.

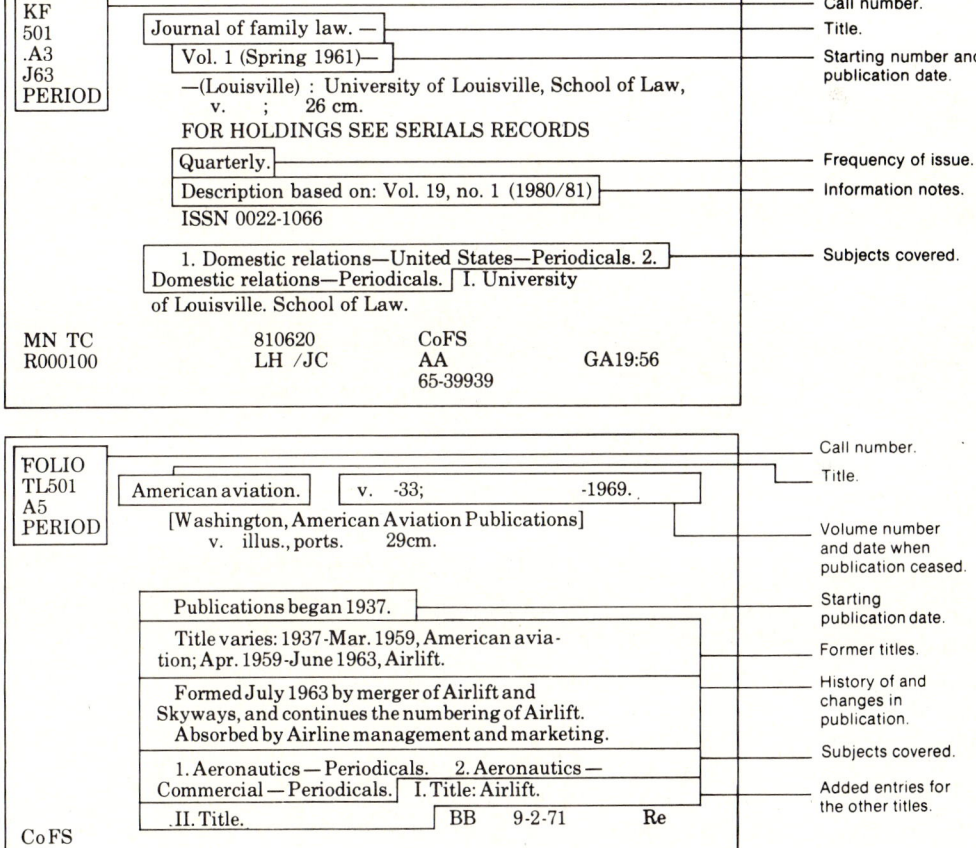

The Serial Record, which is located near the periodical indexes in the South Foyer, also lists which volumes of each periodical are held by the Libraries; numbers of bound volumes are recorded in ink with a dot in front of the number or are typewritten; numbers for unbound volumes are listed in pencil.

(Source: Used with permission of Morgan Library, Colorado State University.)

GATHERING INFORMATION AND IMPROVING CONTENT

System or the Dewey Decimal Classification System; some use both systems. Others are converting from the Dewey Decimal System to the Library of Congress System.

Most libraries have information sheets explaining the classification system(s) used. Some libraries have audio-visual presentations explaining their systems.

General References

The reference collection contains both general information and publications that will help you identify other relevant literature. Reference collections

PANEL 3-3. *Ulrich's* Family

Available for electronic, online searches, *Ulrich's International Periodicals and Irregular Serials Data Base* includes 116,000 active and ceased records of serials prior to the current editions of *Ulrich's International Periodicals Directory* and *Irregular Serials and Annuals.* As a worldwide publication, *Ulrich's International Periodicals Directory* lists 64,800 periodicals in 556 subjects. *Ulrich's Irregular Serials and Annuals* lists some 35,000 serials, annuals, continuations, conferences, and other irregular publications. *Ulrich's Quarterly* supplements *Ulrich's International Directory* and *Irregular Serials and Annuals*, gives new serials, title changes, and cessations and thereby provides current information between editions of *Ulrich's International Periodicals Directory* and *Irregular Serials and Annuals*. *Ulrich's Source of Serials* gives author/publisher information. Each publication includes a detailed user guide. (See below.)

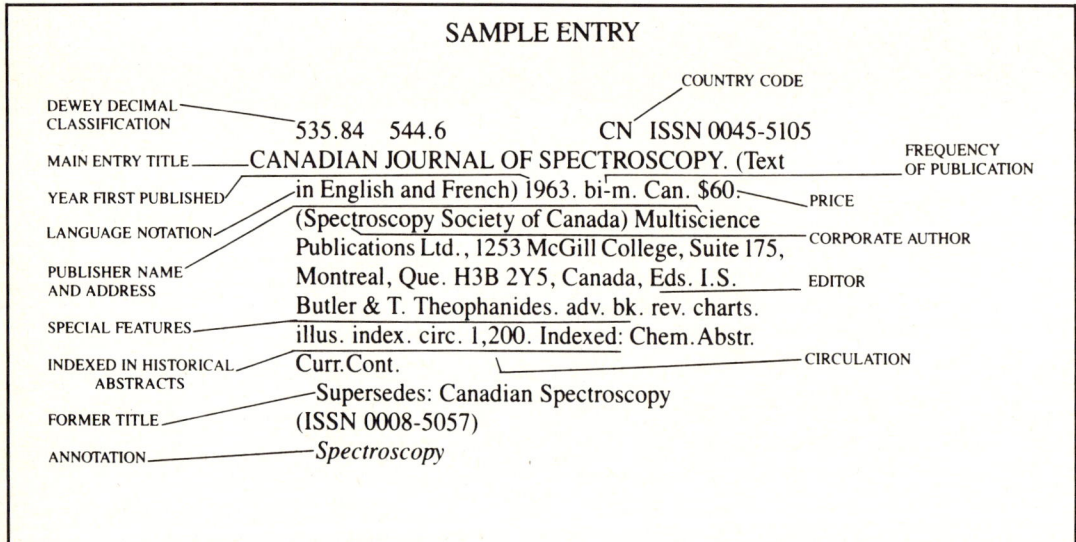

(Source: Published by R. R. Bowker, Division of Reed Publishing, USA. © 1985 by Reed Publishing USA, a division of Reed Holdings, Inc. All rights reserved.)

include statistical abstracts, bibliographies, encyclopedias, anthologies, dictionaries, indexes, abstracts, atlases, gazetteers, and other references. Some libraries have their references in a separate room. Other libraries place the general references near the card catalog or another easily accessible location.

If you're beginning your search, you'll find the *Ulrich's* family of publications (Panel 3-3) helpful in identifying potentially useful periodicals. Eugene P. Sheehy's *Guide to Reference Books* (Panel 3-4) identifies guides, bibliographies, indexes, abstract journals, handbooks, encyclopedias, dictionaries, and other valuable works.

Abstracting Journals and Indexes

Abstracts present a succinct summary of an article's contents, the author(s), publication, date, volume, number, and pages. The *Reader's Guide to Periodical Literature* abstracts from general magazines. Abstract journals cover the thousands of articles reported in technical periodicals.

Indexes, in contrast to abstracts, usually list the titles, author(s), periodical, volume, pages, and date. Some indexes contain abstracts. You'll need to review the technical periodical literature carefully, because even newly published books on rapidly advancing topics may be three to five years behind recent developments. (Most periodicals report developments within 12 to 18 months.)

Abstract journals and indexes provide the keys to finding relevant, recent literature. Panel 3-5 identifies a handful of indexes and abstracts from the hundreds covering the technical and scientific periodical literature. Panel 3-6 lists selected abstracts and indexes. Panel 3-7 explains how to use indexes and abstracts. Various abstracts and indexes have slightly

---

PANEL 3-4. Sheehy's *Guide to Reference Books*

Of its five major sections, the "Social Sciences" and "Pure and Applied Sciences" sections probably will be the most helpful in Eugene Sheehy's *Guide to Reference Books*. In "Social Sciences," the general works and those in sociology, statistics, economics, law, and geography may be of help to you. "Pure and Applied Sciences" covers general works, astronomy, biological sciences, chemistry, earth sciences, mathematics, physics, psychology and psychiatry, engineering, medical sciences, and agricultural sciences. Each section contains several subsections, and these are further divided into disciplines. Under each discipline, Sheehy's work lists guides, dictionaries, encyclopedias, abstract journals, handbooks, histories, films, abbreviations, directories, foreign terms, and other references as appropriate. The categories depend upon the discipline covered.

Before using the individual sections, study "Reference Books and Reference Works" in the preface. "How to Study Reference Books" contains valuable guidelines for searching references.

> ## PANEL 3-5. Annotated Bibliography of Selected Indexes and Abstract Journals
>
> *Applied Science and Technology Index*, Bronx, N.Y.: H. W. Wilson. As a cumulative subject index to English language periodicals, this index covers more than 400 publications in aeronautics and space, atmospheric sciences, chemistry, computer technology and applications, construction industry, energy resources and research, engineering, fire and fire prevention, food and food industry, geology, machinery, mathematics, metallurgy, mineralogy, oceanography, petroleum and gas, physics, plastics, textile industry and fabrics, transportation, and industrial and mechanical arts.
>
> *Bibliography of Agriculture*, Phoenix: Oryx Press. Based on the indexing records of the United States Department of Agriculture's AGRICOLA electronic data base, this bibliography cites journal articles, pamphlets, government documents, special reports, proceedings, and related publications for agriculture and allied sciences.
>
> *National Agricultural Library Catalog*, New York: Rowman and Littlefield. This catalog reports on books and new serials in agriculture.
>
> *Biological Abstracts*, Philadelphia: Biosciences Information Service. *Biological Abstracts*, which began in 1926, reports the world's biosciences research monthly. It publishes semi-annual cumulative indexes on authors, genus and/or species names, biosystematic organization, concepts, and subjects.
>
> *Chemical Abstracts,* Columbus, Ohio: Chemical Abstract Service. These abstracts cover the world's chemical and chemical engineering literature. Users need a strong chemistry background. The introduction suggests using the *General Subject Index*, *Chemical Substance Index*, and relevant years' *Index Guide*. Chemical Abstract Service publishes a workbook to help users learn the system.
>
> *The Engineering Index Monthly*, New York: Engineering Information. This index covers the world's technological literature in all engineering disciplines as published in journals, technical reports, monographs, conference proceedings, and other materials.
>
> *Index Medicus,* Bethesda, Maryland: National Library of Medicine. A monthly bibliography of the serial journal literature in biomedicine and other reports judged useful by librarians.

different organizations. So review the publication's directions on how to use it.

Science Citation Index is probably the primary index for science. Since beginning publication in 1964, when it covered 151,639 source articles, *SCI* has grown to cover more than half a million articles each year. Its cumulative indexes list more than 9 million articles. Such a collection of references requires considerable complexity to organize, so we shall not attempt to explain how to use *SCI* here. Your best bet is a patient science reference librarian, plus actual practice. *SCI* offers quick answers to many questions, including:

Has this paper been cited?
Has there been a review on this subject?
Has this theory been confirmed?

PANEL 3-6. Selected Indexes and Abstracts

The indexes and abstract journals listed below cover national and international technical and scientific literature.

Air Pollution Abstracts
Aquatic Sciences and Fisheries Abstracts
Bibliography of Agriculture
Bibliography and Index of Geology
Biological Abstracts
Chemical Abstracts
Commonwealth Agricultural Bureau
Computer and Control Abstracts
Conference Papers Index
Dissertation Abstracts
Energy Abstracts
Energy Index
Energy Information Abstracts
Engineering Abstracts
Environmental Periodicals Bibliography
EXCERPTA MEDICA
Food and Technology Abstracts
Food Science and Technology Abstracts
Government Reports Announcement
Index Medicus
Index to Dental Literature
International Aerospace Abstracts
International Nursing Index
Meteorological and Geoastrophysical Abstracts
Monthly Catalog of U.S. Government Publications
Oceanic Abstracts
Physics Abstracts
Pollution Abstracts
Psychological Abstracts
Safety Science Abstracts
Science Citation Index
Scientific and Technical Aerospace Reports
Selected Water Resources Abstracts
Toxicity Bibliography

(Source: Based on Colorado State University Morgan Library's leaflet on Computerized Literature Access Search Services for biomedical, engineering, forestry and agriculture, and physical sciences.)

Has this work been extended, the method improved, this suggestion tried?
What are all the current works in which this person is primary author?

SCI also allows you to estimate a scientist's reputation, because you can easily see how often the person is cited by others.

Computer Searches and Electronic Data Bases

Computers are quickly reshaping library use. Through long distance telephone calls and a computer terminal, electronic searches of the more than 1,000 computerized data bases can replace manual searches. Some data bases are computerized card catalogs; others are abstract journals and indexes. Computerized systems and data bases allow you to review quickly tens of thousands, if not millions, of citations of periodical articles. For example, *BIOSIS PREVIEWS* contains more than 3.2 million abstracts of periodical articles. Started in 1969, *BIOSIS* provides semimonthly updates based on its printed counterparts—*Biological Abstracts* and *Biological Abstracts/RRM*.

Electronic data bases have more limited records than printed abstracts

PANEL 3-7. Using Abstracts and Indexes

To use abstract journals and indexes effectively, carefully read and study the instructions or guides to users found at the front of each publication. Some abstracting or indexing services provide special volumes explaining how to use their publications. For example, *Science Citation Index 1984 Guide and List of Source Publications* tells how to use the *Science Citation Index*.

Generally indexes give you limited information, such as an article title, first author, journal or periodical, date, volume, number, and pages. If you looked up "energy conservation" in a 1984 volume of *Applied Science and Technology Index*, you would find the following entries:

Energy Conservation
    See Also
    Electric power—conservation
    Energy consumption
    Energy management
    Energy policy
    Fuel economy

1983 Energy Conservation Awards. il diag *Heat/Pip/Air Cond.* 56:39–40 Ja '84.
Air-to-air energy recovery will recover and expand in 1984. J.A. Parish. *Heat/Pip/Air Cond.* 56:159–60 Ja '84.*

Thus if you were interested in searching for articles on air-to-air energy recovery and its development, you would go to pages 150 to 160 of the January 1984 trade journal, *Heating/Piping/Air Conditioning*.

Indexes give you little information by which to evaluate an article's potential value. Abstract journals usually give you more. To illustrate, let's say you're interested in the feeding habitats of Canada geese, and you want to search *Biological Abstracts/RRM* (1980–) for useful articles. You develop as specific words: "Canada geese," "*Branta canadensis*" (genus, species—the scientific name), "feeding habits," "food habits," "crop damage," "migration," and "seasonal diets." By looking up "Canada geese" in the *Subject Index* (1984, 1051), you find:

```
ASS LARVA PUPA QUEBEC    CANADA /A PHORETIC ASSOCIATON BET    16044
E CONGENITAL ANOMALY            /GENEALOGICAL ANALYSIS OF R   17007
OUNTY NEW-BRUNSWICK             /INCIDENTAL CATCH OF HARBOR   16127
YCUSH HYBRID ONTARIO            /LONGEVITY OF 1ST GENERATIO   16147
SLAND OFF NOVA SCOTIA           /SHORT-TERM CHANGES IN DIST   16153
US-NAMAYCUSH QUEBEC             /USE OF ACCUMULATIONS OF U    16137
ASTERN NEWFOUNDLAND             AREA STOCK ASSESSMENT/REG     16151
UTTERITE BRETHREN OF            DARIUSLBUT SCHMIEDELBUT LE    20107
T/ FALL FOOD HABITS OF          GEESE BRANTA-CANADENSIS NE    16160
GNUS-BEWICKI FEATHER/           GOOSE BRANTA-CANADENSIS NE    15564
ULAR FLORA OF ALBERTA           NOTE CIRSION SPP WEED IDENTI  14461
TION LAKE ONTARIO USA           NUCLEAR EFFLUENT THERMAL P    20060
GENY BRITISH-COLUMBIA           OREGON CALIFORNIA/SIMILARIT   17464
RAIN AT DELHI ONTARIO           TOBACCO SOIL POLLUTION-ACID   19677
SCENS DOSIDICUS-GIGAS           USA MEXICO ATLANTIC OCEAN P   16132
CTIC MIGRATION SEASON           USA SOUTH AMERICA/A NEW FI    15565
ATION DELIVERY MEXICO           USA/ APPLE INTEGRATED PEST    16195
T INTAKE AND CARIES IN   CANADIAN CHILDREN ABSTRACT FRUCTO    15720
TER SURVIVAL 20 YEARS           FORCES AIRCREW EXPERIENCE     20050
HIPPOGLOSSOIDES FROM            WATERS OF THE NORTHWEST       16152
```

To find the abstracts, use the reference numbers, 16160 and 15564, check the Content Summaries section, and you find (1044, 1008):

16160. CRAVEN, SCOTT R.* and RICHARD HUND. (Dep. Wildlife Ecol., Univ. Wis., Madison, WI 53706.) J WILDL MANAGE 48(1): 169–173. **1984. Fall food habits of Canada geese [*Branta canadensis*] in Wisconsin** [USA]./AGRICULTURE, GRIT, LEAD SHOT
CON: Environmental & Industrial Toxicology/General & Systematic Aves/Nutritional Studies/Animal Ecoilogy/Agronomy
TAX: Anseriformes

15564. KUMMER, JOHANNES. (DDR–3500 Stendal, Wahrburger Strasse 88.) BEITR VOGELKD 29(5/6): 311[recd. 1984]. [In Ger.] **Canada goose,** *Branta canadensis,* **near Stendal, East Germany.**/*CYGNUS BEWICKI*, FEATHER
CON: Animal Distribution
TAX: Anseriforms†

The 16160 citation seems to contain useful information, but the 15564 citation looks like a record of sightings in Germany.

To complete your search, you continue to look up terms and record the citations of useful articles.

---

* Applied Science & Technology Index, Copyright © 1984 by the H. W. Wilson Company. Material reproduced by permission of the publisher.

† Reprinted by permission of Biosciences Information Service, Philadelphia, Pa.

and indexes. Check to see when the computerized data base started and whether earlier materials were added to the data base. Few data bases are retrospective; they contain materials published only after the base started. For example, *CA SEARCH*—covering part of the chemical literature—began in 1967 and contains more than 4 million citations. Finding chemical literature published before 1967 requires a manual search of the earlier *Chemical Abstracts* (1907–   ).

Conducting an electronic search involves developing a research statement or question, reviewing the data base's dictionary or thesaurus, and preparing a list of key words and synonyms. In many cases, you work with a librarian who places a long distance telephone call to the vendor—the company that maintains and updates the data base. The call goes to the vendor's computer and logs into the data base. The librarian then enters the key words (descriptors). The computer searches its data base for article titles and abstracts that contain the descriptors and reports the number of article titles containing the descriptors. By combining terms in selected ways, the librarian narrows the search to the more probable articles. At that point, you can request either on-line—shown on the screen—or off-line, printed listings of the citations. Having citations printed at off-peak hours—late in the day—reduces the charges. The vendor mails the citations to you within 24 hours.

Electronic searches are only as good as the descriptors you use. Descriptors that are too general will retrieve too many citations and will cost too much. Descriptors that are too narrow will retrieve too few citations. Until you feel adept at specifying descriptors, let a librarian or other expert help. Many journals and all abstracting services provide guides for designating descriptors. If you use their guides, you'll find much of your work done for you.

ERIC is the database of the Educational Resources Information Center of the National Institute of Education. It indexes nearly 800 periodicals in education. Searching that index involves five steps:

1. Identify your topic in your own terms, as described above.
2. "Translate" your topic into ERIC descriptors or key words. These are contained in the ERIC thesaurus that is shelved with the index itself.
3. List the best descriptors for your topic, and locate them in the main part of the ERIC thesaurus, called the "Alphabetical Descriptor Display." It contains more than 5,000 descriptors. You may find a dozen or more relating to your topic. ERIC instructions help you narrow the list further.
4. For a manual search, take your list of descriptors to the subject index of *Resources in Education* and *Current Index to Journals in Education* to find article titles relevant to your search.
5. For a computer search, follow the directions of your library or information retrieval system.

Key words also come into play at the other end of your project. When you prepare the abstract of your journal article, thesis, or other report that may be included in an electronic data base, you'll be expected to supply key words. That's how the circle gets closed. Key words listed by searchers starting new projects match up against key words listed in data banks by researchers completing old ones. Having used the key word approach to finding relevant articles, you probably have a good list of key words for inclusion in your abstract.

Panel 3-8 explains the Computerized Literature Access Search Service (CLASS) at Morgan Library on the Colorado State University campus. Most university libraries have a similar system. In many libraries, librarians do the searching. But the systems are advancing quickly, and patrons will be doing electronic searches at terminals. In fact, owners of home personal computers were subscribing to selected data bases in the mid-1980s.

Many college and university libraries also belong to national data bases of several libraries' card catalogs, such as the Research Libraries Information Network (RLIN) or the Ohio College Library Center system (OCLC). RLIN member libraries provide tapes of their card catalogs and new acquisitions to a central computer at Stanford University. If your library belongs to RLIN or a similar system, you or a librarian can conduct a computerized search of the card catalog similar to the searches of the electronic data bases of the periodical literature. By identifying your subject and related terms and generating a list of key words, you can search the computerized holdings of not only your library, but member libraries as well. The on-line listings identify the libraries that have the publication. If your library does not have the publication, you can request it through interlibrary loan (discussed below).

Serials—Journals, Magazines, and Periodicals

Serial publications include

1. technical and scientific research journals
2. trade magazines, newspapers, and newsletters
3. popular magazines and newspapers

Technical and scientific research journals report the authors' research. They include such publications as *Wood Science, Agronomy Journal,* and *IBM Journal of Research and Development.* Trade journals, magazines, and newspapers report the practical aspects of professional fields. Examples include *Engineering News Record, Nursing87, Farm Journal, Barron's,* and the *Wall Street Journal.*

Newsletters—usually two- to 12-page mini-newspapers devoted to a specialized, extremely limited topic—report recent advances, often in a

PANEL 3-8. Electronic Searches

1985/86
Engineering Sciences
Computerized Literature Access - Search Service

CSU Libraries has CLASS! Through CLASS, you can have literature searches done on-line with a computer to identify references to magazine and journal articles, books, conference papers, and technical reports.

THE SERVICE: Literature searching once meant spending hours or days manually combing volumes of printed indexes by subject or author for relevant citations. Today the computer has eliminated much of the drudgery. Most of the major indexes to the literature are now available for searching on-line. CSU Libraries has staff members trained to search more than 300 data bases on various subjects. Most searches take from 10 to 20 minutes on-line; citations ordered off-line arrive in about three days.

THE COST: The total cost includes (1) computer connect time charges (based on an hourly rate that varies from base to base); (2) telephone charges; (3) charges for citations printed off-line or online (based on a per citation rate that varies from base to base); (4) University overhead charges; and (5) a small library charge. Searches may be paid for by check, IMO, or cash. The cost cannot be charged to a student account.

HOW TO ARRANGE FOR SEARCHES: Contact <u>Nancy Ellison</u>, Room 214, Morgan Library 491-1877 or Room 208B, Engineering Research Center 491-8475 OR <u>Earlene Bradley</u>, Room 302A, Atmospheric Science 491-8532 or Room 211, Morgan Library, 491-1877. Before the appointment, please prepare a concise statement of the subject of your search, including a list of important keywords and possible synonyms.

SAMPLE SEARCH:
    Statement: Modeling of evapotranspiration in watersheds
    Synonyms and Keywords: model..., mathematical model..., simulation, evapotranspiration, watershed
    Cost: If conducted in WATER RESOURCES ABSTRACTS, requiring 10 minutes on-line, and resulting in 100 citations printed off-line, the cost would be:

$ 7.50	Computer connect time (at $45/hr.)	
$15.00	Off-line prints (at $ .15 each)	
<u>$ 1.33</u>	Telephone & communications network charges	
$23.83		
$ 3.57	*15% University overhead	
<u>$ 6.00</u>	*Library charge	
$33.40	TOTAL COST OF SEARCH	

    *Higher for non-CSU users

LIMITATIONS: Sounds wonderful? It is, if you understand what you're getting and what the limitations are. Some searches must be a combination of computer and manual searching, since entries on the computer go back only a certain number of years; you are limited to what is covered by the data base; relationships cannot always be specified: for example, a search on the "effect of A on B" will also retrieve the "effect of B on A"; searches produce citations to the literature, not the literature itself, which may be difficult to locate and acquire. During the search interview, these and other limitations will be considered before a final recommendation is made.

CONTINUOUS UPDATING SERVICE

    In addition to your initial search, arrangements can now be made for automatic monthly updates based on your established interest profile.

SEE OVER FOR A SELECTED LIST OF ENGINEERING/TECHNOLOGY RELATED AVAILABLE DATA BASES

PANEL 3-8. Continued

SELECTED LIST OF ENGINEERING/TECHNOLOGY RELATED DATA BASES

Name of Base	Corresponding Printed Index	Subject(s) Covered	Years	Conn.	Print	Citations in Base
AEROSPACE DATABASE	International Aerospace Abstracts, Scientific & Technical Aerospace Reports	Aerospace	1984–	$78	$.25	21,000
AGRICOLA	Bibliography of Agriculture	Agriculture	1970–	$35	$.10	2,826,000
FLUIDEX	BHRA – 11 publications	Fluid Engineering	1973–	$69	$.28	158,000
CA SEARCH	Chemical Abstracts	Chemical Sciences	1967–	$76	$.28	6,840,000
CAB ABSTRACTS	CAB – 26 publications	Agriculture, Irrigation, Forestry, Vet. Sci., etc.	1972–	$55	$.35	1,760,000
COMPENDEX	Engineering Index	Engineering	1970–	$99	$.47	1,415,000
DOE ENERGY	Energy Research Abstracts	Energy	1974–	$40	$.15	1,470,500
FOOD SCIENCE AND TECHNOLOGY ABSTRACTS	Food Science & Technology Abstracts	Food Technology	1969–	$75	$.25	281,000
GEOREF	Bibliography & Index of Geology	Geosciences	1927–(N.A.) 1961–(World)	$84	$.30	1,005,000
GPO MONTHLY CATALOG	Monthly Catalog of U.S. Government Publications	U.S. Federal Documents	1976–	$35	$.10	227,300
INSPEC	Science Abstracts: Physics Abstracts, Computer & Control Abstracts, Electrical & Electronic Abstracts	Physics, Electronics, Computer	1969–	$96	$.46	2,494,000
ISMEC	ISMEC Bulletin	Mechanical Engineering	1973–	$84	$.40	174,500
METEOROLOGICAL & GEOASTRO-PHYSICAL ABSTRACTS	Meteorological & Geoastro-physical Abstracts	Meteorology & Geoastro-physics	1972–	$95	$.15	110,500
NTIS	Government Reports Announcements	Technical Reports	1964–	$48	$.20	1,122,000
POLLUTION ABSTRACTS	Pollution Abstracts	Pollution	1970–	$84	$.40	110,000
WATER RESOURCES ABSTRACTS	Selected Water Resources Abstracts	Water Resources	1968–	$45	$.15	176,000
WILSONLINE	Applied Science & Technology Index	Science, Technology	To be announced 1985			

Prices subject to change without notice

(Source: Used with permission of Morgan Library, Colorado State University, Fort Collins.)

matter of days or weeks, in technical, scientific, and business fields. Many newsletters are free, but some cost $100 or more annually. Usually without advertising, often reproduced inexpensively, newsletters report developments and advances of interest to their limited readership. Readership numbers from a few hundred to several thousand. Therefore, libraries often limit their newsletter collections. To find those relevant to your field, check the *Ulrich's* family.

Popular magazines and newspapers often carry articles on technical and scientific subjects. The authors and editors translate and explain the subject for readers with little, if any, specialized background. Such publications include *Today's Health, Omni, Newsweek, Time, Reader's Digest,* and *National Geographic*. When technical and scientific subjects are carried, the stories usually relate the social implications of the subject, such as the impact a technical issue will have on society and the world.

Books and Publications

A major portion of library holdings consists of books and other lengthy publications. Their contents vary widely. Some books consist of compilations of articles. Some carry condensations. Some record a single project in depth. Some present raw data and standards.

Government Documents

Many libraries have collections of government publications either in their regular stacks or in separate stacks. Larger libraries house government publications separately and may have separate card catalogs for them, too.

The United States Superintendent of Documents publishes the *Monthly Catalog of U.S. Government Publications* (Panel 3-9). It lists the federal government's publications and contains both an index section of author, title, subject, series/report, classification number and stock number indexes and a catalog section with a bibliographic description.

The Congressional Information Service, Washington, D.C., publishes the *American Statistics Index* (ASI) and the *Index to Publications of the United States Congress* (CIS). ASI identifies government statistical data, catalogs the publications, provides bibliographic information on the publications, and announces new publications.

Other guides to government publications include John L. Andriot, *Guide to U.S. Government Publications;* Walter L. Newsome, *New Guide to Popular Government Publications;* and Ivan Watters, Jr., *Index to U.S. Government Periodicals*. David W. Parish's *State Government Reference Publications* provides a starting point for identifying literature published at the state level. Government publications also include specialized abstracts such as *Energy Abstracts, Pollution Abstracts, Nuclear Science Abstracts, Scientific and Technical Aerospace Reports,* and *Government Reports Annual Index.*

GATHERING INFORMATION AND IMPROVING CONTENT

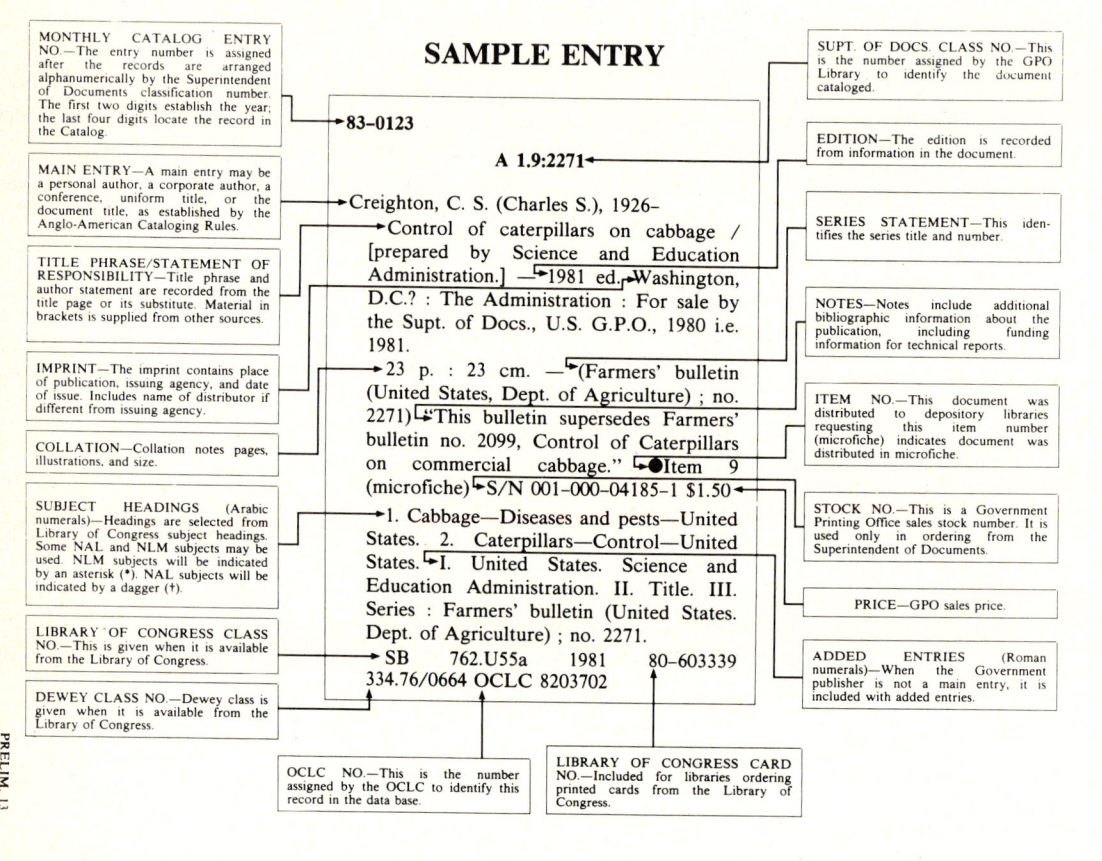

PANEL 3-9. Monthly Catalog of United States Government Publications

If you're searching for Canadian documents, two sources will come in handy. The *Catalogue of Government of Canada Publications* provides the key to locating government publications. *Canadiana*, a national bibliography, lists all works of Canadian publishers and citizens published since 1950.

When you're searching for government publications, remember that different agencies cover the same topics and that agencies change their names. If you're having trouble, seek help from the government documents librarian.

Special Collections

Many libraries maintain collections of maps, musical scores, manuscripts, phonograph records, papers of famous people, and other special materials.

Electronic and Photographic Forms

Advances in electronics and photography and the proliferation of information have encouraged libraries to develop and use alternative storage forms, such as microfiche, microfilm, videotapes, and videodiscs. Such forms need little shelf space. One 4- by 6-inch microfiche card easily holds a 75-page report; a 4.5-inch diameter videodisc can hold the contents of over 300 books.

Advances in personal computers and modems—telephone connections between computers—may even enable you to visit your library through your home computer. But to use the computer effectively, you need to understand your field, its literature, and that literature's organization. You learn that by doing manual searches. Only then can you devise effective strategies for using other information storage techniques.

Interlibrary Loans

Once you've moved from your broad topic to a narrowly defined statement, you may find that your library doesn't have the publication you need or that the publication is checked out or missing. If it's checked out, ask the checkout librarian to have it returned. If it's missing or your library doesn't own it, ask your library to borrow a copy from another library or obtain a photocopy for your use. To get an interlibrary loan, you'll need to fill out a request form.

Be sure to check your library's interlibrary loan request forms to see what information you'll have to provide. As a minimum, record the following:

- Author's or editor's first name, middle initial, last name
- Full title of magazine and journal articles
- Full title of books, journals, magazines, and reports
- Volume, issue, page number, year, month, and date
- Volume number or edition
- Addresses of publishers, if given
- Where you found the reference to the article or publication
- Call number of the publication

If you don't form the habit of recording all this information the first time you encounter it, you'll later find yourself returning to the library to dig it out. In the long run, you'll save time and work by collecting everything the first time.

Interlibrary loans can take from a few days to several months. Your library must first determine which libraries have the publication you want and then request that the owning library lend your library the publication or provide photocopies. Libraries often provide photocopies of single articles for educational purposes, as permitted under copyright law. Your library should be able to obtain a book for one, two, or three weeks. Some libraries charge fees for interlibrary loans; others do not.

DEVELOPING A LITERATURE SEARCH PLAN

Librarians suggest that you develop a plan for searching for the literature you need. Start with a broad focus, and narrow it as you work. To solve a problem, first search the literature to develop an overall understanding of the subject, and then focus on a specific, narrowly defined question. As problems confront you on the job, you probably will not have enough time to acquaint yourself with all the library resources. Learn to use them while you are still in school, and you will speed your problem solving on the job.

After helping students and professionals find literature on thousands of topics, Professor Toni Lueck, Colorado State University's physical science librarian, suggests that you keep a permanent three-ring notebook of your literature reviewing notes. There you can record complete citations and call numbers for general references, abstracts, indexes, government publications, books, journals, magazines, and other relevant literature.

Write Your Question

When you face a specific literature reviewing assignment, Lueck suggests that you write down your research question or problem statement. Then list the broad areas you will find the topic under and the question's specific terms. Next add synonyms and related terms. The specific terms, synonyms, and related terms provide the key words for beginning your search. Try to think of as many key words as possible. As you search the literature and learn related terms, add them to your list.

Once you have your question written down and preliminary list prepared, ask classmates which terms they would add to your list. Ask teaching assistants and professors to review your list and suggest other terms to guide your search. Asking others will turn up useful terms for literature reviewing and will help you to find additional references.

Asking Librarians for Help

If you are doing your first major search, spend some time with your subject librarian to learn how to use specialized publications in your field. Have the librarian review your research question and list of terms. Most librarians can suggest other useful sources for you to search. Keep good notes to guide subsequent searches.

Should you plan a computer search? If you're writing a term paper, Lueck says that a 10- to 20-hour manual search probably will produce sufficient information. If you're a graduate student reviewing literature for your thesis or dissertation, expect to invest at least 80 to 100 hours in both computer and manual searches. For major projects, plan on a computer search to supplement your manual search. If you opt for a computer search, first conduct enough of a manual search so that you understand the scope of your subject. In that way, you can focus tightly on helpful literature.

Plan and conduct your search early in the term so that you'll have enough time to obtain the necessary literature. Browse through the stacks too. Check the titles and subjects near the materials you're seeking. Although not recommended as a primary searching technique, browsing often turns up valuable materials or alternative subjects to review.

EXECUTING YOUR SEARCH

With your question, list of key words and terms, and notebook in hand, review *Ulrich's* and *Sheehy's* publications to determine the references, abstracts, indexes, government publications, and other support literature that you'll need to search. Check *Books in Print* and its supplements to determine the most recent books on your subject. Review the subject section of the card catalog covering your topic. Take detailed notes as you work.

Taking Notes

As you review publications, try using 3- by 5-inch or larger note cards. Punch a hole in one corner of each card, and put them on a ring. The ring keeps your cards together and organized—even if they drop to the floor. Take both bibliographic notes and content notes. For bibliographic notes, take all the information required for the citation style you'll be following. That style will vary according to the fields in which you study. For example, if you write a research paper for a literature class, and your professor is a traditionalist, you will probably use the footnote system, which uses a number raised above the line of text, and the full citation plus any comments inserted at the bottom of the page, the end of each chapter, or the end of the work. To insert footnotes at the bottom of the pages, you must calculate the number of lines the footnote will occupy and leave adequate space at the bottom of the appropriate page.

In the humanities, as in the social, biological, and physical sciences, the more economical author-date system is preferred increasingly. Under that system, the author's last name, followed by the work's year of publication, appear in parentheses—as (Jones 1987), for example. Then the full citation appears at the end of the manuscript. Panel 3-10 illustrates the author-date citation style. Under still another system, footnote numbers appear in parentheses, and the full citation appears at the end of the manuscript under that number. Before you begin collecting data, familiarize yourself with the system your finished paper will require. Study the footnotes, references, bibliographies, and direct and indirect quotes in the material you look at. They will act as reliable guides to citation form and style.

For content notes, you can

1. check out the article or publication and take notes at home,
2. read the article in the library and take the needed notes
3. photocopy the article for personal use, take it home, and extract your notes.

When working in the library, you'll need to take notes in longhand. At home you can take notes by hand, typewriter, or computer. Each note-taking method helps you to learn the material, provides you with details

PANEL 3-10. Author-Date Citation Style

Check carefully for the citation style most common in your field. The author-date style involves listing authors and dates of publication thus:

Popularizing technical topics by using people helps communicate your message. For a television news story on the impact of heavy rains on planting, interviewing farmers and weather experts helps tell the story (Shook, 1982). For print media, Zinsser (1980) recommends using your own experiences.

Then, at the end of your manuscript, list the full citations. Here's a list, alphabetized by the authors' last names:

Shook, F. 1982. *The process of electronic news gathering.* Englewood, Colo.: Morton Publishing.
Zinsser, W. 1980. *On writing well.* New York: Harper & Row.

The complete, alphabetical list may be called, "References," "References Cited," "Literature," "Sources," "Bibliography," or some other term. Specifics of citation styles often depend on the type of sources. Furthermore, some styles group all citations together while other styles group citations by books, articles, technical reports, or other types of sources.

For more guidelines, check the reference desk at your library for two authoritative style guides to technical and scientific writing:

CBE Style Manual Committee. 1983. *CBE Style Manual.* 5th ed. Bethesda, Md.: Council of Biology Editors.
University of Chicago Press. 1982. *The Chicago Manual of Style.* 13th ed. Chicago: University of Chicago Press.

Both manuals give excellent advice on citation styles. The *Chicago Manual* explains another citation style, a numbered system of endnotes.

> ### PANEL 3-11. Paraphrasing Literature
>
> When you take notes, paraphrasing provides an efficient way of condensing essential information. In fact, few technical articles use extended direct quotes. Instead, the authors condense, paraphrase, and summarize information. To paraphrase an article, read it through, and then ask yourself:
>
> - What are the author's major points?
> - What are the study's major findings?
> - What material can I delete and still make my point?
> - What information do I need?
> - How can I best summarize the essential points?
>
> Answer the questions in your own words. Draft a narrative of your answer, and edit it tightly. Avoid, if possible, using the author's style of expression, thinking, or writing. If you don't paraphrase the material in your own words, your report will read like someone else's writing. As Chapter 18 warns, you must always avoid plagiarism and copyright violation.

for future use, and helps you summarize information. When taking notes, paraphrase the original's narrative, as suggested in Panel 3-11.

Experienced library researchers often argue against photocopying material, unless you have some special reason to do so, such as living far from your library. Taking notes forces you to evaluate the material, but photocopying does not. Sooner or later you must decide if the material is really important. Why not do it on first reading?

If you need direct quotes, copy passages carefully. A photocopy can help to reduce transcribing errors. You should have a clear reason for quoting directly. Is the language so precise that you cannot paraphrase it? Will the direct quote convey a special flavor or character?

Work from General to Specific Literature

Implement your literature review plan. When you find a potentially useful article, scan the abstract, title, and headings to determine if it's worth reading. If it is, read it carefully; check its references and citations. By carefully checking cited literature within an article, you may find more useful sources and terms to search out. Be sure to evaluate the article as you read it.

Judy Berndt, social science librarian at Colorado State University, says that students often look up publications in the card catalog, go to the stacks, and find the publication missing. Then they take anything that's available. Too often what they take fits their topic poorly. To evaluate publications initially, Berndt suggests that you ask:

- Is this the best available information?
- What are the author's qualifications?

GATHERING INFORMATION AND IMPROVING CONTENT

PANEL 3-12. Evaluating Literature

When you read literature, ascertain its professional quality by asking:

- Where were the data collected?
- When were the data collected?
- How were the data collected?
- Who collected the data?
- Do the authors place any limitations or restrictions on the data? If so, what are they?
- What are the stated or unstated assumptions behind the study or report?
- How were the data analyzed? What influence does that have on the conclusions?
- Were enough data collected or was the sample size adequate?
- Can you detect any gaps in the data or flaws in the methods or approach?
- Are the conclusions valid based on the overall methodology and analysis?

(Source: Adapted with permission from W. W. Shaner, W. R. Schmehl, and P. F. Philipp, Farming Systems Research and Development: Guidelines for Developing Countries, Appendix 5-N (Tucson, Ariz.: Consortium for International Development, 1982).)

- How much of an expert is the author?
- What is the publication's audience?
- Do I know enough to evaluate the content?

If the publication appears valuable, read it and then use the criteria in Panel 3-12 to determine whether you should use its information in your papers, reports, or presentations.

When you consider literature for citation in communications to experts, use literature of the appropriate level. Mass circulation magazines will help you understand a topic and focus your attention, but they are seldom appropriate for communicating with experts. In preparing an article for engineers experienced in wind electrical generation systems, seldom would you cite general circulation magazines, such as *Popular Science, Reader's Digest,* or *Mother Earth News,* as authoritative sources. To determine the appropriate literature to cite, study the publications that your audience regularly reads and the publications' technical styles.

LIBRARY USE BEYOND GRADUATION

Knowing how to obtain the most from a library becomes more valuable as your career advances. Librarians now estimate that from 1 to 2 million technical and scientific articles, reports, books, and patents are published worldwide each year. Therefore knowing how to find, evaluate, and use technical literature in your field will be extremely important to your career.

USING THE LIBRARY AND REVIEWING LITERATURE

Professors Arthur A. Gerstenfeld and Paul Berger (1980) of the Worcester (Mass.) Polytechnic Institute made a study of 300 scientists working for Hewlett-Packard, Exxon, Xerox, Monsanto, and G. D. Searle. Gerstenfeld and Berger found them relying heavily on their company libraries, or, as they are called in industry, technical information centers (TICs). The libraries at large research and development plants, Gerstenfeld found, perform several functions. They provide weekly literature review bulletins for current awareness, issue monthly technical information bulletins with annual indexes, help professionals carry out literature searches, and assist in retrieving internal reports. Karen Takle Quinn (1982), senior librarian for IBM, calls the library an "intellectual switching center." In her view, libraries are no longer only storehouses, and today's librarians actively participate in helping professionals find answers. Learning your library and knowing your librarian can help you become a more effective professional.

HIGHLIGHTS

1. Library resources include: libraries and reading rooms across campus; subject specialist librarians; general references; abstract journals and indexes; computerized bibliographic data bases; books and publications; government documents; electronic and photographic forms of recording information; and interlibrary loans.
2. When planning your literature search, focus broadly and then narrow down to a specific research question. Write out the question, key terms, and key words. Ask classmates in your major, teaching assistants, and professors to review your question and suggest key terms for your literature search. Allow enough time for conducting your search and obtaining the needed documents.
3. To identify publications relevant to your subject, begin your search by reviewing Sheehy's and Ulrich's references. Take careful bibliographic and content notes. Note taking helps you to learn and summarize material. When you begin to review an article, first review the abstract, title, and headings to see if you should consider it further. If so, read it, and evaluate it carefully to determine its value to your literature review. Paraphrase and quote carefully.

PROJECTS

The following projects will help you to build a reference notebook for library searches. With a three-ring notebook, complete the following projects according to procedures described in this chapter.

1. Prepare an annotated bibliography for at least six major reference documents that cover the field of your future occupation. Include

abstract journals, indexes, and government publications. An annotated bibliography consists of a complete citation to relevant articles and a summary or abstract of the article or publication.

2. Review your library's electronic data bases, and prepare an annotated bibliography of the relevant electronic data bases for your field.

3. Prepare an annotated bibliography covering at least three major research journals, three trade publications, and three popular publications that carry articles in your field.

4. Select a research journal article appropriate to a major report or paper on which you're working this semester. Using the guidelines provided for evaluating literature, analyze the article, and write a description of your evaluation.

5. Using the guidelines provided in earlier chapters, prepare an analysis up to 1,000 words long of one article from a trade publication appropriate to your field. Specify your audience, such as members of an introductory or advanced course.

FOR MORE HELP

GATES, J. K. 1983. *Guide to the use of books and libraries.* New York: McGraw-Hill.
Gates provides details on using libraries in dozens of references.

4
Collecting Information in the Field

OVERVIEW

In Chapter 4, we discuss data collection in the field from the communicator's perspective, including

☐ Observing
☐ Interviewing
☐ Conducting surveys of large groups

The library and the controlled laboratory experiment form two great sources of information. But another source provides still further information: the world at large, peopled by experts, by those who use or are affected by your technology, and by those who contribute to the quality of your communication.

Observing, interviewing, and surveying techniques can supplement your library use. Observing, interviewing, and surveying, as this chapter shows, will help you gather information from people. Direct observation will help you to gather information about many technical and scientific topics. By using these techniques in your research, you can get reliable and useful information for your reports.

LEARNING BY OBSERVING

From December 1831 to October 1836, Charles Darwin served as naturalist on the H.M.S. *Beagle* and studied at firsthand plants, animals, geology, and natural history. His trip took him from the South American coast and nearby islands to Australia, New Zealand, and the Azores. He observed carefully, took detailed notes, returned to England, and reflected on his data for some 20 years before publishing his *Origin of Species*. Historian of science George Sarton (1948) tells us that Leonardo da Vinci—the great Renaissance artist, engineer, and scientist—would begin to read and an idea would cross his mind. He would abandon his reading for observation, experimentation, and design. Sarton says that people have not yet found a better way than observation to discover nature's secrets.

Observation plays a critical role in nursing, construction, engineering, child development, nutrition, law enforcement, business, and dozens of other professions.

WHAT IS OBSERVATION?

Observation includes all techniques that you use to see, touch, smell, and measure things and concepts. In *Research Methods in the Social Sciences* (1959), sociologists Claire Selltiz, Marie Jahoda, Morton Deutsch, and Stuart Cook point out:

> Observation becomes a scientific technique to the extent that it serves a formulated research purpose . . . is planned . . . systematically . . . is recorded systematically . . . and is subject to checks and controls on validity and reliability. (Selltiz et al. 1959, 200)

Good observation can help you to solve the day-to-day problems in your field.

Too often our observations are flawed by factors we fail to consider. For example, agricultural specialists working with a group of African farmers noticed that cassava—a root crop—had a mosaic virus that reduced yields. Upon questioning local farmers, the specialists learned that the farmers had never seen virus-free cassava. The farmers, having never seen healthy cassava, assumed theirs was. They couldn't observe accurately because they had no frame of reference for virus-free cassava (Shaner, Schmelh, and Philipp 1982).

Test your observational skills by looking at Figure 4-1. What, if anything, do you see? Few people are inherently good observers. But they can become better by recognizing observational barriers and learning how to overcome them.

Distortions

William Rivers summarized the work of the early psychologists Gordon Allport and Leo Postman and reported that distortions include

> *Leveling* . . . [when the] description becomes shorter, more concise, more easily grasped and told. A companion effect is *sharpening:* the selective perceptions, retention, and reporting of a limited number of details from a larger context. Then comes *assimilation:* absorbing new information in a way that will not disturb the habits, interests, and sentiments in the subject's [viewer's] mind. These three processes occur simultaneously and represent an effort to reduce a larger amount of information to a simple and meaningful structure . . . this is . . . *embedding.* (Rivers 1975, 60)

FIGURE 4-1. Test your observational skills.

Do you see the Dalmatian in the top illustration? If you have never seen a dog, would you be able to see the Dalmatian? Probably not—you would not have a visual frame of reference. Do you see a vase or two people in the bottom illustration? Concentrate on the white and you see a vase. Concentrate on the black and you see the profile of two people. What you look for determines what you see. *(Source: R. L. Gregory,* The Intelligent Eye *(New York: McGraw-Hill, 1970) pp. 14, 16.)*

Thus we tend to find what we are looking for and to look for what we can find. We tend to pay *selective attention*. Thomas Kuhn (1970) points out that what an individual brings to a problem often determines how he or she proceeds to work on the problem.

A fifth factor, *familiarity,* distorts observations, too. For example, community development specialists know that some small town residents have this observational problem. To counter it, the specialists assign an out-of-towner to photograph decay in the town—broken streetlights, ragged awnings, and so forth. At a town meeting, specialists show the photographs, and the residents react with dismay. The camera's fresh eye has given them a new, and undistorted, look at familiar surroundings.

A sixth factor, *reactions to observing,* may change behavior. For example, some years ago two small boys watched a shoplifter filling her purse with cosmetics in a supermarket. Then she turned and saw a 7- and 8-year-old watching her intently. She returned the cosmetics to the shelves. Similarly, an intently watched dog or zoo animal reacts to observation by watching its watchers. To observe subjects in their natural state, you must find a way to blend unobtrusively into the surroundings.

Developing Good Observation Techniques

Observing well requires knowing what you're looking for, recognizing common distortions, and overcoming them. Begin with a clear problem statement, definition, and description of your variables or what you want to observe. Then carefully describe how you'll observe the variables.

Overcoming Observational Barriers

When you observe, develop a questioning attitude toward what you observe. To overcoming leveling, ask:

- What details am I leaving out?
- What details should be included?

To overcome sharpening, ask:

- What details am I stressing?
- Should I be stressing other points as well?
- What am I overlooking?

To overcome selective perception, ask:

- What details am I recording?
- What details am I deleting?
- How does my personal bias reflect my observations?

To overcome familiarity, ask:

- Can I observe from different perspectives?
- How would someone with a different perspective observe the same thing?
- Can I focus on different elements to improve my observations?

To overcome reactions to observation, ask:

- How can I observe unobtrusively?
- Can I observe artifacts that a subject has left behind?

INTERVIEWING

Although documents may be missing from the library or just plain wrong, and observations may be subject to misperceptions, the interview can be even more perilous. But professionals must use the personal interview even so. The interview, say Eugene Webb and Jerry Salancik (1966), is like the roulette wheel in a western town. "Isn't it crooked?" asks the newcomer. "Oh sure," replies the old-timer. "Then why do you play?" "Because it's the only game in town." (p. 1)

The interview is sometimes the only game in town. Often it's the quickest way to gather information. It can tell you if your new computer manual is helpful to your customers, if your solar energy panels are economically competitive, if a hundred other aspects of your research and development programs are feasible.

But interviews do take time, and taking people's time in an interview diverts them from their own work. Therefore, one assumption underlies the advice that follows: when you interview colleagues, you owe them the most efficient possible use of their time and brains.

Ten Steps to Productive Interviews

After 20 years of reporting and writing, Ken Metzler went to the University of Oregon to teach writing and reporting. He quickly found that his students were intimidated by people in positions of power, such as police chiefs, city managers, and other officials. So Metzler wrote *Creative Interviewing,* one of the best books on interviewing techniques. To research the topic, Metzler interviewed more than 150 public figures in Hawaii and Oregon on their impressions of interviewers. He talked with many professional interviewers. Most said they had learned the craft by trial and error.

"Experience may be a good teacher," said one, "but it's a damned slow one" (Metzler 1977, xi).

Although Metzler wrote his book for young journalists, his points apply equally well for technical and scientific professionals. Metzler identified 10 phases in the interview process, which we paraphrase:

1. *Define the Purpose*. Knowing the information you need guides your preparation for the interview. Write down what you want to know.

2. *Prepare Yourself as Thoroughly as Possible*. Thorough preparation lets you ask good questions, and your respondent, favorably impressed, will help you rather than brushing you off.

3. *Ask for an Interview Appointment*. Because people are busy and may resent unannounced visitors, phone ahead and request a time to interview the person with whom you need to talk. Explain the details of your topic so that the person has time to consider your probable questions. Doing so increases the chances that your interview will provide you with useful information. When asking for an interview, indicate how much time you will need.

4. *Plan Your Interview in Advance*. At a minimum, list the detailed topics you want to cover in the interview. If you feel more comfortable writing out and ordering your questions prior to the interview, do so. It will help to structure your interview.

5. *Let the Respondent Get to Know You*. According to Metzler, the person being interviewed judges the interviewer's sincerity, trustworthiness, and professional competence in the first few minutes of the meeting. Small talk, says Metzler, identifies one "as human rather than mechanical." A few minutes' talk about the surroundings or about finding your way to the place for the interview will help put you both at ease. But after a few minutes, begin asking questions.

6. *Ask Your Questions*. This segment is, of course, the body of your interview. At first you may find things do not go smoothly, at least from your perspective. Your respondent may still be sizing you up and may be uncertain about exactly what you want. Nevertheless, proceed with your questions. As time passes, you will probably establish rapport.

7. *Collaborate in the Interview*. Establish a sense of traveling together down the same path, and your interview will be productive and comfortable. It's natural to feel pressure when interviewing a professional. If you've prepared adequately, you'll soon be into the critical questions. Do not ask elementary questions that your preparation should have answered. But do ask for a clarification if you do not understand something. Asking questions about things you do not understand helps establish rapport with an expert. As the Chinese proverb puts it, "He who asks is a fool for five minutes. He who does not is a fool forever." Do not become so intent on your next question that you lose track of the present one. You may overlook natural follow-ups. By all means, use your list so that you'll know when you've finished your interview.

8. *If you Have to "Drop a Bomb," Do It in Stages*. If part of your interview focuses on a view not held by your respondent, or your evidence does not support your respondent's approach, you will have fewer problems if you have established rapport early in the interview. Whatever the information you have to impart, let it down a piece at a time instead of all at once. Professionals know that not everyone agrees with a particular approach. If your research requires you to tell your respondent that you give an opposing view equal weight, try to break the news diplomatically. Instead of saying, "Professor X disagrees violently with you, and I tend to think he's right," try a different approach. "I know you're aware of Professor X's views," you might say. "And I wonder what you think about them?" The open-minded approach works best.

9. *If You Drop a Bomb, Try to Recover*. Seldom will you need to "drop a bomb," but when you do, you will have fewer problems if you managed the earlier phases well. You need not apologize, nor must you agree, but do use a sympathetic approach. You should be able to part on reasonable terms.

10. *End the Interview*. When your time is up, say something like, "I see my time is up. May I take a moment to check my notes?" Stop and think whether you have asked all your questions and whether you need points clarified. Ask any final questions you may have. If you've been offered any documents during the interview, ask for them now. Ask if you can check back after you've reviewed the material or if you need to clarify a point. Then give a simple thanks, and be on your way.

Note Taking

When you begin interviewing, take notes with pen and paper, and keep your pen moving while the respondent is talking. That method allows you to take notes matter-of-factly when the respondent says something of importance. If you let your pen remain idle and something noteworthy comes along, you risk startling the respondent. Your respondent may retract the statement. Note taking helps you commit what you hear to paper and reinforces what you are learning. You begin to evaluate what you are being told, and that saves time later on.

You may feel tempted to use a tape recorder rather than take notes. Some professionals who interview regularly argue that tape recorders capture a complete and accurate record of the interview. If you want to use a tape recorder, ask permission when you call for an appointment. Tape recorders have their drawbacks. Some professionals who interview regularly argue that tape recorders inhibit respondents. When you tape record an interview, you'll probably need to transcribe it, a time consuming process. Units break down, batteries die, and tapes run out. Such disruptions intrude on the interview's continuity and, possibly, on your rapport with your respondent. Some interviewers find it difficult to pay attention to the tape recorder, to ask questions, and make backup notes at the same time.

GATHERING INFORMATION AND IMPROVING CONTENT

> ### PANEL 4-1. Common Barriers to Interviewing
>
> From their review of 150 studies of conducting interviews in law, journalism, and social science, Eugene J. Webb and Jerry R. Salancik produced *The Interview, or the Only Wheel in Town.* They identified factors that inject bias and negate information obtained through interviews:
>
> - Location, which can distract interviewer or respondent
> - Presence of other persons, who may inhibit responses
> - Interviewer characteristics (such as competence in the respondent's field)
> - Precision of questions asked
> - Selective memory of the respondent
> - Knowledge and willingness to answer of the respondent
>
> To avoid such pitfalls, interview respondents in a place where neither of you will be distracted, be well prepared, ask precise questions, probe, and clarify answers.
>
> *(Source: Adapted from Eugene J. Webb and Jerry R. Salancik, The Interview, or the Only Wheel in Town, Journalism Monographs no. 2, ed. B. Westley (Columbia: University of South Carolina College of Journalism, 1966).)*

If you opt to use tape recorder, practice with it beforehand, make sure its batteries are fully charged, and take along enough tape to complete the interview. Use a tape recorder with a counter, take backup notes, and record the counter number on key points on your interview notes. That trick may save you from transcribing the whole interview.

After you've left the interview, review your notes as soon as possible. If you interviewed someone in your facility, spend the first few minutes after you return to your desk reviewing and expanding the notes. If you've driven somewhere for the interview, review and expand your notes before starting the return journey. If your respondent refused to let you take notes or use a tape recorder, spend a few minutes after the interview committing to paper what you heard.

As you review your notes, whether during the interview or afterward, list new questions that arise. Then call the respondent and ask for clarification. Such a call can reinforce your reliability. The respondent also may have thought of something further to tell you.

Panel 4-1 identifies common interviewing problems. When used carefully and skillfully, interviews provide good information that sometimes can be gotten in no other way.

SURVEYS

Salomon Brothers, a New York investment banking firm, surveys 22 major oil and gas companies and 64 independent producers about their exploration plans for the coming year. The University of Vermont's Agricultural Experiment Station mails a questionnaire to learn the characteristics, attitudes, and participation levels of resident and nonresident hunters. Honeywell, Inc., of Minneapolis, Minnesota, surveys production engineers about needed improvements in maintenance for textile plants. A forester surveys landowners about their timber management practices. A graduate student in nutrition mails a questionnaire to nutritionists on their educational needs. A graduate student in wildlife surveys local Audubon Club members about how they learned of a film series. A senior animal science major surveys stable owners about their horse training needs.

Surveys abound. Professionals and students use them. Robert Feber (1980), chair of the American Statistical Association's subcommittee on survey research, defines a survey as "a method of gathering information from a number of individuals, a 'sample,' . . . to learn something about the larger population from which the survey was drawn" (p. 3). We're speaking here, not about the type of surveys civil engineers do with level and transit, but about attitudes, opinions, and information that people may have.

Without a basic understanding of surveys, you can easily go astray. Done improperly, surveys waste time, money, and effort. If you understand the survey process, you can conduct useful surveys and evaluate data from other surveys.

Should You Survey?

Some years ago, a group of public relations majors planned a public relations campaign for a university department. The students drafted a questionnaire about the professors' teaching loads, sizes of classrooms, equipment, facilities, and department needs. The students distributed the questionnaire to the faculty, but most questionnaires were returned with terse comments. Why? All the answers to the students' questions had been succinctly presented in the department's 5-year self-evaluation, which the faculty had completed the previous semester. Before you survey, ask, "Is the information I need available from other sources?" Often it is. When you examine existing information, evaluate its quality according to the criteria in Panel 4-2. If the information proves reliable, you need not survey. Alternatively, existing information may give you insight into material you'll need to obtain in your survey.

Survey costs vary depending upon methods used, length of questionnaire, required support staff, computer analysis, and other expenses. Recent costs for professional surveys have been $15 per respondent for self-administered and mail questionnaires, $25 for telephone interviews, and $40 for personal interviews (O'Keefe 1985). If you, as a student, use a survey, you can cut costs by doing the work yourself, obtaining free assistance from professors and graduate students, applying for free computer time, and requesting support from your department.

> ### PANEL 4-2. Evaluating Polls and Surveys
>
> Whenever you review a study or report based on surveys, carefully evaluate its content. Professors G. Cleveland Wilhoit, of Indiana University, and Maxwell McCombs, of Syracuse University, recommend asking:
>
> - Who sponsored the survey?
> - Who was interviewed? (Which population is being described?)
> - How were people selected for interviews? (What was the sampling design?)
> - How many people were interviewed?
> - How accurate are the results? (What is the estimated size of the sampling error?)
> - Who were the interviewers?
> - When were the interviews conducted?
> - What were the actual questions asked?
> - How were data tabulated and analyzed?
>
> (Source: Maxwell McCombs, Donald L. Shaw, and David Grey, Handbook of Reporting Methods (Boston: Houghton Mifflin, 1976), p. 86. Copyright © 1976 Houghton Mifflin Company. Used with permission.)

Conducting the Survey

Conducting a survey requires preparing a plan and executing it as outlined in Figure 4-2. Before you begin, familiarize yourself with each step in the process.

Establish Your Objectives. Begin by asking, "What questions or problems do I want to answer when I complete my survey?" List them. Then ask, "What information or data do I need to answer the questions?"

Determine the Type of Survey. You can use self-administered questionnaires, mail questionnaires, telephone interviews, or personal interviews. With self-administered questionnaires, respondents answer questions at a meeting, in class, or at some other gathering. But this form limits the generalizability of your data, a point discussed in more detail later. Mailed questionnaires are sent to the persons sampled; they fill out the questionnaire and mail it to you. With telephone interviews, you or other interviewers call the respondents, ask questions over the telephone, and fill out the questionnaire. With the personal interview, interviewers visit the respondent at work or at home, ask questions face-to-face, and complete the questionnaire.

Each approach has its advantages and disadvantages. Although mail questionnaires are usually the least expensive way to survey a population, some survey researchers reserve mail surveys for populations that have a special interest in the topic. They say that mail surveys are inappropriate for the general public. Not so, argues Donald Dillman, a sociologist at Washington State University. Dillman (1978) says that a proper design

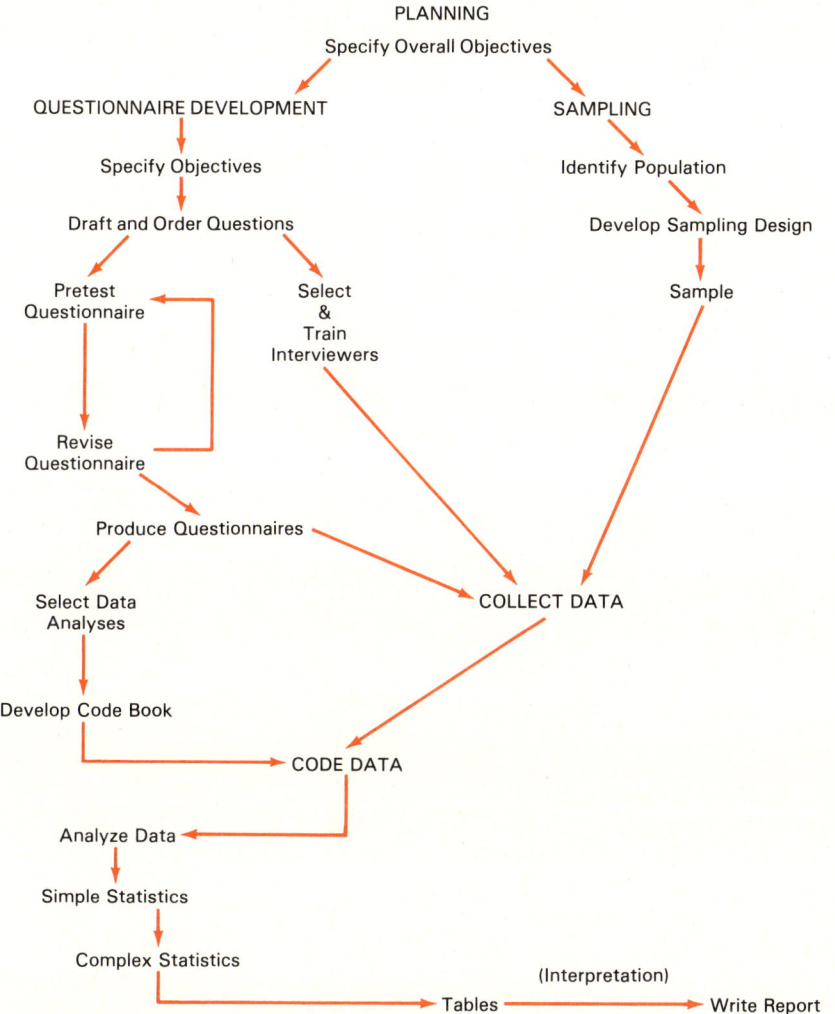

FIGURE 4-2. The flow of survey activities.
(Source: Adapted with permission from presentations by Harry Sharp, Wisconsin Survey Research Laboratory, University of Wisconsin at Madison, August 1974; based conceptually on B. Benjamin, "Statistics in Town Planning," Journal of the Royal Statistical Society 132 (ser. A, pt. 1, 1969), 7.)

overcomes many of the problems associated with mail and telephone surveys. These problems include short questionnaires, low response rate, and slow responses.

Telephone surveys work well if a population has telephones, if you know their numbers, if you can reach them, and if they don't suspect you of selling something. Keep telephone interviews to less than 20 minutes, or the respondents may hang up on you. Interviewers must quickly establish and maintain rapport with respondents. A pleasant and understanding approach with clearly worded, succinct questions encourages rapport.

Although more complex and detailed, the personal interview often produces better data and allows interviewers to observe respondents directly. Interviewers can ask detailed and probing questions. But the expense of hiring interviewers, the high cost of travel, and the limited number of interviews completed per day may make this prohibitive. Furthermore, interviewers may encounter difficulty with would-be respondents who refuse to be interviewed, citing the time and invasion of privacy an interview costs.

Development of the Questionnaire. Some researchers call the questionnaire a "survey schedule." Questionnaires typically have three sections: the first sets the stage and establishes rapport with the respondent; the second gathers the needed data; and the third probes sensitive areas and collects demographic information.

Questions can elicit

1. knowledge, information, and misinformation
2. attitudes, opinions, viewpoints, and prejudices
3. past, present, or possible future behaviors

When writing questions, start with specific objectives. Draft your questions to meet those objectives. Use a mix of open-ended and close-ended, or structured, questions (Panel 4-3). Avoid complex terms and jargon. Keep your questions short and relevant.

Complex, involved questions create problems. When questions run on and on, with multiple variables, they confuse respondents, who cannot give a clear, singular response.

Most respondents won't tell you that they haven't thought about your question's topics. They are more likely to answer your questions and tell you what they think you want to hear. To overcome this problem, first ask respondents if they've thought about the topic. If so, continue asking questions. If not, skip to other questions or end the interview.

In wording your questions, avoid those that are leading, double-barreled, loaded, or that permit overlapping responses. "Is there too much waste in government?" is a leading question. Perhaps the all-time example of the loaded question is, "Have you stopped beating your spouse?" The question is "loaded," because either a "yes" or a "no" answer confesses a crime. On the television program *You're Not Elected, Charlie Brown* (Schulz 1976), Lucy gives a prime example of the leading question. "You're going to vote for my brother, aren't you?" she asks, as she grabs a student around the collar and shakes him vigorously. Lucy's leading question seems obvious, but others aren't so obvious:

"Do you believe that good communication skills are necessary in your profession?"

COLLECTING INFORMATION IN THE FIELD

PANEL 4-3. Question Formats

In preparing questionnaires, you can use either open-ended or close-ended questions.

Open-ended questions provide respondents the opportunity to say what they think, but the resulting answers present coding difficulties. To code responses, you must place them into categories. Close-ended questions are easier to code, but the response categories must be exhaustive and mutually exclusive.

Open-ended Questions
1. What problems do you face in becoming a more effective writer?
_____
_____
_____

Closed-ended Questions
2. In the last month or so, how often have you thought about the role communication skills will play in your career?
( ) very often   ( ) often   ( ) not very often   ( ) not at all
- -

Instructions:
Please rate how strongly you agree or disagree with the statements given below. Let "6" mean you strongly agree and let "1" mean that you strongly disagree. Circle your rating.

	Strongly Agree					Strongly Disagree
1. I have been adequately introduced to the scientific method.	6	5	4	3	2	1
2. I have been adequately introduced to problem solving methods.	6	5	4	3	2	1

- -

Few people would answer no to a question like that. Rephrase the question to:

> "Some professionals tell us that communication skills are very important; others tell us that they are not very important. Generally, would you say that communication skills are very important, important, not very important, or not at all important in your profession?"

Such questions tell the respondent that no one response is better than others and allow a respondent to admit that communication skills are not important in his or her profession.

"Do you drink alcoholic beverages or smoke cigarettes?" is a double-barreled question. Respondents may do one or the other, both, or neither, but the interviewer can't tell which. Avoid double-barreled questions by asking each point individually.

Avoid questions with overlapping answers, such as:

"Would you say your income is between $10,000 and $20,000, between $20,000 and $30,000, or . . . ?"

Respondents making exactly $20,000 would not know how to answer the question. Use discrete categories such as "between $10,000 and $19,999, between $20,000 and $29,999," and so on.

When you write your questions, select your words carefully. Stanley Payne (1980) reports that the Opinion Research Corporation of Princeton University ran studies on the effect of varying the words *might, could,* and *should* in a yes/no question. In three matched samples, 82 percent responded "yes" when the question used the word *should,* 77 percent when the question used *could,* and only 63 percent when the question used *might.*

Finally, avoid jargon or words that only a few people understand. Use the language of the respondents; talk their language.

Ordering Questions. As you arrange your questions, keep asking, "Will putting one question ahead of another question influence subsequent responses?" One question often influences subsequent responses. Some respondents are sensitive about divulging personal information about their income, age, education, and other matters. Pretest potentially sensitive questions, and place them late in the survey.

Cover Letters and Instructions. Too many businesses and salespeople pretend that they're conducting surveys when they're really making a sales pitch. As a result, some people are so wary of interviews that they refuse to answer survey questions. Often you can overcome this type of resistance by sending a cover letter announcing the purpose of the survey and providing an address or phone number that respondents can use to verify its legitimacy. The cover letter should explain the survey, how the respondent was selected, should ask for cooperation, and promise confidentiality. Print cover letters on letterhead stationery, and have a director, chairperson, or other authority sign them. For personal and telephone interviews, send the letter out 7 to 10 days before the interviewer will contact the respondent. Always enclose a cover letter with mail and self-administered questionnaires.

Pretesting Questionnaires. To pretest your survey, select two or three people. Have them complete the questionnaire, or interview them. Afterward, ask if they found any questions confusing, awkward, or difficult.

For mail and self-administered questionnaires, ask if they had any trouble following the directions. Revise the questions and directions that caused difficulty. Pretest your questionnaire and directions again, and revise as needed. Finally, run a pretest on a dozen or so people from the population you'll interview later. Analyze the responses, questions that confused people, caused them problems, or that were worded ambiguously. Look too for discrimination among responses—*everyone should not answer the question in the same way*. Why? In most surveys, you're looking for differences in the population being surveyed.

Producing the Questionnaire. Once you have a final draft, have the questionnaire typed or printed. Proofread it carefully. A printed questionnaire gives the impression of being shorter than a typed one. Use a neat, clean, and attractive layout. Strive for balance between blocks of print and white space, and avoid crowding questions together.

Print more questionnaires than your sample size. Mail surveys often require two or three follow-up mailings to achieve a high response rate. You'll need extra questionnaires for training telephone and personal interviewers.

Sampling. Try to use a random sample—random in the statistical sense—of the population. When you use a random sample, you can generalize back to the original population. You can also determine the sampling error or how closely your sample's responses represent the population's knowledge, attitudes, or characteristics. If you don't use a random survey, you can only talk about the specific respondents; you cannot generalize to a larger group.

For many surveys, random sampling from a *current and complete* list of the entire population is best. But some lists are neither. For example, student directories published at most universities and colleges may not list all students. Some students enroll late; others request that their names not be included. A more accurate list would be the registrar's records of student enrollments. But you may have to use the student directory. If so, you must indicate that your population is drawn from students listed in the directory and not a random sample of all students at your school. Avoid "dirty" lists, those that are incomplete, old, and contain names of people who have moved or died. When you run across a dirty list, or if no list exists, seek help from a statistician.

To sample a list randomly, use a systematic random sample. In theory, you should specify the sampling error you want, that is, how closely you want the response to represent the population. From this sample error, you would calculate the sample size you would need to obtain valid, reliable results. But for many surveys practicality dictates sample size. To illustrate, say you want to survey the 17,000 students on campus, and you've defined a student population as all individuals listed in the student directory. Let's say that your budget allows a sample size of 100. Systematic random sampling requires sampling every *kth* individual. To determine *k,* divide the population, 17,000, by the sample size, 100. You get an interval of every 170th name in the directory. To begin at a random point, use a table of random numbers, and select a random number from 1 through 170. If your random number turns out to be 87, start with the first name listed in your directory and count to the 87th name. That's the name of the first person you'll interview. Then you count 170 names and select the second person to be interviewed. You keep counting by intervals of 170 until you've sampled 100 names. Systematic random sampling works fine as long as the list has no cycle, pattern, or periodicity in its organization.

An alternative approach, but one that does not allow generalizability, is called a *purposeful sample*. You would select individuals according to a characteristic relevant to the survey's purpose. Thus you might select all seniors in electrical engineering or all students attending an environmental organization's monthly meeting. But such samples do not allow you to generalize to a larger population. Why? You have no way of knowing how representative the individuals you sampled are of the larger population.

Selecting and Training Interviewers. Prepare careful instructions for interviewers and respondents to follow. Pretest the instructions when you pretest the questionnaire. If you're conducting telephone or personal interviews and you'll have others helping you, you'll need to select and train your interviewers carefully. Look for people who establish rapport quickly with others, who follow directions explicitly, and who will complete the questionnaires accurately.

All interviewers must ask questions and conduct their interviews in the same way or the data will be faulty. Everyone must read the questions in exactly the same way. If they don't, the responses may not be comparable. You'll need to familiarize interviewers with the survey's details and have them practice interviewing until they're competent.

Fieldwork. When you begin collecting data for a random survey, strive for an 80 percent response rate. For every 100 people contacted, you want at least 80 usable questionnaires. If your response rate drops below 65 percent, researchers caution that respondents may differ significantly from nonrespondents. The group answering your questions may not be representative of the population itself. The closer to 100 percent you can push the response rate, the less likely are your nonrespondents to differ from the respondents and the population. If your survey produces a low response rate, describe your results only in terms of the respondents.

Incomplete questionnaires pose another problem. For telephone and personal interviews, encourage the respondents to answer all questions. If you have unanswered questions, check with a statistician.

Data Analysis. Don't wait until you're ready to analyze your data before seeking a statistician's help. Instead, coordinate your data analysis when you plan your survey. If you don't, you may collect data you can't analyze.

Analyses can be simple or complex. For small samples and short surveys from which you need only frequencies, or the percentage of people responding to various questions, you can tally responses by hand. For larger, more complex surveys, seek the help of a sociologist, journalism professor, or statistician.

Drawing Conclusions. Let your research questions direct your interpretation of the data. Limit your interpretations to your data. If you drew a purposeful sample or you had a low response to a random sample, limit your conclusions to those who answered your questions.

Writing the Report. As you draft your report, keep the list of questions in Panel 4-2 clearly in mind. Others may ask these questions of your work, so include the information. Remember to tailor the report to the audience for which you're writing. A well written and clearly interpreted report using frequencies or percentages poses few problems for people familiar with data. For more general audiences, you'll need to interpret the statistics.

HIGHLIGHTS

1. Observations should be planned, systematic, recorded accurately, and subject to checks and controls. Distortions include leveling, sharpening, assimilation, embedding, selective perception, and overfamiliarity with the subject.
2. Successful interviews require defining your purpose; conducting backgrounding; asking for the interview ahead of time; planning the interview in advance; letting the respondent get to know you; asking your questions; treating the interview as a collaboration; dropping any "bombs" in stages; recovering from any such "bombs"; and ending the interview.
3. Before you survey, check for existing information. If it's available, evaluate it carefully to see if you even need to survey.
4. Begin a survey with a detailed set of objectives, draft your questions, order them, and pretest your questionnaire. Revise it as needed.
5. Define your population, obtain a clean list, and use a random sample, if possible, so that you can generalize to the entire population. If you use a purposeful sample or have a low response rate to a random sample, limit your generalizations.
6. Carefully collect, code, and analyze your data. Generate the necessary tables and illustrations, and write your report.

PROJECTS

1. Select a busy spot on campus, such as the entrance to the student center. Place yourself at a good observation point, and for 30 minutes, observe who goes in and out. What categories of age, sex, and apparent occupations emerge from your observations? How many cases in each category do you find? What conclusions can you draw about who uses the building? Are your observations valid for the whole day?

2. Select the company, agency, or organization for which you'd like to work after completing your degree. Prepare a set of interview questions on the job you'd like to hold. Interview a person who holds that position.

3. Critique and then rewrite the following questions:
   a. Do you believe that computer literacy is a necessary skill for technical and scientific professionals?
   b. Would you prefer a beginning salary between $20,000 and $25,000; between $25,000 and $30,000; between $30,000 and $35,000; or between $35,000 and $40,000?
   c. Do you want a job where you can write, or do you want a job where you can program computers? [ ] yes [ ] no
   d. Have you stopped speeding and driving recklessly?
   e. Do you believe that an exercise program and proper diet are necessary for maintaining your health?

4. Let's say that you want to sample 50 students from your school's student population. What's the sampling interval for a systematic, random sample? Report your calculations.

FOR MORE HELP

BABBIE, E. 1973. *Survey research methods*. Belmont, Calif.: Wadsworth.
Babbie tells how surveys actually are conducted rather than how they should be. He provides useful techniques for minimizing survey difficulties.

DILLMAN, D. A. 1978. *Mail and telephone surveys: The total design method*. New York: Wiley.
The "total design" method overcomes many of the problems commonly associated with mail and telephone surveys.

METZLER, K. 1977. *Creative interviewing*. Englewood Cliffs, N.J.: Prentice-Hall.
Metzler provides an excellent, in-depth treatment of interviewing based on solid research.

RIVERS, W. L. 1975. *Finding facts: Interviewing, observing, using reference sources*. Englewood Cliffs, N.J.: Prentice-Hall.
Rivers details guidelines for gathering information from various sources and discusses techniques for reporting research in various disciplines.

PART THREE
Creating Drafts

P art III begins by outlining ways to identify audiences and to learn more about each audience's characteristics. The balance of Part III discusses components of the technical communication process (Steps 4 through 7), including what to communicate, selecting the proper channel, and drafting the copy.

The chapters review audience analysis techniques, how selected professionals write, and various writing strategies.

Part III addresses the problem of writer's block. It can result when people stick to narrowly defined, rigid rules instead of trying to express their ideas on paper. Sticking to rigid rules stifles the ability to write. Part III suggests alternative ways of approaching writing and acknowledges that more than one "right way" exists when it comes to technical communication.

5
Audience Analysis: Improving the Odds for Effective Communication

OVERVIEW

In Chapter 5, we
- ☐ Define audience analysis
- ☐ Explain why audience analysis is essential
- ☐ Suggest what you need to know about your audience
- ☐ Explain Level I and II audience analysis techniques
- ☐ Introduce Level III audience analysis techniques
- ☐ Discuss how to use information about your audiences

As a major in a technical or scientific field, you probably do not think of yourself as primarily a writer or speaker. Yet, as the professionals testified in Chapter 1, success in science and technology depends heavily on communication skills.

TO UNDERSTAND AUDIENCES, APPLY COMMUNICATION THEORY

In communicating about your profession, you usually want an audience to know or do something. Important as good organization, proper grammar, and clarity of expression are, they do not guarantee that your audience will understand you. Even the simplest organization and shortest words make assumptions about those who read or listen to a message. The most dangerous assumption you can make is to assume that your audience has the same frame of reference—knowledge, experience, opinions, and attitudes—as you.

Consider again the basic principles of communication theory. Because an audience seldom has the same frame of reference as a communicator,

> ### PANEL 5-1. Tailoring Communication for Audiences and Channels
>
> In an audio-visual handbook for United States Fish and Wildlife Service biologists, Carol Boggis explains the concept of designing different scripts for different audiences.
>
> Script style departs drastically from style of technical reports or any other written medium. Scripts are written for the *ear*, while print is written for the *eye*. Your listeners have only one chance to hear a *spoken* message and must listen at the pace you read it; but they can read and re-read *print* at their own pace. Here's an example:
>
> *Technical report:* Standard measurements of length of six newborn swift foxes (*Vulpes velox*) taken in the study area were 15.5 cm, 14.7 cm, 16.5 cm, 14.9 cm, 15.0 cm and 16.2 cm (Smith, 1965).
>
> *Script for wildlife biologists*: Newborn swift fox in the study area averaged 15 and a half centimeters in length.
>
> *Script for television audience:* Newborn swift fox pups would fit in the palm of your hand.
>
> (Source: C. Boggis, An Audiovisual Handbook for the Scientists: A Link Between Research and Application, Master's Thesis, Colorado State University, 1980. Used with permission of C. Boggis.)

the communicator must adjust his or her message. The great orators of ancient Greece, for example, saw their audiences face to face. They learned to read those faces, and that immediate feedback allowed them to adjust their messages accordingly. Similarly, you intuitively apply elements of Aristotelian rhetoric in choosing certain words to explain to a three-year-old why fish live in water and different words to explain the same phenomenon to a college zoology class. You use one approach to borrow $5 from a friend and quite a different one to ask a bank for money. As a professional, you need to adjust your communication both to your audience and to your mode of communication, as shown in Panel 5-1.

In dealing with the complex subject matter of any professional field, it is easy to overlook the need to consider audiences. Many professionals fail to communicate because they proceed as though talking to themselves, with little or no consideration of who receives and must act on their messages. Thus a chemical engineer in charge of quality control writes what to the engineer seems a logical, clear request for new testing equipment. But the person two or three steps up the organizational chart who must act on the request may be a business school graduate to whom chemical engineering remains a mysterious, though wonderful, field. The chemical engineer knows that the new equipment will improve quality control, but the reader wants to know more:

- How much will the equipment cost?
- How will the new equipment make products better?
- What is the financial return on the investment?
- How long is the financial pay-off period?

- How will the investment relate to the company's overall mission?
- Will the equipment increase profits? If so, by how much?
- Will the equipment improve quality enough to change company image?

Even within a single company or department, different readers look at a message from different perspectives.

You may argue that the decision maker owes it to the company to grasp what the quality control engineer is trying to say. True, but in that case the decision maker must be an expert in every field represented in the company. That's not likely to happen. So, in the practical, everyday, intense competition of seeking to be heard, the chemical engineer who fails to consider the audiences risks wasting their time and even losing out. The professional who understands and addresses audience perspectives practically guarantees success.

We are discussing here the principle of *equivalent enlightenment* or *understanding*. If the engineer spells out the rationale behind the request, the business school graduate will not need to master the technical intricacies. The engineer will have created equivalent understanding between sender and receiver of the message. Knowing how the business manager makes decisions will help the engineer frame the request. The engineer does not need a master of business administration degree. Putting oneself into another person's position requires equivalent, not complete, understanding.

To achieve equivalent understanding, select your audiences carefully, if you have a choice, and then learn as much as possible about them. What you learn will help you take positions, explain decisions, make requests, and recommend actions. By considering the sender–receiver communication model and what you know about your receiver's communication habits, you can select the appropriate channel—report, memo, telephone call, meeting, technical article, speech, or slide presentation. If you use a channel your audience uses regularly, you increase your chances of being heard. What you know about basic communication principles and what you know about your audience should influence every phase of your communication.

DETERMINE YOUR AUDIENCE

Have you ever received a letter, memo, or some other message and not known why you received it? Often writers and speakers fail to target their audiences. Their shotgun approach scatters their messages to everyone, even to people with no use for the information.

Not long ago one of the nation's leading companies asked the authors to conduct a communication workshop. Managers were frustrated, we were told, because of badly written reports that they could not understand.

Could we possibly teach the production and testing unit supervisors better grammar and usage so that their documents would be more understandable?

We could try, we replied, but first might we look at the supervisors' work? Management eagerly gave us copies of the reports, memos, minutes of meetings, internal letters, and other materials. Almost at once, the problem began to emerge—in our minds, anyway. True, many documents were not grammatical, but they were not especially poorly written. But they had one commonality: almost all were written exclusively from the author's perspective and showed little regard for the reader's.

When we first met the supervisors for the communication workshop, they were frustrated. Nobody up there was listening, they said. Document after document went up the chain of command, and nobody answered them. Those at the top needed a course in how to read.

We had found a classic case of failure to define the audience. But we had to let our clients discover that for themselves. So we asked each workshop participant to take a recent memo, letter, or report and to visit each person who had received it. A week later, the clients reported their results. They were shocked; some were even outraged. Many recipients on the routing lists had filed the message without reading it. Others had discarded it. Still others said they didn't know what they were to do with the document but had kept it, thinking it important. Few recipients clearly understood why they had received the message or what they were to do with it. From then on, we had our students' attention.

Our training exercise illustrates a major communication problem facing our society—information overload. Every day you and your audiences are bombarded by hundreds, if not thousands, of messages. That bombardment creates a serious problem for you as you try to be heard. Simply, your messages must compete with other messages. Your audience may not be in the most receptive mood for your message.

Unless you are sending good news—a salary increase, a holiday schedule, increased benefits—others will receive and handle your message on a scale of enthusiasm ranging from the dutiful to the joyless. The press of other messages, all demanding attention, often fosters a hurried, grudging reception. Competing messages may be the noise, described in the sender–receiver communication model, that interferes with your message.

Therefore, as a professional, begin your audience analysis by asking:

- Who should receive my information?
- Why do I want them to have my information?
- What do I want the recipients to do with my message?
- How will the information help the recipients to do their jobs better?
- What impact will the information have on the recipients?
- Will recipients ask, "Why did I receive this information? What am I to do with it?"

These questions will help you to identify your audiences and why they should receive and act on your message.

Realizing (and acting on the realization) that different audiences often need the same information is one of the most valuable lessons you can learn from this book.

Sam Brown, a recent mechanical engineering graduate who now works for a small manufacturing company, adapted a personal computer to run a mechanical spring-testing machine. He wrote a computer program to complete five tests. By using the computerized spring-testing machine, a technician could test twice as many springs an hour. At least six audiences within the company were interested in knowing about different aspects of Sam's latest development:

1. Sam's supervisor needed to know how the system worked, how accurate and efficient it was, how the new tests compared with manual tests, how well the computerized system was documented, and how long it took Sam to develop the system.
2. Technicians needed to know how to operate the system, how many springs they should test each hour, possible problems, required set-up and start-up procedures, and safety precautions.
3. The company's training officer needed to know the information necessary for training technicians to operate the system.
4. The personnel officer needed to know about changes in the technicians' job specifications, job description, and additional training. Would future technicians need computer training? If so, at what level?
5. Other engineers would be interested in learning the engineering principles Sam used in adapting the personal computer to the mechanical testing equipment. Could the company's engineers adapt personal computers to other testing equipment? If so, how?
6. The accounting division chief would need to know the savings the computerized spring-testing machine produced. Would there be hidden costs? If so, what would they be?

In some organizations, an employee in Sam's position would merely prepare a technical report and distribute it to everyone, leaving to the recipients the task of digging out the information they needed. A more enlightened approach would be either to direct the recipients to the relevant sections of the report or to provide them with specific information. The supervisor would recieve a cover letter with the report. The cover letter would direct her to the abstract or executive summary and to those sections of the report she needed to read. The engineers would receive the report with a cover letter directing them to its theory and methods sections.

For the training officer, a meeting would be scheduled to discuss and illustrate the needed training for technicians. The training officer or the originating employee then would prepare a training manual and give a short course to the technicians. Based on their discussions, the originating employee and the training officer would prepare a memo for the personnel officer, explaining the computer background that technicians would need. If the employee had reported a cost-benefit analysis in his technical report, his letter to the accounting division chief would refer to the relevant section of the report, or the employee might provide a memo summarizing the key accounting information.

Throughout your career, you'll find yourself in positions like these. You'll need to communicate to many different audiences. Once you have identified your target audiences, your communication work has just begun.

ANALYZE AUDIENCE CHARACTERISTICS

Audience study methods range from the simple to the complex. A fellow student who asks you, "What does the professor want?" engages in audience analysis. So does the manufacturing company that hires a research firm to conduct a market study. Somewhat arbitrarily, in this chapter we divide audience analysis techniques into three levels of intensity. Level I primarily involves thinking and asking logical questions. Level II involves more systematic efforts to learn audience characteristics, mainly by examining materials you know the audience uses. Level III analysis requires using scientific techniques, often at considerable expense.

Having identified your audiences, ask yourself, "What do I need to know about my audiences to help me do a better job of communicating with them?" Start by listing each audience separately and noting what will influence how you communicate with that audience. What you need to know depends upon your message's complexity and the consequences should your communication fail. The greater the stakes, the more you need to know about your audience.

If you've been working for a supervisor for a year, and you are drafting a memo to explain your activities for the coming month, you probably won't spend too much time analyzing your audience. If, on the other hand, you're trying to convince a business to buy $1 million worth of your company's services and products, you'll want to invest more time studying the company, its needs, and how best to obtain the contract.

Because the questions you ask about a particular audience depend upon your message and the desired outcome, you might ask:

- What does the audience know about the topic?
- What misinformation may the audience hold on the topic? On related topics?

- Which points will interest the audience most?
- Will the audience want to read the document or hear a presentation?
- How visually oriented is the audience? Will data, photographs, line art, tables, graphs, or other visuals be more effective than a narrative?
- What do I want the audience to know or to do after receiving the message?
- Does the audience understand and use terms the way I do? If not, how does the audience understand and use terms?
- Which newsletters, magazines, journals, and other publications does the audience read regularly?
- Do members of the audience skim a document, read every word, or read only the summary and introduction?
- How much time does the audience spend reading?
- Does the audience suffer from information overload?
- Of the communication forms available, which ones are most appropriate for presenting my message?
- Do members of the audience have any physical or mental problems that hamper their reading, listening, or seeing?
- Do audience members have prejudices, attitudes, or perceptions that may complicate how I present my message?
- Will demographic information on age, sex, and education be useful?

You'll not always find answers to all these questions, and with practice, you'll develop a sense for which are important most of the time in your particular situation. When you seek to reach a new or unknown audience, review the list above from beginning to end.

Level I Audience Analysis

Begin by considering what you know about your audience in the context of the sender–receiver communication model and the audience's frame of reference. Your experiences with your classmates, professors, professionals in your field, and individuals with less technical background will help you design communications for them.

Step 1. Describe the Audience.   As a minimum, answer the following:

- Who are my audiences?
- What are their educational backgrounds and experiences?
- What do they know about my subject?
- What do they expect of my communication?

Name the specific groups or individuals with whom you'll communicate. Then consider their educational backgrounds and experiences. Are they college graduates? High school graduates? What kinds of technical or scientific backgrounds do they have? For individuals with no technical background in your field, you'll need to translate the information and explain basic terms. Are your audiences up to date on advances in the field? Have they been following your work? Finally, what does the audience expect from your communication? An audience's past experiences with you, your peers, and your predecessors will mold its expectations of your message. Your communication will mold its expectations of your future communications.

Step 2. Consider Yourself.   As a sender, answer the following:

- Who are you? What is your professional position relative to that of your audience?
- What is your educational background and experience? How do they differ from your audience's?
- What do you know about the subject of your communication? How does your level of information differ from that of the audience?
- What expectations do you have of your message?

Although simple, such questions and your answers to them can help you when you compare your frame of reference with those of your audiences. Such a comparison quickly shows you differences between how you approach the topic and how your audiences will.

Step 3. Describe the Potential Message.   To help formulate your communication, consider the following:

- What are the key points you want to make?
- What supporting points do you want to make?

Try talking through your summary out loud. Doing that a couple of times and then listing your points on paper should help you to identify your main points.

Step 4. Determine How to Design Your Communication.   With your audience in mind, your position, and your main points, consider how you will need to slant your message to communicate your objectives. Answer the following:

- What should your key points be?
- How should you organize the points to produce the desired result?
- How should you select the language and design the visuals to help you achieve equivalence of understanding?

Consider a hypothetical example of how Ken Reed, a soil engineer, planned to report the soil test results for a home site.

Nancy and Joe Johnson had purchased a lot and wanted John Homer, a builder, to construct their house. When they applied for a mortgage loan, the mortgage company required soil tests, because the company did not lend money for houses that might be damaged by high ground water or poor soil conditions. Nancy and Joe contracted with Ken for the tests. After the tests, Ken had to communicate with three different audiences—the bank's loan officer, the builder, and Joe and Nancy. John Homer, the builder, had taken soil and foundation classes in college, had built houses in the area for 20 years, understood soil types and accompanying problems, knew how to build homes to overcome soil problems, and read soil reports regularly. Furthermore, he had visited the home site, examined the soils, and expected problems.

Though not an expert on soil tests, the bank loan officer similarly was familiar with soil reports and could definitely read the "bottom line"—that is, whether ground water or poor conditions would or would not cause trouble.

In contrast, Joe had an accounting degree and Nancy a degree in nursing. Neither knew the details of home construction nor soils, soil testing procedures, or the associated technical terms.

Ken, an experienced soil engineer, knew the details of conducting soil tests and the implications of different soil conditions for building. His engineering background and five years of conducting tests in the area gave him a highly technical background on soils and foundations. He knew that his message would have to provide the information needed by all three audiences. He realized, too, that if his report did not document any soil conditions that might damage the house, he could be held liable if the house were damaged by a soil problem.

Ken's tests showed no ground water in the two test holes drilled to a depth of 14.5 feet. Field and laboratory tests revealed silty top soil and a sandy, silty clay subsoil with a swelling problem. The soil contained montmorillonite, a clay that expands when wet and that can buckle foundations, floors, walls, and other concrete structures. When dry, such soil contracts and presents no problem.

Ken's technical report to the builder would need to explain procedures, findings, and recommendations. Ken would need to explain carefully the special conditions and building required. He thought that the house could be built, but the best ways were expensive and difficult. Ken would recommend a structural floor with an empty space underneath it. He would then point out that if the owners were willing to risk subsequent damage, possibly years later, they could use concrete slabs independent of the foundation and slip joints. Any water or moisture that moved under the foundation might damage it and the house. He would provide technical details to the builder.

Joe and Nancy received two copies of the technical report, one to keep and one for the bank. In a cover letter, Ken translated and summarized his findings and their implications. Joe and Nancy could build their home, but the special building techniques might be cost-prohibitive. Ken wrote that John would explain the technical aspects of the problem, the construction procedures, and estimated costs. He suggested an alternative: use a floating floor, and avoid landscaping and planting landscaping near the house. They could not let moisture from runoff, lawn watering, or plumbing wet the soil under the foundation. If the soil became wet, it would expand and possibly crack or destroy the foundation. Their insurance would not pay for repairs.

Ken thus analyzed his audiences and their needs correctly and gave each the information required for a decision. Joe and Nancy built their house properly, and Ken continued to have all the business he could handle, both from lending agencies and from home owners and builders.

Level II Audience Analysis

In settings where you will communicate with large and diverse groups, you will need to know as much as possible about their backgrounds. For these settings, we suggest a second, somewhat more detailed level of audience analysis for gathering information that can help you formulate your message.

As a student, you can learn more about your audiences by examining publications, observing professionals, and studying biographical information. We will discuss each of these possibilities in turn.

Examining Publications. By studying diverse publications in your field, you'll gain insight into how publications tailor information to their readers' needs, and you'll better understand how you can tailor your communication to diverse audiences.

Publications fall into three general groups:

1. technical and scientific articles in academic and research journals, which usually report empirical information
2. trade publications, which report applications of research and explain how to solve day-to-day technical problems
3. general publications, which present information for individuals who have little technical background

Examples of technical, research, and academic journals include *Advances in Water Resources, Health Physics, Journal of the Geological Society, Genetics, Technology in Society, Biometrics, Biological Conservation, Journal of the American Chemical Society, Weed Science, Journal of Sound Vibration, Journal of Medical Engineering and Technology,* and *IBM Systems Journal.* Technical and scientific fields have thousands of other journals. Try to find several in your field.

Examples of trade publications include *Nursing87, Farm Journal, American Forests, Engineering News Record, Control Engineering, Highways and Public Works, Water Engineering and Management, Western Fruit Grower, Wood, World Oil, Supervision,* and *Solar Age*. Although they cover specialized topics, most trade publications use a format similar to that of general circulation magazines.

Examples of popular magazines and general publications include *Time, Newsweek, Popular Science, Scientific American, Psychology Today, Popular Mechanics,* and *Reader's Digest*. Your school's library probably carries a selection of general publications that serve your major.

A few fields, such as industrial science and construction management, may not, upon first examination, appear to have research publications. So you'll need to look in related areas. For example, the fields of wildlife and natural resources have few trade publications, but forestry and wood sciences have many related publications. If you're unsure about how to find the publications, study the chapters on using the library and gathering information.

Although various gradations of research, trade, and popular publications exist, these three broad categories are important for you, because they say a lot about how professionals address audiences. One of the most difficult decisions you face in technical communication, day in and day out, is to decide which terms you can expect your audience to know and which you must define. Familiarity with different levels of professional publications in your field gives you a good feel for what may and may not be commonly understood. But you'll learn much about formatting, too, by examining how the publications proceed.

Panel 5-2 illustrates an informal analysis of research, trade, and popular publications. What should you be looking for as you analyze your field's publications to get a better understanding of your audiences? Consider the citation styles, article organization, table and figure use, article length, writing quality, and publication breadth.

Research publications usually cite literature with either the author-date, numbered reference, or footnote citation style. Although trade journals occasionally use citations, most use either direct or indirect quotes, as popular magazines and newspapers do. Another citation style is to include full titles within the narrative.

Most research journals organize articles under headings such as Introduction, Methods, Results, Discussion, and Conclusions. Although the headings vary, contents are comparable. Research articles also use succinct, descriptive titles and an abstract, which is a summary of the article.

In contrast, trade journal articles usually state a problem and then suggest solutions. The content dictates the article's organization. Popular publications either use a news story or feature article organization. News stories answer some, if not all, of the following questions: Who? What? When? Where? Why? and How? They do so in the lead, the first paragraph or two. Then the article covers key points in descending order of importance.

> ## PANEL 5-2. Comparing Readerships
>
> For an informal audience analysis, consider: the *Journal of Computer and Systems Science,* an academic journal; *Creative Computing,* a specialized or trade journal; and *Newsweek,* a general circulation magazine. The publications' titles suggest that the longer the title, the more education its readers will need.
>
> The *Journal of Computer and Systems Science* has readers with master's or doctoral degrees in mathematics, computer science, and engineering. Its writing uses complex scientific terms—"three-dimensional tenor," "matrices," "substitution," and more—with special meaning to the readers. (It also uses lots of the passive voice, a writing problem identified in Chapter 10.) The *Journal* uses a straightforward layout and design for historical accounts of scientific studies. Illustrations consist primarily of tables, mathematical expressions, derivations, and figures. This publication requires a highly motivated and educated reader.
>
> In contrast, *Creative Computing*'s readers own and use personal computers. But its readers have less extensive computer backgrounds than the *Journal*'s readers. Some readers may be high school students, but most readers probably are technical, scientific, and business professionals. Topics include: product evaluations, profiles of software and hardware, applications, and other practical information. Readers use the publication to learn more about computer applications and to use their computers more efficiently. Illustrations include tables, drawings, photographs, and diagrams. The writing is more readable than that of the *Journal of Computer and Systems Science.* Our review suggests that *Creative Computing*'s readers are strongly interested in computers, knowledgeable in some terms and jargon, but not as sophisticated as the *Journal of Computer and System Science*'s readers.
>
> The February 22, 1982, issue of *Newsweek* carried "To each his own computer," a 4,500-word article on personal computers. *Newsweek* defined and explained technical terms for its readers. Rather than provide details on equipment operation and its internal workings, the writers explained the industry's start, development, and emergence into a billion-dollar-a-year industry in less than 5 years. The writers carefully avoided jargon and gave readers a general overview of the personal computer's potential uses. Readers with no computer background learned about bytes, ROM, RAM, and other terms; all readers learned about the personal computer's history, economics, and potential. *Newsweek* illustrated the piece with photographs, cartoons, sidebars—short, boxed stories—and a table. *Newsweek* added human appeal to the article by reporting on specific individuals within the industry. The smooth writing made the article extremely readable. Of the three publications, *Newsweek* had the least computer-sophisticated readers.

Features begin with an enticing lead; the topic and major points guide the article's organization.

Research publications use highly technical tables and figures. Trade publications translate the information into more easily understood visuals. The reason why, of course, has to do with the assumed level of education or sophistication of the audience. Popular publications use the least complex visuals. When they use tables and figures, the illustrations are quite simple.

Article lengths vary, too, among these sorts of publications. Generally, research publications have the longest articles. To estimate an article's length, count the number of words in 5 column-inches (one column wide and 5 inches deep), and calculate the average number of words per column-inch. Measure the number of column-inches in the article, and multiply by the average word count per column-inch.

Examine the writing. Are the sentences short, direct, and to the point? How would you assess the quality of the writing? Look through several issues, and examine how the publications use technical terms. The complexity and use of terms give further insights into readers' backgrounds.

Consider how the publications treat breadth of content. Research articles usually focus on a narrowly defined problem of limited interest. In contrast, trade articles treat topics with wider appeal. Popular articles cover topics that appeal to large numbers of people.

In all, each article's style, content, and approach give you insight into the readers and their understanding of technical topics. Over time, publications either mirror the technical sophistication of their readers or go out of business.

Observing Professionals. If your school has a student chapter of a professional organization for your major, attend its meetings. Frequently professionals give presentations on campus, and you'll have the opportunity to learn more about your profession and its audiences. If you can attend local, state, regional, or national meetings of your profession, do so. Observe the presentations, and talk with participants.

Take advantage of your school's internship, practicum, and shadow-day programs—one-day visits with professionals—to broaden your professional experience and communication skills.

Using Biographical Information. Professional directories and similar publications, accreditation and certification requirements, and biographical data also will help you to understand your audiences. Some professional organizations publish directories that report each member's name, age, education, areas of specialization, research interests, professional experience, personal data, and other relevant information.

Check your library or ask your adviser if your professional field has a directory. If so, study it for general characteristics of your field's members. You may need to turn to publications such as *American Men and Women of Science*. To see which "Who's Who?" publications cover your field, check your library's card catalog, *Books in Print,* or *Biography Master Index,* a computerized data base.

If you're writing assignments for classes, you can learn something about your professors' and instructors' educational backgrounds by reading your college catalog. Some departments provide students with advising manuals that contain brief biographies of professors.

Level III Audience Analysis

When you have the opportunity to work with technical communicators, public relations practitioners, and other professional communicators, you'll find that they sometimes use sophisticated techniques to gather information about their audiences.

Subscription Forms. Many trade publications use subscription forms, such as that shown in Panel 5-3, to control who receives the publication, to learn more about readers, to sell advertising, and to determine editorial practices. But subscription forms, like any analysis technique, have limitations. Drawing conclusions about readers from the data on subscription forms may, on occasion, lead to erroneous conclusions, as Panel 5-4 illustrates.

In-depth Interviews. Besides subscription forms, professional communicators use in-depth interviews of audience members. The interviewing

PANEL 5-3. Controlled Circulation Magazines

The publication *Research & Development* controls its circulation to professionals in chemistry, aeronautics, geology, electrical engineering, physics, and related fields. To get a free subscription, an individual must clip a subscriber service card from a current issue, fill it out, and submit it. To qualify, readers must meet the criteria identified on the subscription form. Those who do not qualify for a free subscription must pay an annual fee. Here's an example of *Research & Development*'s subscription card.

(Source: Research & Development. *Used with permission of* Research & Development.)

PANEL 5-4. Natasha Crowe

Subscription forms, commonly known as "bingo cards," suggest an element of chance. Useful as they are, bingo cards can produce misleading information, as Steven J. Marcus, managing editor of the Massachusetts Institute of Technology's national magazine, *Technology Review,* reported in the "My Turn" section of *Newsweek:*

> Natasha Crowe, a close acquaintance of mine, recently received an unsolicited invitation from Joanne Black, senior vice president of American Express Company's Card Division. "Quite frankly," the letter began, "the American Express Card is not for everyone. And not everyone who applies for membership is approved." Tasha (as she is affectionately called) ignored the letter. A few weeks later she received a follow-up from a different vice president, Scott P. Marks, Jr. "Quite frankly," Mr. Marks reminded her, "not everyone is invited to apply for an American Express Card. And rarer still are those who receive a personal invitation the second time."
>
> Despite the honor, Tasha has continued to disregard this and similar invitations she has lately been receiving. For one thing, she has no job. Her savings are minimal. Her credit history is essentially a vacuum and therefore her credit rating, I'd imagine, is lousy. She doesn't speak English. She's my cat, and I love her.
>
> American Express is not alone in wanting to do business with my cat. And I am most impressed that her suitors are not what anyone would call, if you'll pardon the expression, Mickey Mouse. For example, Diners Club International has offered her "preferred customer" status. She has been invited to subscribe to the *Wall Street Journal, U.S. News & World Report* and *Fortune.* . . . Tasha is a bright cat, but why is she suddenly irresistible to all these heavy-hitting enterprises? . . .
>
> All I did was subscribe to a single magazine—a technical journal that covers the chemical industry—in her name, including a position in a fictitious company to comply with the subscription form. From this, someone who markets mailing lists concluded she is a female engineer-executive—a rare combination . . . with considerable commercial appeal, to judge by the letters that have been sliding through our door slot ever since. . . .*

Marcus admitted elsewhere in his piece that it was a "cheap shot" for him to put his cat's name on the subscription form of the chemistry publication. And no doubt the overwhelming majority of subscribers to that publication were qualified scientists, engineers, and professionals. The incident does show, however, the caution editors must use when analyzing mailing lists.

(*Newsweek, March 22, 1982, p. 3. Copyright 1982 by Newsweek, Inc. All Rights Reserved. Reprinted by permission.)

---

techniques discussed in Chapter 4 work well in uncovering facts about audiences. Simply focus your interview questions accordingly.

Surveys. For gathering information on larger audiences, professional communicators often use mail, telephone, and personal surveys of a random sample of the audience members. The Gannett Company—which owns more newspapers than any other company in the United States—surveyed 20,000 readers and nonreaders in communities where it publishes newspapers. Gannett staff members then used the information to change the content, focus, and appeal of its newspapers, both to maintain current readers and to win new ones.

Content Analysis. As a long-established form of communication analysis, content analysis provides an objective, structured way of quantifying information about the audiences. John Naisbitt, coauthor of *MegaTrends* (1983), uses content analysis of some 6,000 newspapers to assess current social and business trends and to project future ones. With the information Naisbitt collects, he prepares a newsletter and conducts seminars for corporations about their audiences and clients.

Communication, marketing, and social science research provide additional techniques to help professionals better understand their audiences and how those audiences react to communication. The "For More Help" section at the end of this chapter lists publications covering the more commonly used techniques: focus group, nominal group technique, and the Delphi process.

Using Audience Information

Once you've learned all you can about your audience, you'll feel better prepared to plan, organize, tailor, and produce your communication at an appropriate level. Having studied documents the audience commonly reads, and after learning something about members' backgrounds, you'll be more comfortable dealing with those big questions of which terms to use and when to define them.

Notice how different levels of publications titled similar stories about a common but little-known venereal disease:

> *Journal of the American Medical Association,* April 4, 1986: "Criteria for Selective Screening for *Chlamydia trachomatis* Infection in Women Attending Family Planning Clinics"
> *Science News,* April 12, 1986: "Chlamydia Detection and (Maybe) Protection"
> *Harper's Bazaar,* June 1985: "Chlamydia Epidemic: The Campus Bug That Wrecks Your Sex Life"
> *Jet,* December 23, 1985: "Chlamydia—Nation's No. 1 Sexually Transmitted Disease"

Of these publications, only two were certain enough of their audience to assume considerable scientific knowledge. *Journal of the American Medical Association* knew it was speaking to physicians, including gynecologists and others familiar with the diagnosis and treatment of venereal disease. *Science News* knew that its readers had sufficient scientific training to cope with technical terms, even though they were not physicians. *Harper's Bazaar* and *Jet,* however, made no assumptions about what readers might know and used definitions even in the headlines.

Panel 5-5 shows part of the *Journal of the American Medical Association* article. The writers were physicians speaking to other physicians, and

PANEL 5-5. Reporting Chlamydia Screening Methods to Physicians

Criteria for Selective Screening for *Chlamydia trachomatis* Infection in Women Attending Family Planning Clinics

H. Hunter Handsfield, MD; Lora L. Jasman, MD; Pacita L. Roberts, MS; Vivien W. Hanson, MD; Robert L. Kothenbeutel, MD; Walter E. Stamm, MD

• Clinical and epidemiologic factors associated with *Chlamydia trachomatis* infection were examined in women attending two family planning clinics in order to develop criteria for selective screening. *Chlamydia trachomatis* was isolated from the cervix of 98 (9.3%) of 1,059 women. Five demographic, behavioral, and clinical characteristics were independently predictive of chlamydial infection by stepwise multivariate logistic-regression analysis: aged 24 years or less, intercourse with a new partner within the preceding two months, examination results showing purulent or mucopurulent cervical exudate, bleeding induced by swabbing the endocervical mucosa, and use of no contraception or a nonbarrier method. A screening program that tested women with two or more of these risk factors (65% of the total) would encompass all who had a 4.7% or greater predicted risk of chlamydial infection and would detect 90% of all infections. Selective screening of sexually active women for chlamydial infection is advocated as a necessary and cost-effective public health measure.
(*JAMA* 1986;255:1730–1734)

Chlamydia trachomatis is among the most prevalent sexually transmitted pathogens and has been shown to be an important cause of nongonococcal urethritis, epididymitis, and proctitis in men[1-3]; mucopurulent cervicitis, urethritis, and salpingitis in women[1,2,4,5]; and various infections in newborn infants.[1,2,6] In addition, *C trachomatis* has been linked with infertility in women,[1,2,4,7] and with complications of pregnancy, such as premature rupture of the fetal membranes, premature delivery, and postpartum endometritis.[1,2,4] Because chlamydial infections commonly are asymptomatic or cause mild or nonspecific symptoms and signs, screening of sexually active women has been advocated as a potentially effective public health measure.[8-11] Screening has not been widely implemented, however, partly because cultures for *C trachomatis* are expensive and not widely available. Recently developed immunological methods to detect antigens of *C trachomatis* in genital secretions[12,13] may make screening more practical, but cost still limits the implementation of screening programs, especially in public clinics that serve the populations at highest risk. However, selective screening of sexually active women might offer a less costly alternative to universal screening. We undertook the present study in two family planning clinics to identify easily applied demographic and clinical criteria that would facilitate selective screening of sexually active women.

Materials and Methods

Patient Population

The study population consisted of consecutive women aged 14 years or older attending either of two urban family planning clinics....

(Source: Journal of the American Medical Association 255 (April 4, 1986): 1730. Reprinted by permission.)

> ## PANEL 5-6. Reporting Chlamydia Screening Methods to Scientists Generally
>
> ### Chlamydia detection and (maybe) protection
>
> Even its victims may never have heard of it, but *Chlamydia trachomatis* is the most common sexually transmitted disease agent in the United States. It is also one of the most damaging. The severity of the problem has prompted guidelines for its detection, published in the April 4 JOURNAL OF THE AMERICAN MEDICAL ASSOCIATION, as well as investigations into possible defenses against infection. A report at the recent meeting of the American Society of Microbiologists (ASM) confirms that, in the laboratory, an ingredient in most spermicides prevents chlamydial infection of cells.
>
> The problem with detecting chlamydia is that it is often painless: Asymptomatic in up to 70 percent of the women it infects, it can cause infertility before infection is discovered. "Ideally, you'd like to do universal screening," says researcher H. Hunter Handsfield of the Seattle-King County Department of Public Health, who helped formulate the detection guidelines. But until recently, the only dependable test for chlamydia was expensive, difficult and largely unavailable outside of major medical centers.
>
> New immunologic tests that detect the *Chlamydia trachomatis* bacteria in genital secretions (SN: 5/7/83, p. 296) are better suited to clinic use, but the cost of the tests is still a stumbling block—especially, Handsfield and his colleagues note, "in public clinics that serve the populations at highest risk."
>
> In a study of 1,059 women at two family-planning clinics, the researchers found that five characteristics could identify women likely to be infected with chlamydia. The characteristics: no more than 24 years old; having had a new sex partner within the preceding two months; having a purulent cervical discharge; bleeding during parts of the vaginal exam; and using a nonbarrier method of contraception, or no contraception at all. Testing any woman with two or more of these risk factors, they say, would have caught 90 percent of the chlamydial infections at the clinics while reducing the number of women tested by 35 percent....
>
> *(Source:* Science News, *April 12, 1986, 31.5–6. Reprinted with permission from* Science News, *the weekly newsmagazine of science, copyright 1986 by Science Service, Inc.)*

assumed that the list of terms in the first sentence required no definitions. Panel 5-6 gives an excerpt from the *Science News* article. Note especially the first sentence, which reveals the writer's assumption that, scientific training or not, readers needed to be informed of the wide spread of the disease. Physicians, on the other hand, would already know that if they were interested in screening procedures.

Panel 5-7 explains how a group of writers at the Document Design Center in Washington, D.C., tailored information to a particular audience. The approach helped readers understand the employee handbook and saved the personnel staff time and the company money.

PANEL 5-7. How Professional Communicators Use Audience Analyses

The Document Design Center of the American Institutes for Research of Washington, D.C., uses a variety of audience analysis techniques to communicate specialized information to selected audiences. In the October 1981 newsletter, *Simply Stated,* the editors, Robbin Battison and Joanne Landesman, explained how the center's staff improved an employee benefits handbook:

When we were first consulted about this project, we were told simply that the benefits handbook was "in bad shape." Employees couldn't understand it and administrators of the benefit plans had a lot of headaches as a result. Could we help?

The first thing we had to do was to analyze *why* employees had such a hard time understanding the handbook. As one of our pre-design steps, we interviewed 50 employees (in a stratified random sample ranging from file clerks to vice presidents) about how they used the book. We asked them specific questions about where they keep the book, whether they have read it, and what happened the last few times that they had to refer to it. We then asked them to look through the book to find answers to specific questions that we asked them. The results, predictably, were miserable. These employees did not know what their benefits were nor how the program operated. They could not find relevant information in a reasonable time period; when they could find it, they often could not interpret it accurately. When interviewed, they tended to *underestimate* what their benefits were. When actually filing claims, however, practice showed that they tended to *overestimate* them.

This particular book had grown by amendment over several years, and most people commented on the complex legalisms, jargon, and extensive cross-references. As one woman put it, "I know the answer's in there somewhere—if I look long enough." Many of the people interviewed would consult the book briefly, then bypass it entirely by putting routine questions to administrators of the benefit plans or to the claims processing office (which had to assign some staff to the phones full-time). Senior employees and employees who had worked there for a long time would ask their knowledgeable or well-placed friends for the answers to questions.

Why were people bypassing what should have been their primary source of information? The handbook demanded, as it turned out, *recodification* and *reorganization*, as well as rewriting. It needed to be recodified because the benefit plans had grown by amendment and the cumulative amendments had not been smoothly incorporated into the text. Information on a given topic was distributed throughout the book. An employee could never be sure that complete and accurate information on a topic would be found in a single section.

The handbook had to be reorganized because its complex structure did not match employees' perceptions of what was important information and what was incidental. It needed an organization that directly addressed the concerns of the employees.

How could we correct the book to make it more useful and effective? One of the major problems concerned the section headings that supposedly grouped information for the reader. First, there were too many of them. In some chapters, we found five different levels of unnumbered headings, which made it difficult for people to perceive the organization of the sections they were scanning or reading. The headings were also graphically indistinguishable, which made it difficult to perceive how different sections related to each other. In our revised version, we used just two levels of headings within each chapter, and we numbered the lowest level. We used indented bullets to set off more detailed points within the numbered sections.

Second, the employees never really read the handbook cover-to-cover, but they did use it for reference when they had specific problems or questions. However, there was little correspondence between the content of the headings and the questions that people wanted answered. We set about correcting this with a principle we frequently invoke: using informative, personalized questions as headings.

For example, we rewrote "Coordination of Benefits" as "What if someone in my family is

(continued)

PANEL 5-7. (continued)

covered by two group health plans?" "Special 100% Reimbursement Feature" became "Is there a limit to how much I must pay out of my own pocket?" "Conversion Privileges" became "Can I convert my health coverage to an individual policy?" Every numbered section was written as a personal question.

These revisions were intended to anticipate the readers' needs for information and to break up the information into accessible, manageable chunks. Moreover, they helped make the handbook a less formidable document, because readers can now scan a heading and make more confident decisions about how relevant a particular section is to their questions. We predict that they will spend less time searching through the document to make sure they catch every last detail.

Recodification helped to sort out the layers of amendments and make the benefit plans more unified and coherent. Reorganization helped to put information in the right place, where it would address the kinds of questions that employees had told us were important to them. Both of these types of changes in turn led to other decisions about how and where to place definitions and whether to repeat certain provisions verbatim at different points in the handbook.

With these basic changes mapped out, and with the decision to write headings as personal questions, we were ready to address problems at the paragraph and sentence levels. It turned out, of course, that many of the complex legalisms, cross-references, and convoluted syntax of the older version stemmed from traditional writing techniques, not from any inherent technicalities or resistance to clear writing. We were able to move passages around, condense them, and in some cases eliminate them altogether. Thus, portions of the new handbook that precisely parallel the old handbook are rare. One example of prose that has been rewritten, but not reordered, is:

BEFORE

Benefits are paid if an insured employee or eligible dependent incurs covered charges because of pregnancy. Reimbursement for hospital and out-of-hospital maternity charges will be made on the same basis as for any non-maternity condition covered under the plan.

AFTER

If you or one of your insured family members becomes pregnant, the Plan will pay for medical care in the same way that it pays for any other medical condition.

The point of this case study is that we were not able to start revising at the level of individual sentences, rewriting them one by one, until we made basic decisions about the overall organization and approach of the text. Like the mechanic, we knew we could always change the spark plugs when it was a convenient step in the complete tune-up.

(Source: Simply Stated, no. 21, October 1981. Reprinted with permission from Simply Stated, the monthly newsletter of the Document Design Center, American Institutes for Research.)

HIGHLIGHTS

1. Audience analysis begins by identifying readers, listeners, and viewers.
2. Identifying readers begins by asking: Who should receive my message? What do I want the recipients to do with my message? Why do I want them to have it? How will the information help the recipients do their jobs better? What impact will the information have on recipients? Will the recipients ask, "Why did I receive this information? What am I to do with it?"

3. To determine what you need to know about your audiences, begin by asking: What does the audience know about the topic? What about the topic will interest the audience? Why will the audience read, listen, or watch what I have to present?
4. Level I audience analysis entails describing the audience, considering yourself, describing the potential message, and then determining how to tailor your communication to the audience.
5. Level II analysis includes evaluating publications, observing professionals, and using biographical information.
6. Level III analysis, used most frequently by professional communicators, includes a variety of communication, marketing, and social science research techniques.

PROJECTS

1. Select a research or academic journal or publication in your field. Study at least two issues. Prepare a short description of the publication, its audience, and readers, based on the following questions:
   a. What is the education of the articles' authors?
   b. What is the education of the publication's readers?
   c. What types of experiences have the readers had?
   d. What would you estimate as the age range of readers?
   e. Do their education and age range influence how readers see particular topics?
   f. What is the knowledge level of the readers on the topics covered?
   g. How are the articles organized?
   h. How familiar are readers with the jargon used in the publication?
   i. What types of illustrations—tables, line drawings, graphs, photographs, or others—are used? Why?
2. Select a trade publication for your profession, and study the two most recent issues. Answer the questions posed for Project 1.
3. Select a popular publication that sometimes carries articles about your profession. Answer the questions posed for Project 1.
4. Your adviser has scheduled a meeting on campus for local professionals in your field. You have been assigned to prepare a presentation that explains a development in your major, such as course changes, over the last two years. Prepare a list of questions to guide your audience analysis of the professionals you would expect to attend the meeting. Gather the information, and then write a 600-word narrative describing the professionals. Turn in both your questions and your narrative.
5. Assume that your school has a magazine and career day for students who have not selected a major. Your adviser has asked you to

prepare an article describing your major and your intended career. Before you begin, you need to know more about your audience. Develop a set of questions to ask these students, and then complete an audience analysis. Write a 600-word narrative about the students. Turn in both your questions and your narrative.

6. Select a recent development in your major that has been reported in the popular media. Then trace the reporting of that development through the trade and technical or scientific publications. Photocopy at least one article from a scientific or technical journal reporting on the development, one article from a trade magazine, and one article from a popular magazine. Analyze the differences in the use of terms, the slant, organization, and details reported in three articles. Write a report of 750 words or less describing the audiences based on your publication analysis.

7. If you are to prepare a major project or technical report for the semester, complete an audience analysis of the people who will read your report. Follow the guidelines provided in this chapter. Prepare a 500-word report describing your report's target audience.

FOR MORE HELP

CAERNARVEN-SMITH, P. 1983. *Audience and response*. Pembroke, Md.: Firman Publications.
The author discusses considerations of audience for professionals who write. The book has many examples of products written for different audiences.

DELBECQ, A. L., A. H. VAN DE VEN, AND D. H. GUSTAFSON. 1975. *Group techniques in program planning: A guide to nominal group and Delphi processes*. Glenview, Ill.: Scott, Foresman.
The authors provide a detailed, step-by-step approach to both the nominal group technique and the Delphi process, based on their own experience.

KEENE, M., AND M. BARNES-OSTRANDER. 1985. Audience analysis and adaptation. In M. Moran and D. Journet (eds.), *Research in technical communication*. Westport, Conn.: Greenwood Press.
The chapter reviews audience analysis and the role it plays in effective technical communication.

KRIPPENDORFF, K. 1981. *Content analysis: An introduction to its methodology*. Beverly Hills: Sage Publications.
Krippendorff provides specific steps and guidelines for people planning to use content analysis.

6
How Professionals Write: A Look at Writing Processes

OVERVIEW

In Chapter 6, we
- Look at how technical and scientific professionals write
- Consider how professional writers work
- Consider how instructors and professors write
- Summarize approaches to writing

If, as you proceeded through school, you were told there's but one way to write, read on and you'll discover what researchers and writers themselves tell us about how students, professionals, and professors write. If you've struggled hard when you try to write, you'll see why you're not alone as you try to commit words to paper. If you've thought writing easy for almost everyone but yourself, you'll soon learn why few people find it easy. If you've become frustrated when trying to write, you'll begin to understand why. If you've thought writing a mysterious, abstract, and confusing process, you'll soon discover that writing can be less mysterious and that it can be concrete and clear.

HOW TECHNICAL AND SCIENTIFIC PROFESSIONALS WRITE

Realizing that little is known about how technical and scientific professionals write, English professor Jack Selzer (1983) of Pennsylvania State University tracked an engineer through many kinds of writing. Selzer asked the engineer to answer on tape a series of questions each time he completed a letter, proposal, report, or other document. Later the professor interviewed the engineer at length.

The engineer, Kenneth E. Nelson, of Chicago, works for one of the nation's largest design and systems firms. He holds a B.S. in urban systems engineering from the University of Illinois and an M.S. in civil engineering from Northwestern. As manager of his firm's Chicago office, he produces many different proposals, most of which follow a format imposed by clients.

Professor Selzer found that Nelson spends 80 percent of the time he devotes to "writing" doing what Selzer describes as "inventing" content—deciding what should go into the document. To decide, Nelson first establishes his purpose and then analyzes his audience. Usually he calls key people at the organization that is to receive the proposal or people who have dealt with the organization before. He also interviews his colleagues, brainstorms with them, reviews similar documents previously produced by his firm, and reviews the general field literature.

Once Nelson has chosen his content, he turns to the actual writing. This activity occupies about one fifth of the time he spends preparing a document. He begins with extremely detailed outlines, complete with headings, subheadings, and indentations.

As Selzer told the Modern Language Association in reporting his study:

> For everything from important letters to lengthy final reports, Nelson commits himself to a detailed plan based on a tidy outline. If the outline is for a proposal, letter, or progress report, he completes the outline before he begins. If the outline is for a final report, he begins it early in the engineering project and then revises it as necessary later in the project. (Selzer 1983, 14–15)

Professor Selzer observed Nelson writing. The engineer worked straight through, from start to finish, and produced 1,000 words in just under two and one half hours. In the piece of writing Nelson produced under Selzer's gaze, he scratched out one paragraph, three sentences, and three other single words. Clearly, Nelson's research, outline, and other preparation had left him ready to compose. Nelson's writing process worked quite well for him.

Selzer could not say whether Nelson's method is the same as that other engineers use. But Selzer concluded that Nelson is an efficient and effective writer, whose proposals are promptly completed and swiftly accepted. Furthermore, Selzer observed that Nelson's way of writing followed a process similar to that used by some technical writers—a point covered later in the chapter. But first, let's consider how other technical and scientific professionals write.

Matt Young (1982), a physicist at the National Bureau of Standards in Boulder, Colorado, says that he usually completes his research and data gathering before he begins writing. He is not conscious of the writing process and does not use an outline, but when writing technical reports and journal articles, he first pulls information from past research and literature reviews and then explains his methodology, reports the project's results, discusses its implications, and summarizes the key points. Then he writes the abstract.

In contrast, Dr. Hassel Ledbetter (1982), chairman of the editorial review board at the National Bureau of Standards in Boulder, says that he starts thinking about where the article will be published at least a year prior to writing the draft. Only after he spends considerable time thinking about the topic and reviews sufficient literature, develops an adequate theory, and conducts experiments does he begin writing. He completes his calculations, develops illustrations and tables. Then Ledbetter organizes the article in his head and writes some sections in longhand. He explains that writing is in fact a thinking process. He starts at the article's beginning and moves through to its end. If he finds information missing, he stops, gathers additional information either from literature reviews or a return to the laboratory, and only then returns to his writing. Once he starts putting words on paper, he drafts a 20-page, typed manuscript in one and one half to two days.

Surveys of Technical Professionals

Professors Nancy Roundy of Iowa State University and David Mair of the University of Oklahoma reported their survey of 70 technical writers in the *Journal of Advanced Composition* (1982). Thirty were technical, scientific, engineering, and assorted majors enrolled in a technical writing course, eight were students working part-time in industry, two were university professors, and thirty were engineers and researchers in industry.

The writers surveyed used several strategies. All use some prewriting—planning, thinking, and inventing—and rewriting activities—revising and editing. For manuscripts longer than 10 pages, they spend more time in prewriting and rewriting as clearly separate activities. In contrast, when writing short manuscripts, the professionals often merge prewriting with rewriting. They seem to spend a smaller percentage of their time on pre- and rewriting activities for short manuscripts than for longer ones.

Furthermore, the writers do not work straight through from beginning to end, as engineer Nelson does. Instead, they loop back to prewriting as they proceed. All the technical writers revise—adding, rearranging, substituting, and deleting materials—during and after the initial composing process. Professors Roundy and Mair found that writers often break their writing, rewriting, and editing activities into subsets of activities. They rewrite throughout the composing process at three levels: content, structure, and style. First, they solve the major content and structural problems of their drafts before working on the paragraphs and sentences. They focus on making style changes after considering the audience and after completing their drafts. In all, the writers use content and a logical development to guide their revision.

In *College English,* Lester Faigley and Thomas P. Miller (1982), of the University of Texas at Austin, reported their survey of 200 professionals from agriculture, mining, construction, manufacturing, transportation, utilities, sales, finance, insurance, real estate, services, and government.

Respondents told the researchers that they create a variety of written products, in a variety of media, with a variety of composing processes. They rarely revise memos and letters but frequently revise reports, especially reports going outside their company.

Lee Odell, Dixie Goswani, Anne Herrington, and Doris Quick (1983), of the Document Design Center in Washington, D.C., have studied professionals' writing processes in diverse settings. They reported that professionals who write in nonacademic settings have detailed knowledge of their audiences and have developed a repertoire of strategies for dealing with the audiences. Specifically,

> These writers can usually assume that their writing will actually be read by someone who can, in fact, be informed or persuaded and that the reader may very well be dependent upon the writer's ability to inform or persuade. Finally, these writers know that their writing will have important consequences; if they do not write reasonably well, they may not obtain what they want; they may have the additional trouble of dealing with an angry or confused reader; they may not be promoted or given merit raises; or they may lose their jobs. (Odell et al. 1983, 19)

Linda Flower, of the Department of English, and John Hayes, of the Department of Psychology, at Carnegie-Mellon University have completed several studies of writers in laboratory settings (Flower and Hayes 1980, 1981, 1984; Hayes and Flower 1983). The researchers asked both experienced and inexperienced writers to think aloud as they completed a given assignment. The researchers found that experienced writers spend more time planning before beginning to commit words to paper and that they have a richer collection of cognitive maps, schemata, or plans to select from in organizing the assignment.

In diverse studies of scientific and technical problem solving, researchers have found that successful problem solvers know many different cognitive strategies for dealing with the problem they face. Their training and past experiences have helped them build their knowledge of problem solving strategies. Experienced problem solvers, whether writers, technical, or scientific specialists, have developed an understanding of many different problem solving strategies. When faced with a particular problem, they first search their mental bank of strategies. If they've faced and solved similar problems, they use that experience to help guide their solution of the problem they face.

HOW PROFESSIONAL WRITERS PRACTICE THEIR CRAFT

The growing body of research about how technical and scientific professionals write suggests that they use many different but related writing techniques. Let us explore how several technical writers and a popular writer go about their craft.

At the IBM Santa Teresa Laboratory a few miles south of San Jose, California, five people sit talking about how they write (Allard et al. 1982). At this program development center for IBM's General Products Division, lab workers design, develop, test, and maintain a variety of programs for IBM customers throughout the world. Probably the most advanced programming lab within IBM, Santa Teresa ranks among the highest of high technology centers. Programmers there create and maintain applications for computer language programs such as APL, FORTRAN, BASIC, COBOL, and PL/1.

The discussion deals with how the five go about writing. They belong to a group of about 50 people working in system information, meaning they prepare the written documents that accompany programs. They are, in effect, go-betweens for programmers and designers and program users.

Harriet Duzet, longtime member and national officer in the Society for Technical Communication, is a "planner/writer." Her job involves planning the "library" of user manuals that accompany a program and then writing the individual books. She interviews programmers, asks them thousands of questions, and makes a master plan or outline of what must be included and in what order. Like engineer Nelson, she employs a detailed outline. Unlike him, though, she may write sections out of linear order. She composes on the latest model IBM word processor, but at day's end she sends for a paper printout, and she may use pencil and scissors to edit further.

Jim Overholt composes in longhand. Although he used a typewriter when working in advertising before joining IBM, he found that his present job demands a different approach. He describes his longhand system as an "outline of sorts," into which he puts important points in no particular order. When satisfied, he transfers the outline to a word processor and expands it into a formal piece of writing.

Jim Vreeland begins on a word processor, randomly listing words to build an outline, and expands the outline into the section he is writing.

Fidel Salinas, who worked in marketing for IBM before becoming a writer, composes on a word processor. He confesses to having felt nervous "about wasting computer time" when he first started writing that way. Often he found himself staring at the screen, waiting for the words to come. Later he picked up speed.

Terry Allard, supervisor of the group, majored in English and later edited a newspaper. He still needs deadlines or "official dates" to meet before his writing comes easily.

Although each person uses highly personal methods of linking words together to create "writing," each keeps the copy flowing from pencil or word processor. None waits for every bit of information before beginning to write. "Write what's available," Overholt stresses. "Write up what you have, and leave what you don't have for later."

Chris Morgan (1980), former editor-in-chief of *Byte,* a computer magazine, makes the point that you can write in increments that you assemble later, somewhat in the way that motion pictures are made. Morgan, befitting the editor of a computer magazine, uses his computer to shuffle parts of a document into final order. E. B. White, of the *New Yorker,* was fond of scissors and paste, which he used to assemble and reassemble his work. John McPhee, best-selling author of *Coming Into the Country, Levels of the Game,* and other books, uses index cards and a large bulletin board, on which he mixes and matches directions to himself. All these writers recognize a central point about writing: *you must write what you can when you can.* You cannot afford the luxury of waiting for inspiration.

William Zinsser, the author of *On Writing Well,* explained his writing process in *Writing with a Word Processor* (1983). From his chapter explaining his way of committing words to paper, consider the following:

> Writing is a deeply personal process, full of mystery and surprise. No two people go about it in exactly the same way. We all have little devices to get us started, or to keep us going, or to remind us of what we think we want to say, and what works for one person may not work for anyone else. The main thing is to get something written—to get the words out of our heads. There is no "right" method. Any method that will do the job is the right method for you. (Zinsser 1983, 96)

HOW PROFESSORS AND INSTRUCTORS WRITE

Two influential technical writing teachers, Lois DeBakey and F. P. Woodford, themselves differ on approaches to writing. DeBakey, professor of scientific communication at Baylor College of Medicine in Houston, has written science articles, books on science writing, and has taught writing to doctors and scientists. She stresses the importance of preliminary planning and of not composing too soon:

> A frequent, and grievous, error among novices is to begin writing too soon. The result is a diffuse, rambling, confused [piece] that lays bare the incertitude and chaotic thinking of the [writer]. An obscurely defined, vaguely described, sloppily developed project is bound to suggest murky thinking and is likely to portend haphazard research. (DeBakey 1978, 28–29)

Clear thinking leads to clear writing, DeBakey says, and the opposite to the opposite. She suggests making a sentence outline, which is like an abstract. From that the larger paper can be written. Her method runs from idea, to outline, to brief synopsis, to full paper.

Woodford, who chaired a committee on graduate training in scientific writing for the Council of Biology Editors, endorses the idea that writing can help you organize your thoughts. As editor of the influential writing manual that the council produced, Woodford argued for writing a 150- to 250-word synopsis *first*, before making an outline. "Writing such a synopsis sharpens the writer's thoughts," Woodford says. "Outlines often allow them to remain vague" (Woodford 1972, 10).

In trying to learn more about the overall writing process, Tom Waldrep, of the University of South Carolina, asked 30 composition, writing, and English teachers to explain for him how they go about writing. From their explanations, Waldrep compiled *Writers on Writing* (1985). Below, excerpts reflect the different approaches used even by those who teach writing.

William Lutz, associate professor and chair of the Department of English at Rutgers University-Camden, holds a law degree from Rutgers. He explains his approach to writing:

> Before I write, I write in my mind. The more difficult and complex the writing, the more time I need to think before I write. Ideas incubate in my mind. While I walk, drive, swim, and exercise I am thinking, planning, writing. I think about the introduction, what examples to use, how to develop the main idea, what kind of conclusion to use. I write, revise, rewrite, agonize, despair, give up, only to start over again, and all this before I ever begin to put words on paper. (Lutz 1985, 189)*

English professor Donald Murray, at the University of New Hampshire, writes fiction, poetry, and nonfiction. He won a Pulitzer prize for newspaper editorial writing. Here Murray explains his writing process:

> *Immersion*. I am involved with the subjects I write about long before I know what I am going to write about them. And I am on duty 24 hours a day, reading, absorbing, connecting, thinking, rehearsing. . . . Writing is . . . an essential kind of talking to oneself.
> *Concentration*. I need to be able, at the time of prewriting and writing, to concentrate on one task over all others—at least for an hour, an hour and a half, two hours, 15 minutes, 10, less but still a moment when I fall out of the world, forgetting time, place, duty and listen to the writing flowing through me to the page.
> *Deadlines*. I have to have deadlines that are self-imposed or imposed by others and, I confess, the deadlines of others are more powerful than my own.
> *Planning*. I spend most of my time planning what I may write, making lists, taking notes, making notes, making more lists, talking to myself in my head. . . . I try not to be too formal about how I plan—planning should be, above all, play—and I try not to write too early.
> *Drafting*. I write fast; I rush forward writing so fast my handwriting becomes incomprehensible even to me, typing beyond my ability so that the letters and words pile up on the word processor like a train wreck, or dictating so fast I can produce 500, 1,000 in an hour; 1,500, 2,000, 3,000 in a morning.
> *Rewriting*. I'm doing it less and less. Rewriting means the creation of a new

---

* W. Lutz, How I Write, in *Writers on Writing*, ed. T. Waldrep (New York: Random House, 1985). © 1985. Used by permission of W. Lutz.

draft with major changes in subject, focus, order, voice. These days I plan more and rewrite less. But when I rewrite, I start back at the beginning, seeing the subject anew. . . . Rewriting is mostly replanning.
Revising. Of course these first drafts—or third or fourth drafts—will have to be fussed with, cut, added to, reordered, shaped, and polished. . . . That's fun, once you have a draft in hand. (Murray 1985, 222–5)*

Ronald F. Lunsford, associate professor of English and director of composition and rhetoric at Clemson University, describes his approach:

My writing process varies according to situation and purpose. . . . In responding to a letter of inquiry about our freshman English program, I usually know pretty well what I need to say before I begin. The same is true when I write a memorandum to members of our composition staff. In these situations, I often find myself rehearsing in my head what I am going to say. By the time I sit down to write, I know what is going to be in each paragraph, and the writing consists of fleshing out, in words, the thoughts I have already formulated. (Lunsford 1985, 179)†

On and on go the explanations of how composition and English teachers themselves go about writing. The points they, technical and scientific professionals who write, and professional writers make reflect many different approaches. Yet commonalities do emerge.

COMMONALITIES ACROSS FIELDS

From the foregoing discussion and examples of writing processes, the following points emerge:

- Experienced writers have developed and use various writing strategies.
- Experienced writers vary their writing strategies with what they're writing.
- Although the writing process may appear linear upon first examination, a closer look shows that writers constantly return to earlier plans, recast segments, and reconsider organizational approaches, their audience, purpose, and the potential reaction.
- Experienced writers spend time planning, thinking about, and developing the content of their written communication before committing words to paper.

---

* D. Murray, Getting Under the Lightning, in *Writers on Writing*, ed. T. Waldrep (New York: Random House, 1985). © 1985. Used by permission of D. Murray.

† R. Lunsford, Confessions of a Developing Writer, in *Writers on Writing*, ed. T. Waldrep (New York: Random House, 1985). © 1985. Used by permission of R. Lunsford.

- Experienced writers carefully consider their audiences, the audience characteristics, the environment and conditions under which the audiences will read the documents, and the desired impact of the documents on the audience.
- Experienced writers carefully consider the purposes and functions of their products before, during, and after committing words to paper.
- Experienced writers have developed a wide variety of cognitive schemes for planning and organizing their writing. The many different planning and organizational patterns they know often speed their writing.
- For short written products like memos and letters, many writers do not use outlines on paper but plan the structure and content and then commit words to paper.
- For longer manuscripts, some writers use detailed outlines, but the structure and form vary widely. Few use formal outlines; instead they use private outlines. The private outline may be a general listing of ideas, topics, phrases, or words. A few writers use elaborate outlines, much like the formal outlines you may have learned in high school.
- Some writers carefully draft and build their manuscripts word by word, sentence by sentence, paragraph by paragraph, page by page, refining as they work.
- Other writers dump their copy, rushing quickly through the manuscript, to revise and polish later. They hurry through the drafting stage to get their words onto paper or screen.
- When writers polish their manuscripts, they use a variety of techniques to delete unnecessary words and tighten their copy.
- Some writers write in longhand, some use manual, electric, or electronic typewriters, and a growing number work directly on word processing units.

The implications for your transition from in-school to on-the-job writing are considerable. Writing at work will be easier if you learn different approaches to speed up your writing process. That is the purpose of the next five chapters. Study and practice the techniques given and you can develop a process that helps you write.

HIGHLIGHTS

1. Professionals who write regularly use a variety of techniques and strategies for writing.
2. Professional writers develop approaches to writing that are not unlike those of scientific and technical professionals who write.

3. Professors who both research writing processes and teach writing courses use many different writing strategies and approaches adapted to their individual writing styles.
4. There are commonalities to the writing strategies of professionals who write, professional writers, and writing teachers and researchers.

PROJECTS

If you analyze your writing process, you can better understand what you do and enhance your writing skills. Spend some time thinking about the questions below:

a. Do you use a prewriting technique, such as thinking about the audience, the topic, the points you want to make?
b. Do you know your readers' characteristics? Do you anticipate reactions? If so, what are they?
c. Do you identify a specific purpose for your writing? If so, what is it?
d. Do you use an outline or list to guide you through your writing? If so, briefly explain and describe your aid.
e. Do you write with pen and paper, typewriter, or a word processor?

Apply the foregoing questions to the projects below.

1. Consider the last business letter you wrote seeking information or applying for a job. If you have not written a formal business letter recently, consider a letter to a friend or relatives. Prepare a 500-word narrative about your writing process.
2. Consider your last major paper or project requiring a finished manuscript of at least 10 pages. Prepare a 500- to 1,000-word narrative about your writing process.
3. In class, discuss how you go about writing long reports or papers. Take notes on the processes that your classmates use. Prepare a 500- to 750-word narrative comparing and contrasting these approaches.
4. In class, discuss how you go about writing short letters, memos, and reports. Take notes on the processes that your classmates use. Write a 500-word narrative comparing and contrasting these approaches.

5. In class, discuss the problems you face in becoming a better writer. Take notes on your classmates' problems. Write a 500-word report on these problems, offering suggestions for their solution.

FOR MORE HELP

WALDREP, T., ed. 1985. *Writers on writing*. New York: Random House.
    In Waldrep's book, 30 professors explain how they write. The explanations show the diversity of approaches to finished products.

7
A Closer Look at Drafting Manuscripts

OVERVIEW

In Chapter 7, we

- ☐ Suggest differences between professional and student writing and explain how you can begin the transition
- ☐ Discuss process writing models
- ☐ Expand on planning strategies for writing documents
- ☐ Suggest organizational strategies for structuring your writing
- ☐ Discuss selected techniques for developing paragraphs
- ☐ Suggest ways of committing ideas to paper and disk

The previous chapter explored how various writers write. Now we elaborate on different writing strategies and processes. As you read, compare each strategy with the one you have been using. Then try out new strategies. Your objective is to find techniques that improve your writing and that ease your transition from student to on-the-job writing.

Writing in school differs in at least six ways from writing on the job. First, writing on the job functions to inform, instruct, persuade, and document. In contrast, most in-school writing functions to instruct, with you doing the learning. You study a topic, and your writing reflects what you have learned. Your professors usually evaluate your writing according to this criterion.

Second, in-school writing assignments are designed to help you learn to express yourself—professionals are assumed to have mastered that skill.

Third, most school assignments have but one reader, the instructor. In contrast, most professional writing has many different readers: supervisors, their bosses, peers, customers, clients, and maybe hundreds of others.

Fourth, in most cases, your teachers know more about your subject than you do. But as a professional, you probably will know more about your subject than most of your readers. Therefore, you must consider your readers' frames of reference more carefully than you have for your classroom assignments.

Fifth, as a student, you may have greater leeway about waiting until the spirit moves you to write. As a professional, you'll find yourself writing day in and day out, working under tight deadlines, and seldom allowed the luxury of waiting until that inner voice tells you to begin. You must write when and where you can.

Sixth, although sometimes you may have had difficulty choosing topics to write on in school, you'll seldom have problems finding topics to write about on the job. Your day-to-day professional projects and assignments will provide ample topics.

WRITING PROCESSES

By studying writing processes, you are likely to become a better writer. The following discussion reviews "free writing," the Document Design Center's model, and a model based on Linda Flower's and John Hayes's research on problem solving techniques in writing.

Free Writing

Peter Elbow (1973) describes "free writing" as "automatic writing," "babbling," or "jabbering." The basic strategy is to begin writing and continue writing without stopping, thinking, or worrying about details.

Rush your ideas down. Don't worry about what you're saying or how you're saying it. Don't worry about the order of your points. Don't worry about ideas, how you express them, your grammar, spelling, word choice, or any one of a hundred other factors that may slow you down. Don't worry about editing the document when you're writing. Don't evaluate your writing. Don't worry about mistakes. Simply capture your ideas on paper.

When you use free writing, separate the creative activity—the writing or committing of words to paper—from editing—the judgmental, often destructive activities. Don't edit in any way. Elbow argues that over the years, most people have learned that they make mistakes when writing, so they unconsciously try to avoid making mistakes when they write by constantly editing their writing. And that, he says, is a mistake. It hampers the free flow of ideas. Turn off that inner voice that keeps telling you you're making mistakes, and you avoid what Elbow calls "premature editing." *Edit after you've committed your ideas to paper.*

To begin free writing, Elbow recommends exercising for 10 to 20 minutes at least three times a week. Keep your free writing exercises short. Set passages aside, and return to them later. Using this method, you will be able to spot incoherent passages. When you spot one, try free writing again. And again. It will help you to think through what you're trying to say and how to organize more logically. Elbow declares:

> Writing things down can accelerate growth. When you write things down—as long as you don't write them down with too much commitment—you are able to see them in perspective. It is as though holding onto that thought or perception were a burden to your mind. Writing is a setting down of that burden, and it lets the mind take a rest from it. Now the mind can better see what is limited about it and take up a new thought or perception. (Elbow 1973, 46)

Once you have your thoughts on paper, you can begin the editing process. Elbow says that the essence of editing is "easy come, easy go." You must cut deeply and critically. Don't be too tied to what you have said and how you said it. As Elbow explains:

> Editing is almost invariably manipulative, intrusive, artificial, and compromising: red-penciling, cutting up, throwing away, rewriting. And mostly throwing away. For this process, follow all standard advice about writing: be vigilant, ruthless; be orderly, planned . . . try to think of a better word, struggle for the exact phrase, try to cut out the "dead wood," make up your mind what you really mean. (Elbow 1973, 38–39)

Chapters 9, 10, and 11 explain editing and revision techniques that will help you tighten your drafts. Elbow's free writing approach provides but one writing strategy. Let us consider another.

The Document Design Model

The process model in Figure 7-1 evolved from the work of the Document Design Center of the American Institutes for Research under a project funded by the National Institute of Education. Researchers Dixie Goswami, Janice Redish, Daniel Felker, and Alan Siegel (1981) investigated writing in many different fields. Their model provides strong guidelines for technical and scientific writing. To paraphrase their assumptions:

- Writing is a dynamic, multi-step process.
- The writing process begins with a determination of who will read the document, why it is being written, and what it will cover.
- After the first draft, the writer must review, edit, evaluate, and revise a document until the intended reader can understand it.
- Writing as a means of communication must emphasize the writer's connection to the reader.
- Writing is a recursive process, with the writer looping back time and again throughout the process.

As illustrated in Figure 7-1, the researchers suggest five separate activities for prewriting. The following discussion adapts their guidelines to technical and scientific communication.

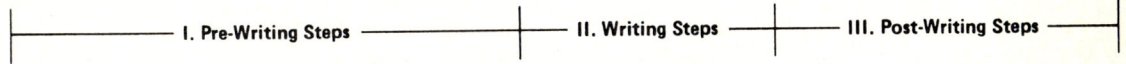

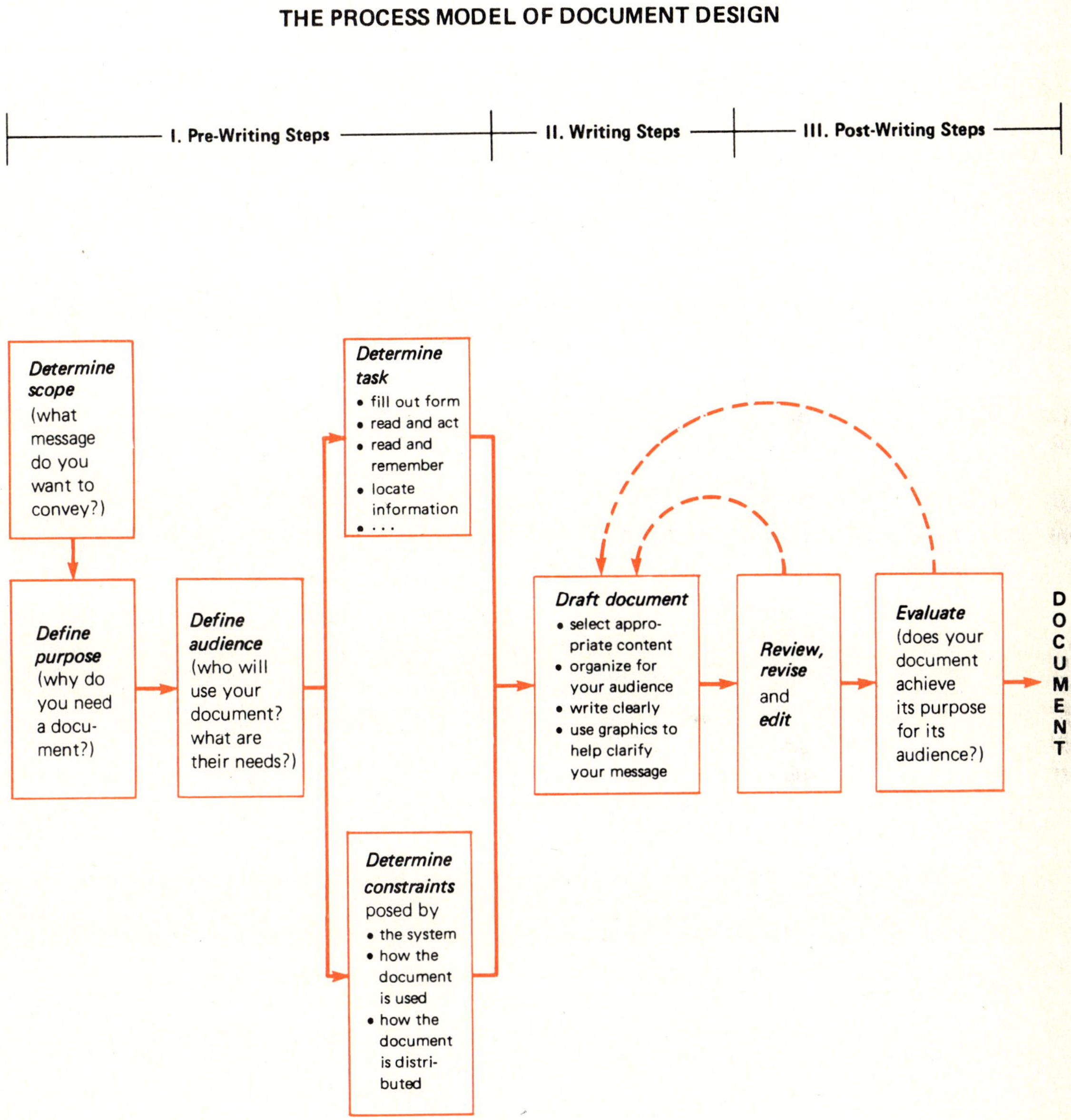

FIGURE 7-1. The process model of the Document Design Center.
(Source: Dixie Goswami, Janice C. Redish, Daniel B. Felker, and Alan Siegel, Writing in the Professions *(Washington, D.C.: Document Design Center, American Institutes for Research, 1981), p. 38. Copyright © 1981 American Institutes for Research. Reprinted by permission.)*

Prewriting: Determine Scope. Ask first, "In general, what is the message I want to convey?" Determining scope in technical and scientific documents requires you to select the relevant from the vast amount of information you have (Panel 7-1). Consider the key points you need to cover, and clearly establish them. Keep in mind that other prewriting activities may cause you to recast and reconsider your scope.

Prewriting: Determine Purpose. Ask next, "Why am I writing the document?" Consider the overall communication functions identified in Chapter 1. Will your written product inform, instruct, persuade, or document? Two questions can help you clarify your purpose:

- What do I want my readers to know?
- What do I want my readers to do?

Specify and refine your secondary objectives or purposes. As you move through the prewriting process, you may change or refine your purpose.

Prewriting: Define Audience. Ask next, "Who will read my document?" Use the guidelines given in Chapter 5 to help identify your audience and learn its characteristics, key steps in communicating.

Prewriting: Determine Task. Ask now, "What do I expect my readers to do after having read the document?" What tasks should your document help readers accomplish? The Document Design Center suggests a variety of tasks for document readers: to understand, locate information, act immediately on the information, fill out forms, and accomplish other objectives. The specific tasks you identify represent a refinement of your document's overall purpose.

Prewriting: Determine Constraints. Ask now, "Which constraints must I keep in mind?" Many different constraints influence how and what you write: deadlines, document length, budget, format, reviewers' roles, printing considerations, distribution, storage requirements, document uses, and use environment. The uniqueness of your organization and the document you're preparing will determine your constraints.

Once the writer begins to draft the document, four considerations arise: content, organization, language, and graphics (Figure 7-1).

Draft Document: Content. You cannot and should not tell all. You must serve as a gatekeeper and select the most relevant information in light of your scope, purpose, audiences, tasks, and constraints.

Draft Document: Organization. The Document Design Center model calls for you to consider how you organize the project, the document as a whole, each section within the document, and the sentences within each paragraph.

PANEL 7-1. Questions to Guide Prewriting

example

Form for Answering Pre-Writing Questions

Pre-Writing Decisions

Ask yourself these pre-design questions about your document. If you don't know the answers, discuss the questions with appropriate others to find the answers.

Brief statement: _____
(type of document and what it is about)

Scope: What is the topic? the message? Where did you go to find out information?

_____
_____
_____
_____
_____

Purposes: What purpose**s** will your document serve?

_____
_____
_____
_____
_____

Audiences: Who will read your document? Who else is concerned about what your document is like?

The (primary) readers who are most important to you are:

1. _____ 2. _____

The (secondary) readers who are also important to you are:

1. _____ 2. _____
3. _____ 4. _____

(Source: Dixie Goswami, Janice C. Redish, Daniel B. Felker, and Alan Siegel, *Writing in the Professions* (Washington, D.C.: Document Design Center, American Institutes for Research, 1981), pp. 68–69. Copyright © 1981 American Institutes for Research. Reprinted by permission.)

(Continued)

PANEL 7-1. (continued)

example

Some important characteristics of the primary audiences that might affect how you write the document are:

1. _____
2. _____
3. _____
4. _____
5. _____
6. _____

For the secondary audiences, you should be concerned about these specific points:

audience _____   concerns _____
                                    _____

audience _____   concerns _____
                                    _____

audience _____   concerns _____
                                    _____

audience _____   concerns _____
                                    _____

Tasks: What does your primary audience have to be able to do to use your document?

_____
_____
_____
_____
_____

PANEL 7-1. (continued)

example

Constraints: What else must you keep in mind when writing? Is it an impediment to your purpose? What can you do about it?

constraint _____ OK? ___yes ___no

If it is not OK, what can you do about it? _____

_____
_____
_____
_____
_____

constraint _____ OK? ___yes ___no

If it is not OK, what can you do about it? _____

_____
_____
_____
_____
_____

When many professionals have data or information to present, they first prepare their illustrations—tables, graphs, charts, and other visuals—and then organize their documents around that data. In addition, your audience, its need for information, and your overall purpose should influence how you organize the document and its components.

Draft Document: Language. Like many writers, the Document Design Center researchers advocate writing clearly, simply, and directly. Achieving clear, concise writing comes with learning basic techniques, as described in Chapters 9 and 10.

Draft Document: Graphics. The Document Design Center discusses graphics considerations such as typefaces, line length, margins, spacing, layout, and use of illustrations. In technical and scientific communication, illustrations play especially important roles. Chapter 8 provides detailed guidelines for developing technical and scientific illustrations.

Post Writing: Reviewing, Revising, and Editing. You improve your writing by setting your drafts aside for a few minutes, hours, or, better yet, days. Then review and edit them yourself. As you make the transition to professional writing, you will find that your peers and superiors review and edit your drafts. Armed with their comments, you'll return to prewriting and revising your drafts.

Not all documents need evaluations, nor do you always have time to evaluate documents as you'd like. But when you do, focus your evaluations on the product or its users. Assess its overall quality and effectiveness in light of its scope, purpose, audience, task, and constraints.

The Problem Solving Model

Viewing writing as problem solving fits technical and scientific communication well, because most fields emphasize such a strategy. Furthermore, professional writers have long viewed writing as problem solving. As William Zinsser comments,

> All writing is ultimately a question of solving a problem. It may be a problem of where to obtain facts, or how to organize material. It may be a problem of approach or attitude, tone or style. Whatever it is, it has to be confronted and solved. (Zinsser 1980, 53)

Over the last two decades, researchers have found that many professionals use problem solving strategies to guide their writing. Foremost among investigators of problem solving strategies are Linda Flower and John Hayes of Carnegie-Mellon University. Their model appears in *College Composition and Communication* (see Figure 7-2). The model shows writing

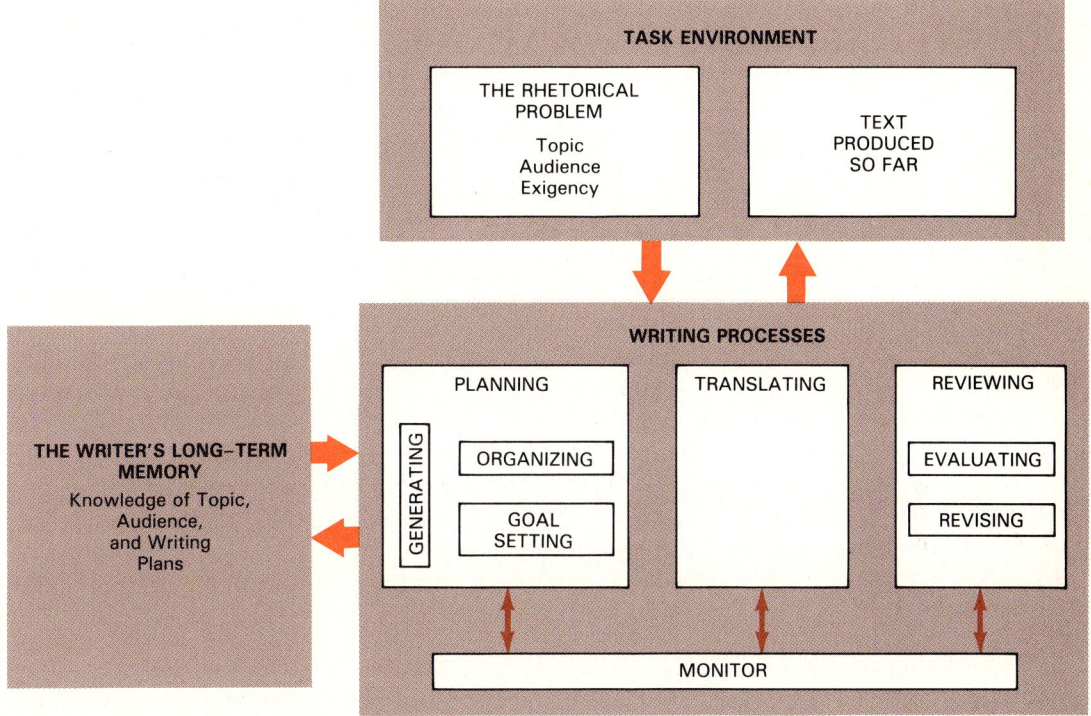

FIGURE 7-2. The problem solving model of writing.

(Source: Linda Flower and John R. Hayes, "A Cognitive Process Theory of Writing" *College Composition and Communication, December 1981.* Copyright © 1981 by the National Council of Teachers of English. Reprinted by permission of the publisher.)

as a recursive process. It can be adapted easily to technical and scientific writing.

The task begins when you solve the technical or scientific problem you face. Flower and Hayes call the assignment to write an article or other document the "rhetorical problem." Your topic, audience, and requirements for writing, or "exigency," make up the components of the rhetorical problem.

In solving your problem up to this point, you may have prepared other text: reports, letters, memos, and documents. You have collected notes, reprints, photocopies, microforms, and other documents. The rhetorical problem plus the text produced so far make up the "task environment."

Now you move to writing processes, themselves. Hayes and Flower (1984) have described planning as generating information, organizing or structuring the information, and setting goals. Generating information requires thinking about the problem, both by searching your long-term memory for information you've accumulated from experience and by considering what you know about the audience. Goal setting helps you select relevant information and influences your organization. Organizing entails grouping your ideas together and structuring your information in a logical pattern.

Translating requires converting your planning—generating, organizing, and goal setting—to symbols on paper. Usually you'll use words as the symbols, but some professionals start with diagrams, illustrations, or notations to express their thoughts. Then they translate those into words.

In the "reviewing" area, as Flower (1985b) points out, expert writers develop plans for evaluating and revising their drafts, and they revise as they refine their understanding of their goals. They review the complete communication, a process larger in scope than just correcting grammatical and spelling mistakes and tightening copy. They monitor themselves throughout the writing process, trying to figure out what's wrong and what can be improved in a draft. First they diagnose problems and then solve them. Monitoring is a process whereby writers reflect on planning, translating, and reviewing. Monitoring tells the writer when to move on to another stage of the writing process.

PLANNING, SELECTING CONTENT, AND ORGANIZING: PREPARATIONS FOR WRITING

When you prepare long manuscripts, such as technical reports, you'll benefit from planning and organizing your ideas first.

Planning: Content, Writer, and Purpose

To communicate your ideas effectively, carefully consider your topic, your audience, and your objectives. By the time you've solved most technical and scientific problems, you'll have more information than you can use in one document. You'll have to select and present different information to different audiences. You'll have to consider the general and specific purposes of your document—whether you want to inform, instruct, persuade your readers or merely want to document your activities. Consider the points raised in Chapter 5 about audiences and their characteristics. Your answers will help you to select information relevant to your readers.

Many things go through experienced writers' minds as they plan their writing. For example, after having reviewed dozens of articles, books, and reports on writing, having talked with professionals, and having practiced professional communication for many years, we faced the problem of selecting appropriate content for this chapter. Two primary audiences came to mind. First, we thought of instructors, the people who select texts for technical and scientific communication courses. Then we thought of you, the students taking those courses. We tried to keep characteristics of both audiences in mind.

We worked under certain constraints. A chapter can have only so many pages and a book so many chapters. Our contract with Random House limits our final manuscript to 756 pages. We had decided on a 20-chapter book and calculated each chapter should average 36 pages of typed, double spaced copy. Another constraint was time. We faced deadlines. Each of us had to average writing and rewriting one chapter every two weeks.

As we thought about this chapter, we considered how illustrations would reduce the copy's length, add visual appeal, and break up the printed page. We also began thinking about what kinds of assignments might help you understand your writing process. Two key questions drove planning for this chapter: what do students who are developing their technical and scientific communication skills need to know about writing processes, and what will help them improve their writing processes?

Several points emerged as we planned the chapter. We thought we should discuss the following topics:

1. how technical and scientific writing differs from composition and literary writing
2. how models of writing processes differ
3. how guidelines can help planning
4. how different organizational techniques and outlining strategies help planning
5. how to commit planning, organization, and thoughts to paper
6. how students can develop a better understanding of their writing processes through assignments
7. how readers can find more information about the writing process

We considered how best to make the points. We drafted an outline (Panel 7-2). From that emerged a 12-page, modified outline. We set it aside to cool for a while, discussed it, and set it aside again. Finally, one of us drafted this chapter.

Selecting Content

As we have seen, generating technical and scientific writing is a selective process. Let us consider several approaches to selecting content: the journalist's approach, "cubing," and brainstorming.

Selecting Content: The Journalist's Approach. Take a lesson from journalists in selecting information. Journalists' primary function is to inform. Occasionally they instruct. Journalists let their answers to six general questions guide their selection process:

- Who?
- What?
- Why?
- Where?
- When?
- How?

Try using these questions to select your content. The questions work well for the content of memos, letters, and short reports and documents. If you're having trouble selecting information, try answering aloud the journalist's questions. When you start talking, you instinctively abbreviate and summarize your information.

Selecting Content: Cubing. Gregory Cowan and Elizabeth Cowan (1980) recommended "cubing" to help writers generate content. Imagine a cube with one of the following directions on each side:

- Describe it.
- Compare it.
- Associate it.
- Analyze it.
- Apply it.
- Argue for or against it.

"It" represents your topic. In cubing, write your responses quickly. Do not spend more than three to five minutes on any activity:

> *You are not hunting for something to say from each perspective;* you are trying a *quick* run into your mind for whatever presents itself on that angle, and the quickness of the run is important. (Cowan and Cowan 1980, 21)

Although cubing helps develop content primarily for essays, it can help you to summarize your content. Cubing is much like Elbow's free writing technique, but it's more directed. Cubing forces you to summarize and abbreviate your points. Once you have the basic ideas down, you can build upon them.

Selecting Content: Brainstorming. Brainstorming, as we've described, involves letting your brain run freely, jotting ideas down in no particular order. You let your thinking motor run full force, without brakes. Let your brain run freely, and don't evaluate your ideas at this stage. After your brainstorming session, you can consider the points in light of your audience, desired outcome, and constraints.

A CLOSER LOOK AT DRAFTING MANUSCRIPTS   131

PANEL 7-2. Segments of a Working Outline

Ideas on Writing Chapter   *16 April '85 — Let cool. Develop detailed outline before writing.*

How writing on the job differs from in-school writing

General observations about writing . . .   *Examples? Needed?*

-- process oriented
-- problem solving   *Keep space limitations in mind.*
-- using process models
-- get something on paper

Key differences between composition and technical writing as a professional?

Are there differences? What do I write about? Readers? Functions --
expressive vs functional . . .   *What else? Expand?*

A look at process models   *Consider your audience*

With which one should you start?

Document design model . . . has potential . . . gives an alternative way
of considering the process

Linear model . . . Should we use this one? . . . in some ways it's   *Delete?*
closer to the Document Design Center's steps.   *Will readers know it?*

Free writing -- Elbow. Should we start here? Dumping style?
contrast to Zinsser's approach

Flower + Hayes — research — influences on process!

Planning considerations

Content's key role in the process

Audience ... consider its needs, etc. . . . desires to reach with your
message. . .

You as the write . . . frame of reference issue

Discovery (Invention) Techniques -- What you know in light of what you'll
use

Cohen & Cohen
Brainstorming
Journalist's 5Ws and How

Ways of organizing information

Pattern notes ... trees ... informal/ formal outlines

Organizing Content

Once you have your ideas in hand, you can organize them into a logical pattern. Eventually you'll need to put the ideas in a linear sequence, because reports, memos, books, manuals, and all other written documents are linear in that they move from point one, to point two, to point three, and so on.

If, when you think of organization, you think first of a formal outline, think again. You need not begin with a formal outline, even though your finished product gives readers the impression that you did use a highly structured outline.

Several techniques can help you organize your content: pattern notes, issue trees, informal and formal outlines.

Pattern Notes. Alan Fields (1982), of Cranfield School of Management in Bedford, England, pioneered the use of pattern notes (Figure 7-3). Pattern notes lend themselves well to organizing points generated during brainstorming, and you can use them in lieu of brainstorming. Pattern notes provide a visual plan—ideal for the visually oriented person—of how to structure information meaningfully.

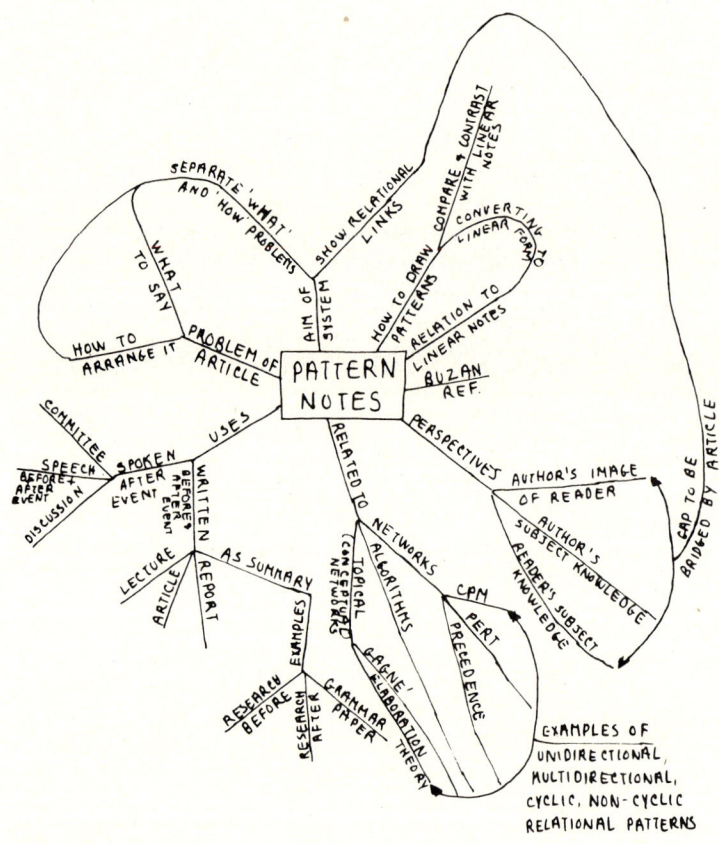

FIGURE 7-3. Pattern notes.

(Source: David H. Jonassen, ed., The Technology of Text *(Englewood, N.J.: Educational Technology Publications, 1982). Reprinted by permission of the publisher.)*

To begin making pattern notes, select your key topic or point, box it in the center of a piece of paper, and then add spokes for subpoints. Finally, link all the spokes together to show relationships and how you might tie them together. If you're dealing with a topic with several key points or "centers of interest," as Fields calls them, you may link together several boxes.

Issue Trees. Linda Flower (1985b) suggests using issue trees (Figure 7-4) to organize your information. Issue trees help writers to sketch out and visualize their ideas. They provide flexibility, because the writer can change, recast, and rework during the organizing process.

To develop an issue tree, write your main point at the top of a page; then list the subpoints under it. Then list subpoints in decreasing order of importance. If you get stuck, ask yourself questions about each branch in the tree:

- What do I mean?
- How so?
- How do I know?
- Such as?
- Why?
- Why not?
- So what?

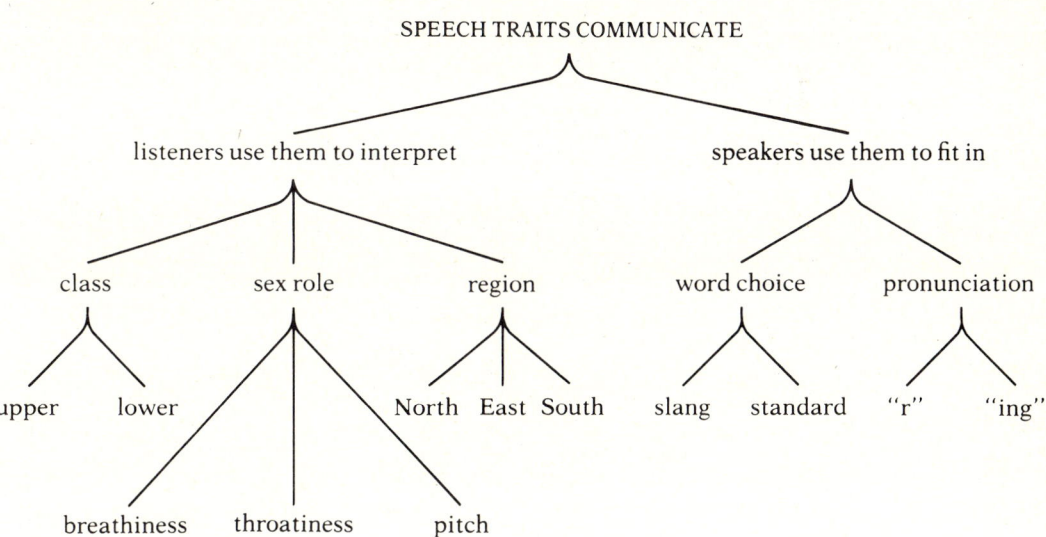

FIGURE 7-4. A basic issue tree.

(From Problem Solving Strategies for Writing, *Second Edition, p. 101, by Linda Flower, copyright © 1985 by Harcourt Brace Jovanovich, Inc. Reprinted by permission of the publisher.)*

Issue trees show you how well you have balanced your treatment of key points. At times, you may find that one branch constitutes most of your focus, almost to the exclusion of other branches. If so, you can balance the branches, before investing hours in drafting your narrative.

Informal Outlines. Informal, working outlines can help you to create and revise a document. They are strictly to help you write and are not for others to see. An informal outline

- organizes your material's order of presentation
- helps achieve a balance among the various parts
- reveals gaps that may exist in material or thought
- helps your writing and helps you to start and finish the project on time
- shows what is to be done

A working outline, such as that shown in Panel 7-2, can help you to organize your thoughts.

To learn how professionals structure their outlines, Blaine McKee (1974), professor of technical communication at Colorado State University, surveyed 180 members of the Society for Technical Communication. Of the 80 who responded to his survey, McKee found that some 90 percent used a topic outline, consisting of words, phrases, and sentences. Only 5 percent each used a formal sentence outline or no outline at all.

As logic suggests, you can organize information in many different ways (see Panel 7-3). Historians frequently outline subjects by time of occurrence. They might list events in the United States from the Civil War to 1900. Historians also outline according to, say, social movements that have affected elections of candidates for office. A salesperson might outline by geographical sales districts or by type of product sold. Someone might create a quantity-based outline, such as one that categorized farms of less than 80 acres, 81 to 160 acres, 161 acres or more. Another farm report might be outlined by type of owner: family, corporation, single operator, or cooperative. The type of outline you select should be guided by the purpose and content of your message and your audience.

Often you'll find that your company or the publication for which you are writing requires certain organizational structures. For example, see Panel 7-4 for a common structure for many technical and scientific reports. To use these organizational patterns, you'll need to organize your content for each section. That's where the logic of your content will influence how you organize it.

Formal Outlines. Formal outlines prescribe specific headings and subheadings, with each major unit indented. A common outline system (dreaded by many a composition student) is this:

PANEL 7-3. Fundamental Organizational Patterns

In his classic, *Basic Technical Writing* (1985), Herman Weisman suggests that writers consider the purpose of their document, their readers' needs, and the nature of the material when considering organizational patterns. Weisman suggests the following as possible organizations:

1. *Chronological Pattern* documents the sequence in which something happened.
2. *Geographical or Spatial Pattern* arranges or presents information based on the physical arrangement of entities—for example, an office's layout.
3. *Functional Pattern* works well for examining how parts of a mechanism or machine work and in what order they work.
4. *Order of Importance Pattern* puts elements in order of decreasing importance.
5. *Elimination of Possible Solutions Pattern* works from the least likely to the most likely or best solution or builds to a climax.
6. *General to Particular Pattern* discusses the general principles and moves to the specific application of the principles.
7. *Particular to General Pattern* presents the specifics first and then derives the overall principles or concepts.
8. *Simple to Complex Pattern* begins with the simple case or topic and moves to the more complex cases or topics.
9. *Pro and Con Pattern* presents the positive and negative points on an issue or decision.
10. *Cause and Effect Pattern* presents the causative factors and then their results.

    I. First level: first point
       A. Second level: first point
       B. Second level: second point
          1. Third level: first point
          2. Third level: second point
             a. Fourth level: first point
             b. Fourth level: second point
                (1) Fifth level: first point
                (2) Fifth level: second point
                    (a) Sixth level: first point
                    (b) Sixth level: second point
   II. First level: second point

Another common outlining system, based on a numerical or decimal system, structures the descending headings: 1; 1.1; 1.1.1; 1.1.1.1; 2; 2.1, etc. The decimal system is commonly used in some companies, government technical publications, and military publications.

    For formal outlines, the general rule is to use only complete sentences, *or* complete phrases, *or* words, not mixing sentences with phrases or words. Each level also should have two or more parts—if you have a first

> **PANEL 7-4. A Basic Technical Article Organization**
>
> **Title**
>
> **Abstract**
>
> Summarizes the methods and results.
>
> **Introduction**
>
> Provides a literature overview that builds the rationale for the research question, objectives, or problem statement.
>
> **Methods**
>
> Explains how you collected the information or went about carrying out a particular activity that provided or produced the data or information to solve the question.
>
> **Results**
>
> Reports the findings of the project. Includes tables, graphs, or other visuals.
>
> **Discussion**
>
> Reports the implications of the findings, what they mean in light of the research question, objectives, or purpose.
>
> **Conclusion**
>
> Provides a succinct summary of the key findings and, possibly, the implications.
>
> **References**
>
> Provides complete citations in the publication's style.
>
> **Appendix**
>
> If needed, includes items, such as detailed tables, notes, additional discussions, explanations, or materials of interest to only some readers.

level, you'll need a second level. All divisions should be treated equally and of similar length. Following these rules will help you gain approval if your formal outlines go out for review and criticism. Keep in mind that any outline should help you write, not hinder you.

Finally, formal outlines often turn up as tables of contents in finished manuscripts. Some writers produce such outlines after they've written, revised, and edited their document. Make sure that your table of contents conforms to the major headings in your manuscript.

DRAFTING PARAGRAPHS

Creating paragraphs may have become an almost instinctive process. You let the logic of your topic guide you. But, on occasion, you may get stuck. The points you want to make just won't go together easily, or when you finally have them on paper, your paragraphs won't flow the way you think they should. By taking a look at paragraph building, you'll gain insight into constructing paragraphs that communicate.

Paragraphs are typographical units, tied together by the logic of their content. They have many characteristics. Frederick Crews and Sandra Schor (1985) suggest that a typical paragraph

1. presents one main idea;
2. conveys thoughts that are connected by logical association and by word signals;
3. often reveals its main idea in a prominent statement, usually but not always toward the start;
4. usually supports or illustrates that idea;
5. may also deal with objections or limitations to that idea, but without allowing the objections to assume greater importance than the idea itself; and
6. may begin or end generally, taking an expanded view of the addressed topic. (Crews and Schor 1985, 246)*

Typically, a paragraph in technical and scientific writing serves many different functions, chief among them being to

- break up copy and make a visually pleasing page
- help the reader
- introduce ideas or concepts
- provide transition from one major idea to another
- summarize information
- emphasize key points

Not only are paragraphs typographical units, they are visually tied to the other parts of the text. In publications, articles are usually introduced with headlines and divided with subheads or headlines. For example, the December 1985 issue of *Photomethods*, a trade magazine for professional photographers, carried an article on photography in Antarctica. Accompanying the article was a sidebar, a three-quarter page, how-to article, "Cold Weather Photography." It was printed in black type with lines around it and a light blue—cold blue—background. A simple headline, "Cold Weather Photography," introduced the article:

> Photography in cold weather is not only unpleasant for the photographer, but also brings up problems in camera and battery operation that do not exist at normal temperatures. To make the work pleasant, dress properly, and wear the type of gloves that allow you to operate the important camera controls with gloves on. (Wildi 1985, 33)

Then the opening paragraph established two problems—those for the photographer and those for the equipment—and began providing solutions.

---

* Frederick Crews and Sandra Schor, *The Borzoi Handbook for Writers* (New York: Knopf, 1985). Copyright © 1985 Alfred A. Knopf, Inc. Used with permission.

Paragraph Length

A paragraph's length, or word count, depends on the column width, spacing between lines, type size, and the author's writing style. An informal review of selected research journals, trade publications, and popular publications showed typical paragraph lengths by word count:

Type of Publication	Range	Average
Research Publications		
Clay and Clay Materials	60–400	200
IBM Research Journal	50–200	90
Trade Publications		
Civil Engineering	30–145	70
Physics Today	50–160	75
Byte	50–120	70
Popular Publications		
Time	70–200	125
Denver Post	5–65	20
Organic Gardening	54–114	71

Scholars studying newspaper design have found that "gray copy," inches and inches without visual breaks, discourages reading. To develop a feel for the grayness of type, view several different newspapers, books, or other publications from a distance of three to four feet (Nelson 1983; Strong 1926). You'll quickly see the print's gray impression. Long paragraphs appear dense and hard to follow. In contrast, short paragraphs add white space by breaking up the copy.

In many technical and scientific publications, paragraphs run from 1 to 3 or more inches long, depending on the publication, authors' styles, and paragraph functions. Paragraphs of instructions, descriptions, and directions may be only one sentence long. Other paragraphs have two, three, or more sentences.

Short paragraphs serve another important function: they help readers to find points quickly. Readers can scan short paragraphs more quickly than long paragraphs. The paragraph breaks give readers visual clues for finding selected information.

Developing Paragraphs

Frederick Crews and Sandra Schor (1985) explain that unity and continuity help build well developed paragraphs. *Unity* means that each paragraph has one leading idea, topic, or concept. That leading idea appears in the main, or topic, sentence. *Continuity* means simply that the writer logically ties each sentence to the previous one. Consider the unity in the following paragraphs:

> The Hessian fly, the most destructive insect pest of wheat in the world, has been controlled on winter wheat in the United States for nearly 30 years. So far, the battle has been won, not with chemical pesticides, but by breeding genetic resistance in the wheat plant.
>
> But while the Hessian fly may be down, it can't be counted out. David Keith, University of Nebraska entomology professor, writes that "the Hessian fly is still lurking around the edges of fields in the American wheat belt. We never know when the fly might change genetically and again become a threat to our wheat crop." As a result, the fly continues to be of intense interest to entomologists in this country and abroad. (Kelley 1985, 10)

The first paragraph introduces the topic—control of the Hessian fly—and then explains how scientists achieved control. The second paragraph says that the Hessian fly is still around, can still threaten the country's wheat crop, and why the fly is still a threat.

To develop paragraph continuity, signal changes for your readers. Crews and Schor suggest using different words to signal changing relationships:

Consequence: therefore, then, thus, hence, accordingly, as a result
Likeness: likewise, similarly
Contrast: but, however, nevertheless, on the contrary, on the other hand, yet
Amplification: and, again, in addition, further, furthermore, moreover, also, too
Example: for instance, for example
Concession: to be sure, granted, of course, it is true
Insistence: indeed, in fact, yes, no
Sequence: first, second, finally
Restatement: that is, in other words, in simpler terms, to put it differently
Recapitulation: in conclusion, all in all, to summarize, altogether
Time or place: afterward, later, earlier, formerly, elsewhere, here, there, hitherto, subsequently, at the same time, simultaneously, above, below, farther on, this time, so far, until now (Crews and Schor 1985, 252–3)*

You can also provide continuity by using pronouns, demonstrative pronouns or adjectives (that, this, those, these), repeating words and phrases, and implied repetitions.

---

* Crews and Schor, *Borzoi Handbook.* Copyright © 1985 Alfred A. Knopf, Inc. Used with permission.

Writing Paragraphs

If you have trouble writing a paragraph, ask yourself a series of questions, such as

- What's the main idea or topic?
- What points must I make in this paragraph?
- What points must I add to expand, support, or modify the main idea?
- What examples can I present that illustrate the main idea?

Think about the questions, answer them mentally, or jot down notes on paper. Then talk your way through your points. Once you've done that, commit your ideas to paper or disk. You can return later to edit and recast the paragraphs.

COMMITTING IDEAS TO PAPER

The time comes when your thinking, planning, and reflecting are ripe, and you must write. You must commit your ideas to paper, whether directly or through word porcessing or dictating. Try to improve your writing speed by experimenting with other systems of committing your thoughts to paper.

Paper, Pencil, or Pen

If you're typical of most students, you probably write most frequently with pen or pencil and paper. When required, you either retype what you've written or have someone type it for you. Some writers argue that they cannot write on a typewriter or a word processor because they must feel their words as they write.

If you're a pen and paper writer and haven't tried writing on a typewriter or word processor, give them a try.

Electric or Electronic Typewriter

Many professional writers compose directly on the typewriter. The machine becomes an extension of their thinking. Touch typing allows a writer to write more quickly and edit more easily than handwriting copy does. When you draft copy on a machine, put your words on paper as quickly as you can. Don't worry about every detail. You can return later, clean up the manuscript, and correct typing, spelling, punctuation, and other errors.

Word Processing Units

Most microcomputers have software programs that permit word processing. Writing directly at a word processor can speed your writing. Many also have programs for checking spelling, grammar, and style. If you touch type, you can commit hundreds of words to disk within an hour. You can return to your copy later and clean up the errors, all without having to retype. Learning to touch type undoubtedly makes writing on word processors easier. If you don't touch type, learn to do so soon. It's a basic skill that more and more professionals need.

To give some idea of a writer's speed with a word processor, *Time* magazine reported that conservative columnist William F. Buckley had shifted to a word processor (Murphy 1985). Buckley dashed off a newspaper column in 20 munutes and a 7,500-word manuscript for a children's book in two hours on his personal computer. Buckley, by the way, claims to touch type at 110 words a minute.

With a word processor, you can delete, add, and revise almost painlessly. The ease of making such changes reduces most writers' dread of having to retype their copy. Furthermore, many professionals report that using a computer for writing cuts writing time by one third to one half compared to writing on electric typewriters.

Because writing on a word processor will become more and more common in technical and scientific fields, we devote half of Chapter 14 to an overview of the process.

Dictating

The old image of a boss dictating a letter to his secretary has given way to a new image of professionals speaking into dictating machines or recorders. The dictation then goes to a secretary or word processing clerk who transcribes the dictation. Developments in computer voice recognition hold promise for dictation directly to a computer. Some day the computer will transcribe spoken words.

If you have access to dictation staff and equipment, try dictating. It too can speed your writing process. First learn how to operate the equipment. Then seek guidance from the person who will transcribe your dictation. To dictate effectively:

1. Plan your document.
2. Visualize your reader.
3. Express your ideas clearly.
4. Listen to, revise, and tighten your draft. (Office of Records Management 1973, 14)

By listening to what you have written, you will learn how to revise your dictation so that it's easier to follow and your writing flows smoothly. When you write by dictating, concentrate on capturing your ideas. Don't let your internal editor create writer's block. Once you have a transcribed document, edit it carefully.

HIGHLIGHTS

1. Professional writing differs from school writing in its function, number of readers, readers' knowledge, working environment, deadlines, and ways in which topics present themselves.
2. Free writing allows the writer to center on putting down ideas on paper as quickly as possible, leaving editing until later.
3. The Document Design Center model focuses on prewriting, writing, and postwriting steps. Prewriting includes determining the project's scope, tasks, and constraints, and defining its purpose and audiences. Writing is producing the document. Postwriting includes reviewing, revising, editing, and evaluating.
4. Using a nonlinear, problem solving approach to writing encourages planning, translating, and reviewing in light of the writer's long-term memory and the task environment.
5. To guide planning, the writer asks: What do my readers need to know? Why? What do I want my readers to know or to do after reading my materials? Why? What does my audience expect of my communication?
6. Strategies for selecting content include posing the journalist's questions—who, what, where, when, why, how—cubing, and brainstorming.
7. Common techniques for committing ideas to paper (or disk) include writing with pen, pencil, and paper; electronic or electric typewriter; word processors; and dictating.

PROJECTS

For the following, select a major project on which you've been working and will continue to work over several assignments. See which of the writing processes and strategies from this chapter help you the most.

1. Try the free writing method for generating a memo to your instructor on the thrust of your major paper for the semester.
2. Using the Document Design Center's process model, prepare a simple diagram on your major topic that answers the questions posed in Panel 7-1.
3. Generate an issue tree on your major topic.
4. Use the journalist's questions as a guide to listing the key elements for a memo to your instructor explaining your major topic.
5. Take the same topic, and use the cubing method. Write quickly to cover the points.
6. Develop an informal outline on your major paper for the term.

FOR MORE HELP

ANDERSON, P. V., R. J. BROCKMANN, AND C. R. MILLER. 1983. *New essays in technical and scientific communication: Research, theory, practice.* Baywood's Technical Communication Series, Number 2. Farmingdale, N.Y.: Baywood Publishing.
  The editors have collected a series of articles on the diverse approaches being researched and the fruits of that research.

FLOWER, L. 1985. *Problem solving strategies for writing.* 2nd ed. New York: Harcourt Brace Jovanovich.
  Flower suggests a problem solving approach to writing that's based on the research she did with John Hayes.

GONZALEZ, J. 1980. *The complete guide to effective dictation.* Boston: Kent Publishing.
  The guide provides a strong, clear, detailed introduction to dictation.

MILLER, C. 1985. Invention in technical and scientific discourse: a perspective survey. In *Research in technical communication,* ed. M. Moran and D. Journet. Westport, Conn.: Greenwood Press.
  Miller provides one of the most thorough literature reviews of invention techniques with application to technical and scientific communication.

8
Illustrations: Technology's Universal Language

OVERVIEW

In Chapter 8, we

- ☐ Explain illustrations' roles and functions in technical and scientific communication
- ☐ Discuss processes for selecting, planning, and producing illustrations
- ☐ Suggest guidelines for producing illustrations

Most technical and scientific fields depend heavily on illustrations to visualize and compress information. The style for most technical and scientific fields is to set off illustrations from the text as either tables or figures. If an illustration isn't a table, it's a figure. Figures include charts, line graphs, and photographs. Some disciplines, such as statistics, chemistry, and mathematics, insert notations, equations, and formulas within the narrative. (Illustration style is not to be confused with graphic design style, which has to do with the layout and typography of a publication and thus goes far beyond the selection and use of figures and tables.)

John S. Harris and Reed H. Blake (1976) argue that tables, figures, and formulas are usually superior to text for communicating technical information. Certainly that is true for mechanical arrangements, such as the parts of an engine or an animal, and for presenting data. Formulas, the shorthand of science, provide extremely concise information.

In solving problems, most technical and scientific professionals use data or visual information. Often their thinking centers on symbols other than words. They think first in their specialized language of symbols and then explain in words what the symbols mean.

ILLUSTRATION FUNCTIONS

In technical and scientific communication, illustrations:

- Summarize key points and reduce narrative length
- Emphasize key points
- Present information visually
- Simplify information
- Enhance understanding
- Add visual appeal

From a functional communication perspective, visuals sometimes supplement the narrative, and sometimes the narrative supplements the visuals. For many technical and scientific fields, visuals carry more information more succinctly than a narrative.

To compare the communication power of tables versus narratives, suppose that we had conducted a survey of hunters and fisherman. We wanted to determine their frequency of reading newspapers and magazines, and we reported their reading frequencies of Sunday sports sections. A narrative might convey the results this way:

> When it came to the frequency of reading the Sunday sports section of their local newspapers, some 66 percent of the fishermen (n = 555)* reported reading the section every week, 22 percent every other week, and 12 percent less frequently. In contrast, some 42 percent of the hunters (n = 575) reported reading the Sunday sports section every week, 16 percent every other week, and 42 percent less frequently.
>
> *n = sample size

Now consider the conciseness and clarity of the same information presented in a table:

TABLE 8-1. Percentage of fishermen and hunters who reported reading the Sunday sports section of their local newspapers

Frequency of Reading	Hunters* (n=555)**	Fishermen (n=575)
Every week	42%	66%
Every other week	16	22
Less than every other week	42	12
Total	100%	100%

*Responses accurate to within 3%
**n = sample size

SELECTING, PLANNING, AND PRODUCING ILLUSTRATIONS

You can put together visuals for your technical and scientific communications in different ways, depending upon your working style. Just as their writing processes differ, technical and scientific professionals differ in their processes of developing visuals.

The following sections explore factors you should consider when producing visuals for your reports, articles, memos, letters, instructions, documents, and other communications.

Selecting Illustrations

To select the appropriate illustration you should consider your audience, its frame of reference, your content, and the outcome you desire from your communication. Consider your audience's ability to comprehend your illustrations. Most technical professionals have little difficulty understanding data presented in tables, complex graphs, and statistics, but general audiences may have some difficulty with these forms.

Choose the simplest visual format that effectively conveys your message. For example, compare the ease of understanding the two components of Panel 8-1. The bar graph quickly communicates the message; the table is less effective and takes longer to interpret.

Most technical professionals understand the tables, data, equations, and illustrations required in their respective disciplines. But they may have difficulty understanding material outside their fields. Thus an electrical engineer easily understands electronics symbols and schematics, but the same engineer may have trouble with biochemical formulas. To use illustrations that your audience understands, you face the task of trying to ascertain its technical skills. That's where your own audience analysis skills become important. For reporting an increase in lung cancer of people living downwind of a city in a medical journal, an epidemiologist probably would use tables. To present the same information to a general audience, the epidemiologist probably would use bar graphs or picture graphs. A company's or a publication's visual style may give you insight into the type of illustrations to select for that audience. A few minutes spent reviewing company reports, journals, and magazines quickly suggest appropriate visuals.

You also need to consider the outcome you want from an illustration. Do you want your audience to learn specific details or to gain an overall impression? *The more general an impression you want, the more general tables, graphs, and line art you can use.*

Content also may influence your selection of illustrations. Consider your problem solving strategy. In many technical and scientific fields, the first result consists of data, usually in tables that are tied closely to the question or problem. If such tables are computer printouts, you'll need to

PANEL 8-1. Compare the Ease of Understanding Table and Bar Graph

TABLE 8-2. Report components most commonly used by National Aeronautics and Space Administration (NASA) managers when reviewing NASA technical reports for possible reading (n = 212)

Component	Percentage Used
Summary	79%
Conclusion	69
Abstract	53
Title Page	46
Introduction	41

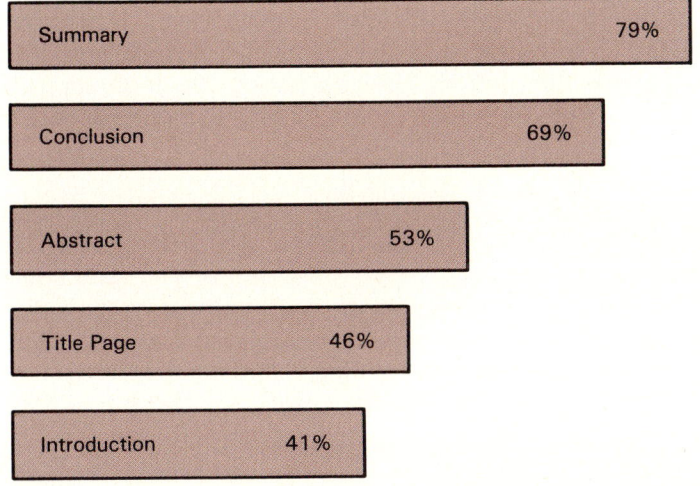

FIGURE 8-1. Report components most commonly used by National Aeronautics and Space Administration managers in reviewing NASA technical reports for possible reading.
(Source: T. E. Pinelli, V. W. Cordle, M. Glassman, and R. F. Vondran, Report-reading Patterns of Technical Managers, Technical Communication 31 (1984) 3: 20–24. Used with permission.)

simplify and summarize them in presenting your data's key points. Later you'll find that you may use the data to prepare line graphs, bar charts, circle charts, or other visuals.

As a professional communicating with many different audiences, you will find yourself recasting your illustrations in different forms for different groups. You'll also need to recast them in the appropriate sizes and proportions for different media. Thus a 3- by 3-inch table in a printed report must be recast to make a legible transparency projected overhead for an audience of 100 people.

Planning Illustrations

You can develop your illustrations at any time, but the earlier you start thinking about and preparing them, the fewer problems you'll have. Consider what types of illustrations you might eventually need, and collect or prepare them as you work through a project. If the project involves physical objects, for example, take photographs as you progress.

Keep in mind that different professionals use different ways of planning and producing illustrations. For problems requiring data analysis, many professionals work methodically through a problem, develop the tables or figures, and then outline and write the results or findings sections around the illustrations. The tables and figures present the specific findings; the narrative reports their interpretation. In other cases, professionals outline their communication and then ask themselves which visuals would help make the points more effectively and clearly. Such an approach requires pulling visuals from files or generating them. Sometimes, both approaches together produce the needed illustrations.

Professionals from different fields take various approaches. Thus if you were an engineer preparing a report on a chemical process you'd designed to treat equipment parts, you'd first prepare a flow chart showing the process. With that flow chart clearly in front of you, you could write the narrative. If you were a nutrition specialist comparing the vitamins in new carrot varieties, you might develop a table summarizing the nutritional analysis of each variety tested. With a clear summary of your results, you then could write your narrative. If you were an agricultural extension agent writing an information sheet for home gardeners about how to identify squash beetles, the damage they cause to squash, and methods for controlling them, your narrative would not be so dependent on visuals. In fact, you could probably write your article without visuals. But if you add close-up photographs or line drawings of the beetles, their damage, and control techniques, you'll communicate more effectively.

How then should you proceed? That depends on your problem and communication task. When planning a *technical report based on data analyses,* consider the question guiding your project, complete your data analyses, and then prepare your tables. Next, around your tables develop your outline, pattern notes, or other organizing device. Finally, write your narrative. By using such an approach, you can let your tables guide your organization and your organization guide your narrative. By developing your visuals before you organize your manuscript, you can let your manuscript support, explain, and interpret the visuals. You may find that writing about the illustrations helps you to clarify your interpretation of the information they present.

For preparing *instructions on how to perform a process,* try planning your illustrations and narrative simultaneously. Suppose that you were preparing instructions for a youth group on how to build a bluebird house. You could plan for a drawing of the birdhouse, a list of supplies, and then you could develop a sequence of illustrations on each step in completing the birdhouse.

For preparing a *progress report* to clients or some other audience, you might use photographs. Let's say that you are a site engineer for a firm constructing a building for an out-of-town client. Your supervisor asks you to prepare a monthly progress report. You outline and draft your report. You report that you're four weeks ahead of schedule. You realize that photographs would help your clients—financial investors—better understand the building's progress. So you shoot a few rolls of 35mm slides of the building. You select and number 20 slides. Then you add a narrative keyed to the slides and insert them in a plastic slide sheet as part of the report.

Whichever approach to planning illustrations you use, check all publication, company, and organization style guides. Many have specific guidelines that you must follow.

Obtaining Illustrations

Once you decide which illustrations you need, you face the question of how to obtain them. You may

- Borrow them
- Prepare them yourself
- Use computer graphics
- Hire a professional artist

Borrowing Illustrations. If you plan to use illustrations from other publications, seek written permission. Copyright law protects most illustrations from use without permission. Chapter 18 on copyright provides further details. When you seek permission, request a high-quality original or a high-quality copy, such as a photo mechanical transfer (PMT). Avoid using photocopies as masters, because they lose quality each time copied.

Preparing Illustrations Yourself. For most class papers and limited circulation professional papers, you can prepare illustrations yourself. Don't let the lack of a studio deter you. With a computer, plotters, and graphics software or with a T-square, triangle, nonreproducing colored pencils, ink or felt-tip pens, and a small supply of press type and symbols, you can prepare many suitable illustrations.

For tables, you need only a typewriter with tab sets. Although a manual typewriter will do, electronic or electric units with film ribbons produce the best tables for photocopies. To prepare a table, first rough out the format and test the format headings. Then judge the data you'll be entering to see if it will fit. With practice, you'll find yourself producing acceptable tables on the first try.

Using Computer Graphics. Computers are rapidly changing the preparation of illustrations for technical, scientific, and other fields. With the proper computer, printer, and software, you can prepare most kinds of visuals—a topic discussed further in Chapter 14. You can use personal computers to prepare tables, line graphs, bar graphs, circle or pie charts, pictorial charts, scientific and technical notations and formulas, flow charts, organizational charts, time charts, maps, line art, designs, and many other illustrations. Computers have the definite advantage of letting you easily change sizes, perspectives, dimensions, and other characteristics of illustrations.

With a computer, you can generate illustrations on screen and transfer them to hard copy—paper, acetate, or film. The quality of the hard copy depends upon the equipment used to print it. Some equipment provides high-quality copy that can be used in professional publications; other equipment is acceptable for student papers, but not for publication.

Compare the quality of the illustrations in Figure 8-2. Some printers do not produce solid lines; their letters and lines are too thin to reproduce well. Some printers produce four-color illustrations. Although visually appealing, four-color illustrations may not be appropriate for a report or publication that will be printed in one color of ink. The four-color illustration may not reproduce well. Let the planned reproduction process influence the form of the original illustration.

Hiring an Artist. Many universities, government agencies, and other organizations have staff artists. You also can find freelance artists in most communities. To find freelancers, check with school art departments, art supply stores, and printers.

Contact the artist long before you need your illustrations, explain what you'll need, and ask if the artist has experience preparing technical and scientific illustrations. Be sure to discuss the artist's charges and the time required to produce the illustrations.

Whether you borrow illustrations from others, prepare them manually, prepare them with a computer, or hire a professional artist, follow the appropriate style manual or guidelines for illustrations. If your field does not have guidelines for illustrations, follow those given later in this chapter.

Combining Illustrations and Narrative

Open many documents, and you'll find such statements as "The data are presented in Table 11," or "Figure 5 provides a line chart showing the

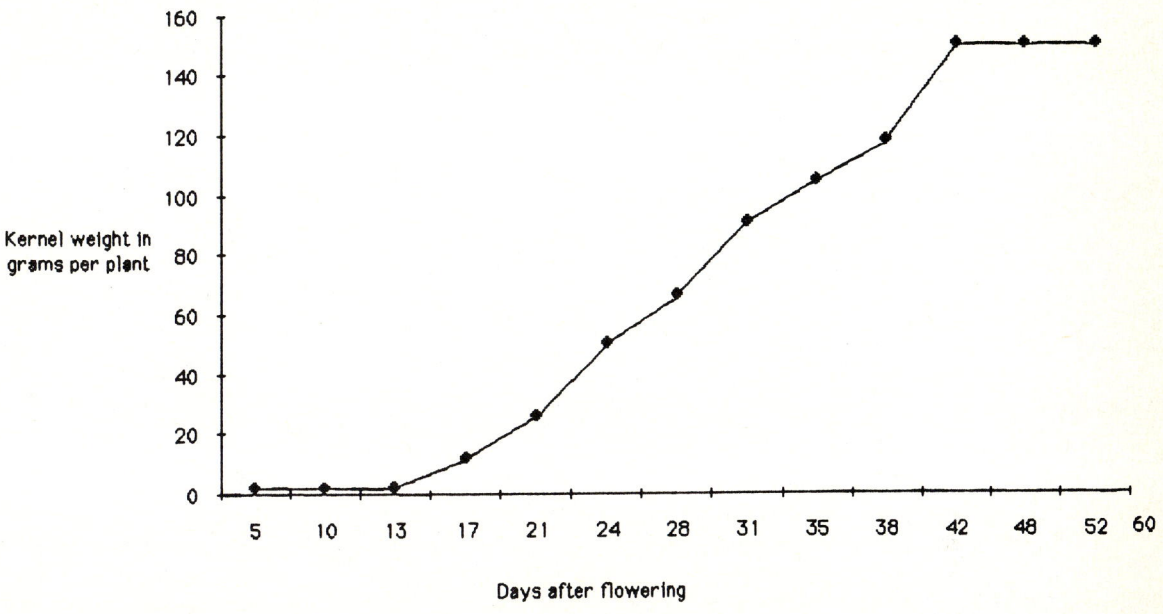

Figure 8-2 A. Growth of corn kernels

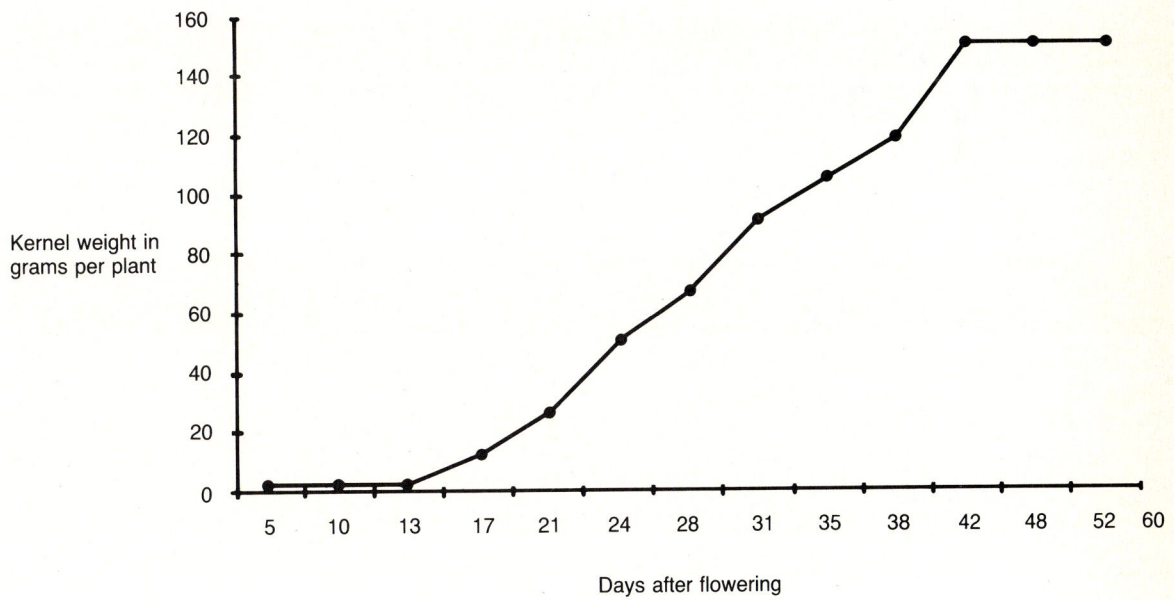

Figure 8-2 B. Growth of corn kernels

FIGURE 8-2. Compare the quality of dot matrix printout and redrawn illustration.

curves." Openings like these waste time and convey no useful information. Your narrative should reflect your interpretation of the data you're presenting. What do the data mean? What conclusions can you draw from the data or information you've collected? Use your technical or scientific background to interpret your data.

When you discuss data tables and figures, explain your interpretation of the data. But don't repeat all details of the illustration. For example, a journal article on otters and mink in the Everglades began with a discussion of results:

> The apparent abundance of otters and mink relative to each other over the total area sampled was not different ($p > 0.05$) over the period of the study (Fig. 2A) or within seasons (Fig. 2B). (Humphrey and Zinn 1982, 378)

The figures gave the data; the narrative gave the interpretation.

Study publications' guidelines for inserting illustrations in the narrative. Usually you indicate the insertion just after the paragraph containing the first mention of the illustration. Sometimes you indicate the insertion on a separate line:

> [Insert Figure 2 about here]

Other publications require the notation to appear in the left-hand margin.

Handling Illustrations

Nothing can be more costly in both time and dollars than losing illustrations. Clearly label your illustrations with your name, address, and their illustration number. Don't write on the illustrations themselves. Instead, on self-adhesive labels, type your name, address, and the illustration number, article or report title, and then carefully place the label on the illustration's back. If adding labels might damage the illustrations, check with your supervisor or the editor to whom you'll submit your manuscript on how to label your illustrations safely. In the margin, use a wax pencil to mark the illustration's top. To protect your illustrations, tape heavy transparent paper to the back of the illustration, and fold it over the illustration's front, or use glassine envelopes.

GUIDELINES FOR ILLUSTRATIONS

The following discussion presents general principles on preparing illustrations and specific techniques for preparing tables, line graphs, bar graphs, circle graphs, isotypes, notations, flow charts, organizational charts, algorithms, time charts, maps, line art, and photographs.

The guidelines do not replace those of the organization whose style manual you may be following. *Follow your organization's style in preparing illustrations.*

General Principles

Let the following points guide you:

1. First determine the final form of the proposed illustrations. Will the information be presented in a technical report, an article, an overhead transparency, a 35mm slide, or a display? Remember: seldom can an illustration for one medium work effectively in another medium.
2. Prepare visuals in proportion to their final size. Thus, for a bar graph to be reproduced as a 3 by 3 inches, make the original 6 by 6 or 9 by 9 inches. Make the visual large, and then reduce it, rather than making it small and enlarging it. Reduction minimizes errors.
3. Make legends and captions legible. When reduced, symbols and letters should be clear and legible. Be sure the letters aren't too small and do not run together.
4. Prepare a separate page for each illustration per page. Insert the illustrations at the end of the report or article. During word processing or printing, illustrations will be inserted in the proper location.
5. Label illustrations fully so that readers can understand the information being presented.
6. Too much information confuses readers. Avoid clutter.
7. Illustrate key points only. Illustrating small points obscures the main points.
8. Include a title for each illustration.
9. Make sure titles and legends are succinct and informative.
10. If you have lots of data, break it into two or more illustrations.
11. When adding labels, legends, and captions, use standard abbreviations. Provide footnotes explaining abbreviations.
12. Provide high-quality copies, such as PMTs. Avoid using photocopies.
13. Proofread and double-check all illustrations before submitting them.

Specific Guidelines: Tables

Tables, a compact way of presenting complex verbal and numerical information, range from the informal to the formal. The complex tables of most technical fields are most appropriate for highly trained audiences. For less experienced audiences, simplify your tables.

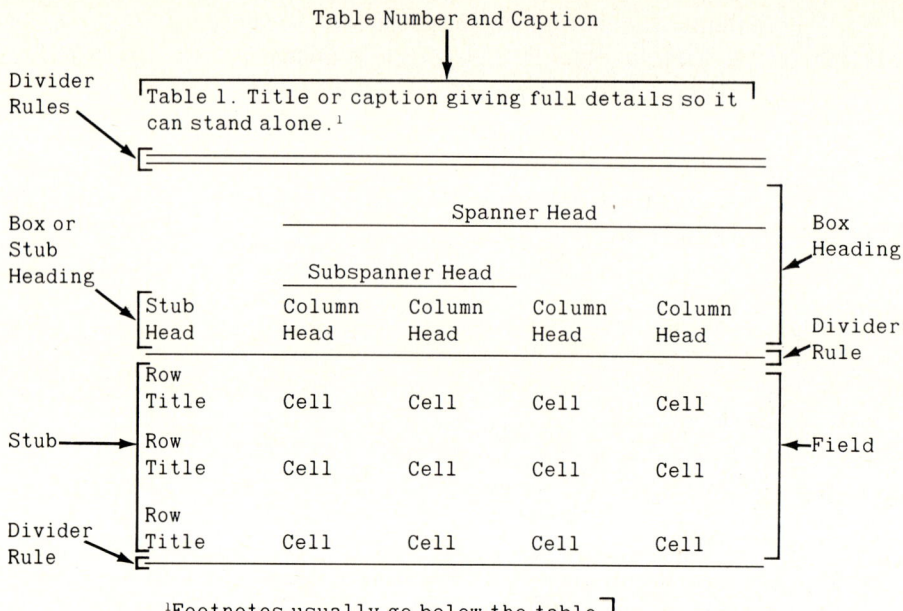

FIGURE 8-3. Basic components of a complex table.
(Source: Adapted from CBE Style Manual: A Guide for Authors, Editors, and Publishers in the Biological Sciences, *5th ed. (Bethesda, Md.: Council of Biology Editors, 1983), p. 76. Reprinted by permission of the Council of Biology Editors; and D. R. Buxton, R. C. Dinauer, D. A. Fuccillo, J. J. Mortvedt, and C. O. Qualset,* Publications Handbook and Style Manual *(Madison, Wis.: American Society of Agronomy, 1984), pp. 61–62 by permission of the American Society of Agronomy, Inc.)*

Tips for Tables. By following the format shown in Figure 8-3 and the points listed below, you will avoid common pitfalls to tables. *Keep in mind that tables work best for audiences trained and experienced in working with tables and data.* When preparing tables:

- Number tables with Arabic numerals: Table 1, Table 2, etc.
- Limit divider rules to between title and spanner head, between box headings and field, and between field and footnotes.
- Avoid broadside tables, tables arranged so that readers must turn them 90 degrees to read them. Reduce large tables on photocopying machines or photographic processes, or recast them to fit vertically on a page.
- Choose clear, uncluttered typefaces to maintain print legibility.
- For most tables, place independent variables in rows and dependent variables in columns. For all tables, let the message dictate the arrangement.
- Indicate units in column heads.

- Report sample sizes: for example, "n = 275."
- Indicate blank cells with a dash or zero.

Michael Macdonald-Ross (1977a) has reviewed the research on illustrations. We paraphrase his recommendations:

- Round numbers to two significant digits to simplify mental arithmetic.
- Provide row and column averages as reference points.
- Use columns to organize and to make important comparisons.
- Order rows and columns by the sizes of numbers to help readers make mental comparisons.
- Don't space out columns artificially to fill a page.

Specific Guidelines: Line Graphs

Simple line graphs provide visual representations for continuous variables such as trends and movements over time, and they give readers an overall impression of data (Figure 8-4). Complex line graphs show advanced mathematical and statistical analyses. They require plotting on graph paper, careful reading, and interpolation. *Complex line graphs are best for highly trained and skilled professionals; simple line graphs are best for readers who lack technical backgrounds.* If necessary for your audience, recast complex line graphs into simple line graphs.

The following guidelines cover only simple line graphs.*

Tips for Line Graphs. Suggestions for preparing line graphs include the following (Felker et al. 1982):

- Show points along the horizontal axis when graphs report time-related data.
- Limit the number of lines in any one graph. Many intersecting lines confuse readers and prove hard to understand.
- Make different lines visually distinct by using different symbols.
- Make each line's points clear.
- Label each line carefully.
- Draw vertical and horizontal axes proportionately.

In one of the few studies of line graphs, they were found to be less effective than bar graphs for general audiences (Powers et al. 1961). Readers need special skills to interpret line graphs (Macdonald-Ross 1977a).

---

*For complex line graphs, we recommend that you see Norbert Enrick, *Handbook of Effective Graphic and Tabular Communication* (Huntington, N.Y.: Robert E. Krieger, 1980).

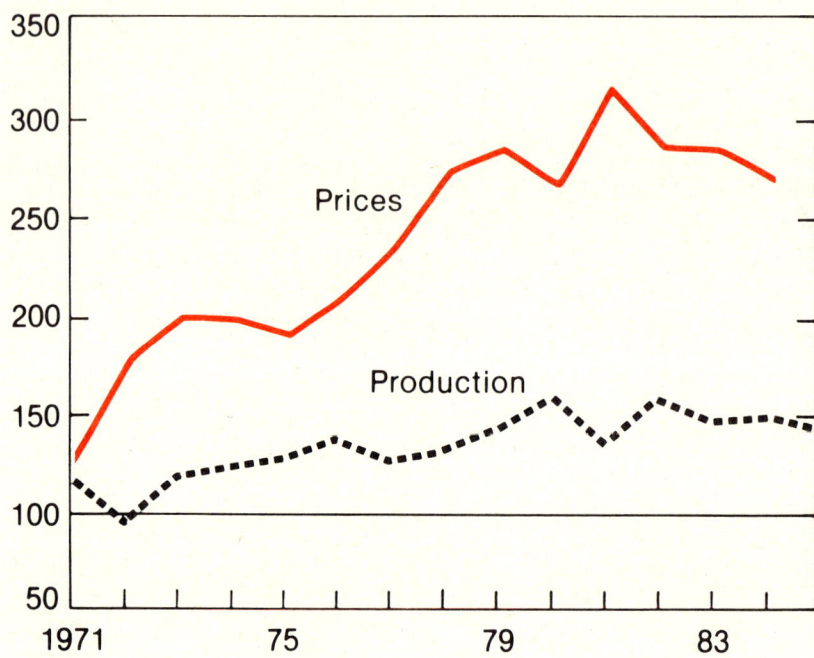

1985 preliminary. Production of 15 major fruits. Season average growers' price weighted by production.

FIGURE 8-4. Production and farm prices of 15 major noncitrus fruits in the United States from 1970 through 1985.
(Source: Agricultural Chartbook (Washington, D.C.: United States Department of Agriculture, 1985), chart 261.)

Specific Guidelines: Bar Graphs

Bar graphs are best used for presenting discrete variables to general audiences who are relatively unskilled in understanding numbers. They are best for showing relationships succinctly, and for showing magnitudes and distributions (Figure 8-5). *Bar graphs usually show general relationships rather than specific points.*

Tips for Bar Graphs. When preparing bar graphs:

- Use the horizontal axis for the independent variable when reporting cause and effect data.

- Use the vertical axis for the dependent variable when reporting cause and effect data.
- Keep the bars the same width in any one graph.
- Start the vertical axis at zero. If necessary, break the axis for values far above zero.
- Keep scales simple and unconfusing.
- Avoid placing two graphs with different scales side by side. Readers may assume that the scales are the same.

Bar graphs are superior to circle and line graphs for presenting information to general audiences (Macdonald-Ross 1977b). Directly labeling

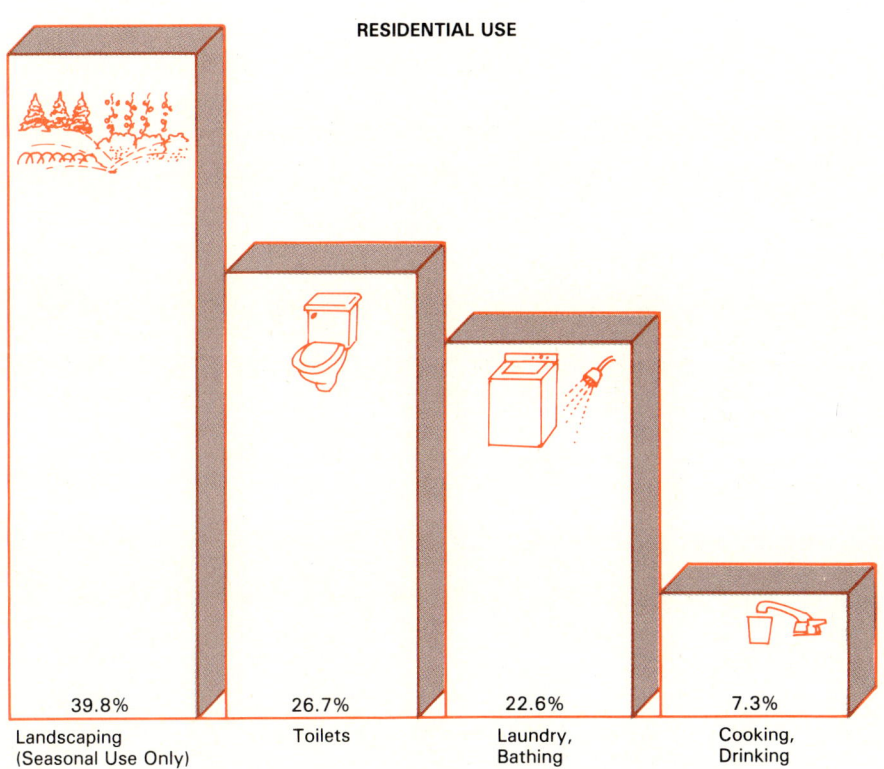

Nearly 40% of the annual water use by the average homeowner in the semi-arid West goes on landscaping around the home.

FIGURE 8-5. Residential water consumption in Denver, Colorado.
An effective bar graph showing key water uses in homes. Note the labels, percentages, and added line art. *(Source: Adapted from a graph developed by the Denver Water Department,* Public Annual Report, Supplement to Empire Magazine, Denver Post, *1978. Used with permission of the Denver Water Board.)*

the bars (Figure 8-5) rather than forcing readers to study legends and axes for information increases understanding (Powers et al. 1961).

Specific Guidelines: Circle Graphs

Circle graphs—sometimes called "pie charts"—commonly show the whole and the relationship of the parts to the whole (Figure 8-6). *They are appropriate for general audiences.*

Tips for Circle Graphs.   When preparing circle graphs:

- Avoid fine segments.
- Arrange segments clockwise from large to small.
- Begin the largest segment at twelve o'clock.
- Place labels adjacent to segments.

Avoid circle graphs of different sizes, because readers have difficulty comparing them (Macdonald-Ross 1977b).

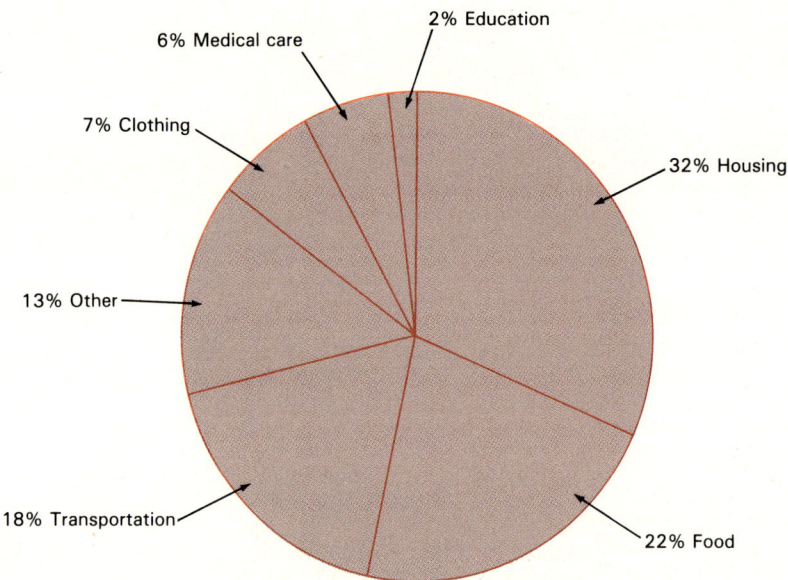

FIGURE 8-6.   A circle or pie graph showing various costs of raising western, nonfarm children.
(Source: 1982 Handbook of Agricultural Charts *(Washington, D.C.: U.S. Department of Agriculture).)*

Specific Guidelines: Isotypes

Isotypes, a special form of picture graphs, were first used in the 1920s. Isotypes—an *International System of Typographic Picture Education*—transform numbers into visual symbols. Isotypes generalize abstract concepts and show relationships and changes over time. *The audience should be able to develop a general understanding of the topic from the isotypes* (Figure 8-7).

Tips on Isotypes. Four principles for preparing isotypes include (Macdonald-Ross 1977a):

- Represent the subject matter with standard pictorial signs.
- Show larger quantities by increasing the number of signs, not by increasing the size of one sign.
- Let each sign stand for a given quantity or percentage.
- Arrange the pictorial signs to present an organized message.

Three additional points on preparing isotypes are (Neurath 1974):

- Use color if possible.
- Arrange symbols so that differences become clear.
- Use horizontal rather than vertical arrangements.

FIGURE 8-7. U.S. bicycle sales 1981–1985 (foreign and domestic makes).
(Source: Based on K. Mayer, Colorado's Bike Psyche, Rocky Mountain News, *April 20, 1986, p. 68.)*

Specific Guidelines: Notations

Communication in chemistry, mathematics, statistics, computer science, physics, medicine, and dozens of other professions depends heavily on notations—symbols, equations, and so on (Figure 8-8). Users must master the symbols and equations, but once they do, the symbols and equations prove efficient communicators, often expressing thoughts that cannot be expressed easily in words. *When you use symbols and notations, be sure that your audiences are familiar with them.*

Notations do have drawbacks. Although they are easily written by hand, notations usually slow up typing and word processing. They also boost typesetting costs. Follow the style manual for your field on handling notations.

Specific Guidelines: Flow Charts

In technical and scientific professions, flow charts communicate procedures, processes, operations, directions, and activities. Flow charts may be abstract symbols—boxes and labels—or labeled drawings of equipment and activities (Figure 8-9).

Tips for Flow Charts. The following suggestions come, in part, from Robert Leffert's *How to Prepare Charts and Graphs for Effective Reports* (1982):

- Be thoroughly familiar with the content.
- Identify the basic components.
- Draw the components in either abstract or picture form.
- Add the links between the components.

Specific Guidelines: Organizational Charts

Organizational charts identify an entity or an organization's major components, or show parts and their relationships. Most organizational charts use simple boxes with connecting lines and labels (Figure 8-10).

Tips for Organizational Charts. Follow the basic principles for flow charts when developing organizational charts. Before preparing organizational charts depicting a company's or entity's structure, check with the administrators to learn if any reorganization plans are pending.

Specific Guidelines: Algorithms

An algorithm is "a means of reaching a decision by considering only those factors which are relevant to that particular decision" (Wheatley and Unwin

Table 6-D-3. Results for a cropping pattern test using a completely randomized design.

	Cropping Pattern	
Traditional	Alternative 1	Alternative 2
	Yield (kg/ha)	
1,800	2,020	2,830
1,760	1,030	3,220
1,510	1,600	3,670
890	1,410	2,940
1,410	2,730	1,980
1,760	1,860	3,720
1,300	1,540	3,210
1,650	1,000	2,140
900	3,100	1,390
1,310	940	2,700

Y_i (sum of treatment yields)	14,290	17,230	27,800
$\bar{Y}_i$ (mean of treatment yields)	1,429	1,723	2,780
S_i (standard deviation)	336	729	751
CV_i (coefficient of variation)	24%(CV_1)	42%(CV_2)	27%(CV_3)

Y (sum of all treatment yields) = 59,320
$\bar{Y}$ (mean of all treatment yields) = 1,977

Mean

The mean is the arithmetic average of a group of values and is usually considered the most representative single value for the whole group. $\bar{Y}_i$ is the mean of treatment i and is calculated as follows:

$$\bar{Y}_i = \frac{\sum_{j=1}^{r} Y_{ij}}{r} = \frac{Y_i}{r}$$

Where: Σ = summation symbol

i = 1, 2, ..., t

t = number of treatments in the experiment (t = 3 in this experiment and we have 3 treatment means: $\bar{Y}_1$, $\bar{Y}_2$, and $\bar{Y}_3$)

j = 1, 2, ..., r

r = number of replications in each treatment (r = 10 in this experiment)

Y_{ij} = individual yield values

$\sum_{j=1}^{r} Y_{ij} = Y_i$ = the sum of individual yield values from j = 1 to j = r in treatment i

As an example, we calculate the mean for the traditional cropping pattern (i = 1):

$$\bar{Y}_1 = \frac{\sum_{j=1}^{r} Y_{1j}}{r} = \frac{Y_1}{r}$$

$$= \frac{\sum_{j=1}^{10} Y_{1j}}{10} = \frac{Y_1}{10}$$

$$= \frac{1{,}800 + 1{,}760 + 1{,}510 + 890 + 1{,}410 + 1{,}760 + 1{,}300 + 1{,}650 + 900 + 1{,}310}{10}$$

$$= \frac{14{,}290}{10}$$

$$= 1{,}429 \text{ (kg/ha)}$$

FIGURE 8-8. Notations.

Notations serve as codes for both chemistry and statistics. *(Source: Reprinted with permission from W. W. Shaner, W. R. Schmehl, and P. F. Philipp,* Farming Systems Research and Development: Guidelines for Developing Countries *(Tucson, Ariz.: Consortium for International Development, 1982), p. 328.)*

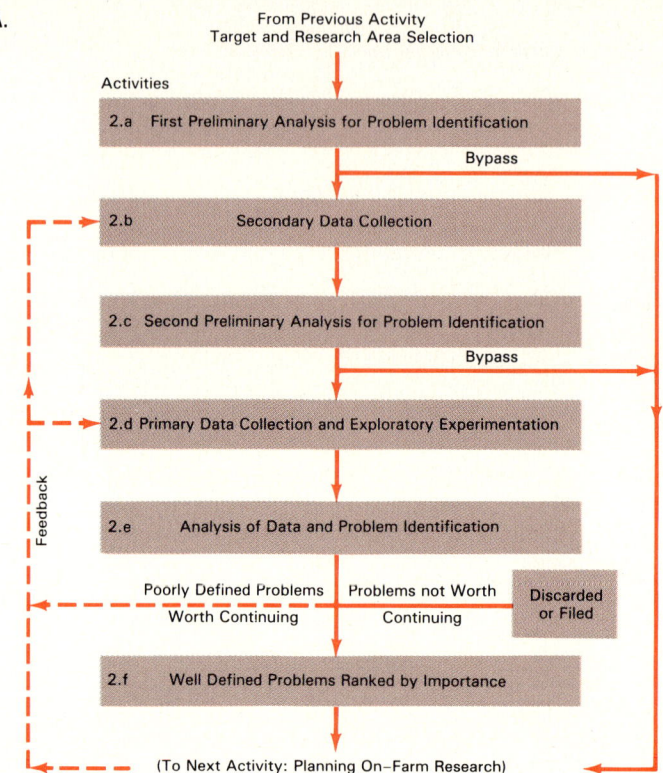

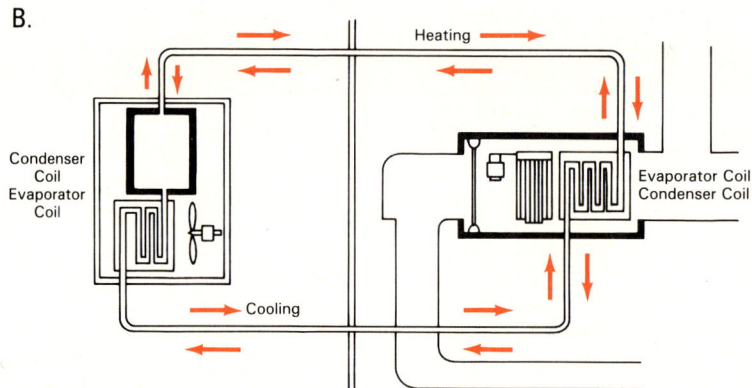

FIGURE 8-9. Flow charts.

A. Flow chart for problem identification in farming systems in developing countries.
B. Heat pump flow chart. *(Source: A. Reprinted with permission from W. W. Shaner, W. R. Schmehl, and P. F. Philipp,* Farming Systems Research and Development: Guidelines for Developing Countries *(Tucson, Ariz.: Consortium for International Development, 1982), p. 28. B. Small Homes Council,* Heating the Home *(Champaign: University of Illinois). © 1971 by the Board of Trustees of the University of Illinois. Reprinted by permission from Circular G3.1, Small Homes Council.)*

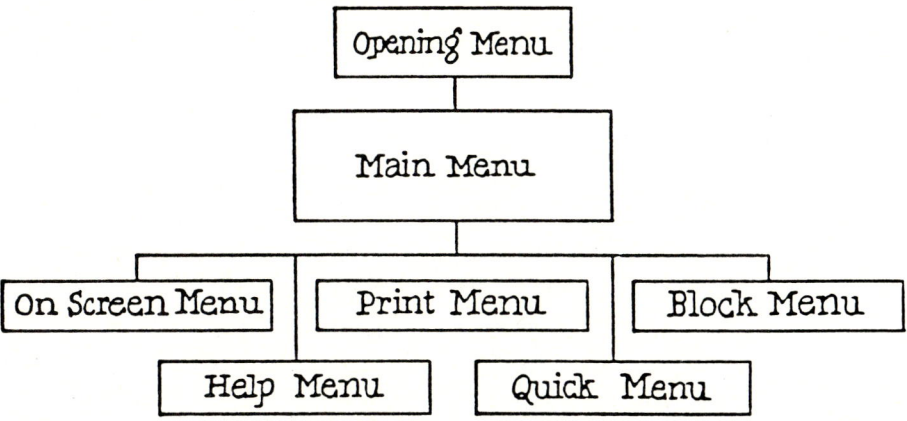

FIGURE 8-10. A basic organizational chart.

The chart shows the menus of the WordStar word processing program. *(Source: WordStar KayPro Reference Manual (1983), pp. 1–5. Used with permission of MicroPro Corporation, San Rafael, Calif.)*

1972, 10). Variations on algorithms include decision charts, decision tables, flow chart algorithms, and tree diagrams.

Algorithms can guide readers in filling out forms (Figure 8-11), performing a process, diagnosing a problem, and many other activities.

Research results are mixed, but they suggest that algorithms have a good potential for technical communication. For example, Melissa Holland and Andrew Rose (1982), of the Document Design Center, found that *users had problems with ordinary language algorithms at first. But once users learned how algorithms work, they found algorithms quicker and easier to follow than narrative instructions.*

Computer scientists use algorithms with specific symbols: rectangles represent activities, diamonds raise questions or identify decision points, circles indicate stopping points, and arrows connect the actions represented by symbols. Once trained, users do not need a narrative to understand algorithms.

Ordinary language algorithms use simple boxes and, if necessary, minimal narrative.

Tips for Algorithms. If you plan to use algorithms, study their use in your respective discipline, and follow style guidelines.

Specific Guidelines: Time Charts

Time charts—visual timetables—can be helpful for developing proposals, scheduling activities, and preparing communications (Figure 8-12). They

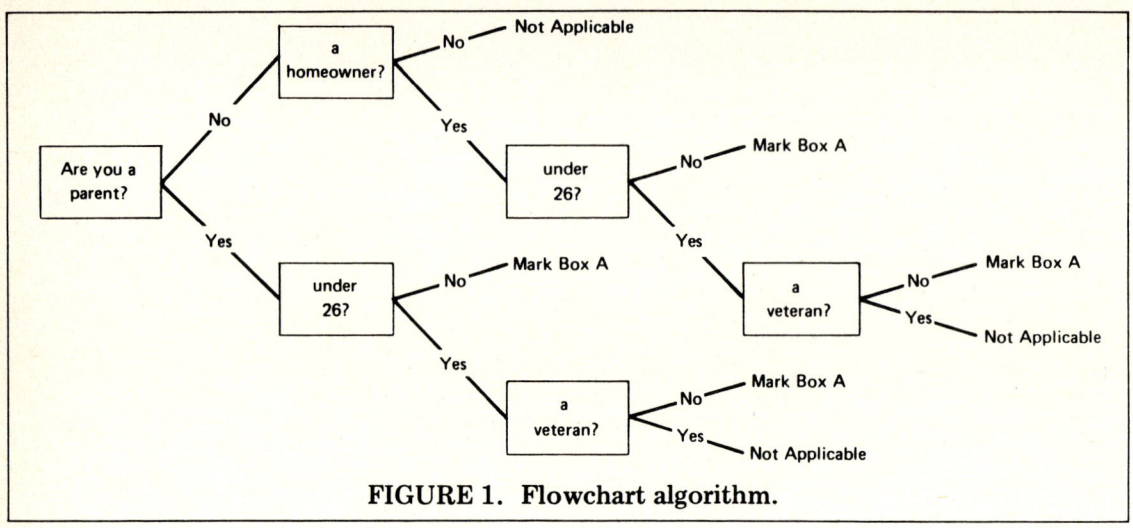

FIGURE 1. Flowchart algorithm.

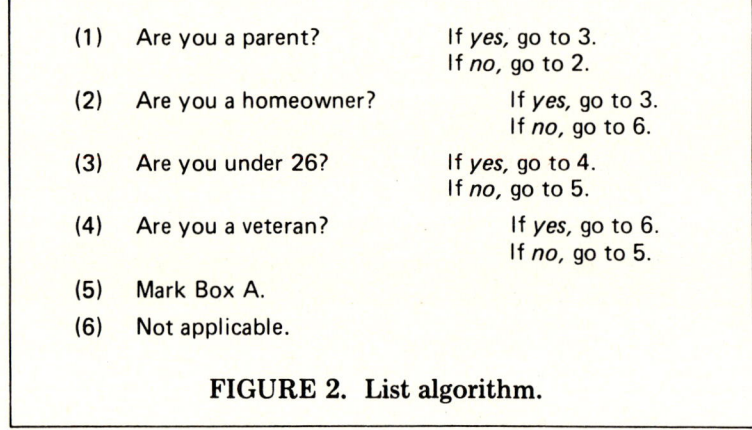

FIGURE 2. List algorithm.

FIGURE 8-11. Flow chart and list algorithms for filling out a form.
(Source: Simply Stated, *no. 23, January 1982. Reprinted with permission from* Simply Stated, *the monthly newsletter of the Document Design Center, American Institutes for Research.)*

can help you to meet objectives, allocate resources, and divide responsibilities among staff. Robert Leffert (1982) divides time charts into timeline and milestone charts and Program Evaluation Review Technique (PERT) charts. PERT charts usually require computer development; they may prove difficult for inexperienced readers.

ILLUSTRATIONS: TECHNOLOGY'S UNIVERSAL LANGUAGE

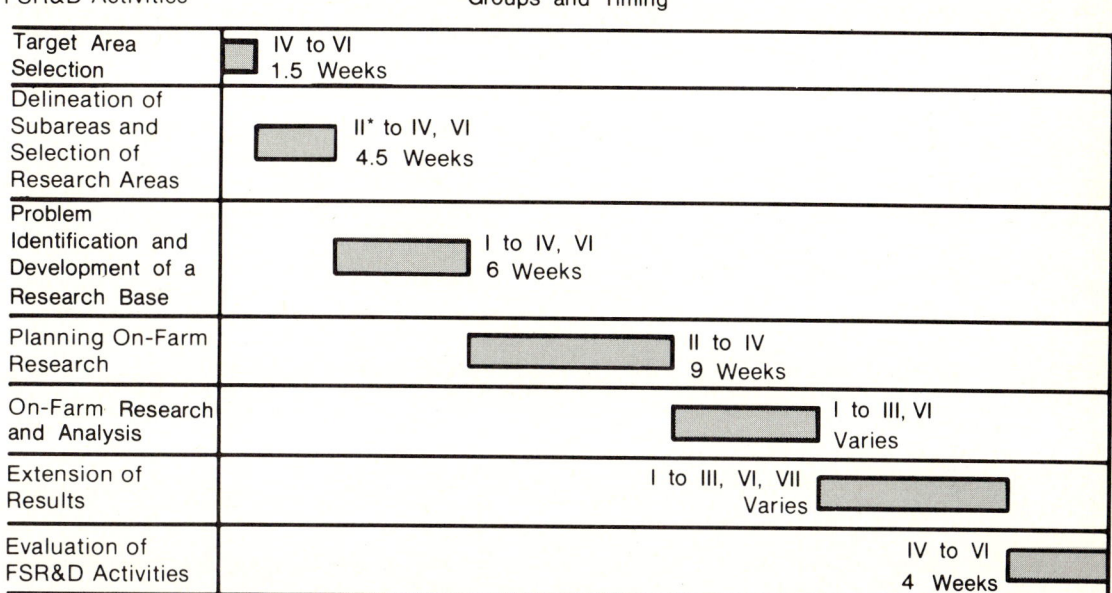

*Provided field teams have been selected

Key: I. Farmers and Farmers' Organizations; II Field Team; III Regional Headquarters Team; IV. National Headquarters Team; V. Key National Decision Makers; VI. Extension Service; VII. Production Oriented Groups.

FIGURE 8-12. A time chart for planning farm research and development.
(Source: Reprinted with permission from W. W. Shaner, W. R. Schmehl, and P. F. Philipp, *Farming Systems Research and Development: Guidelines for Developing Countries* (Tucson, Ariz.: Consortium for International Development, 1982), p. 35.)

Tips for Time Charts. To prepare time charts (Leffert 1982):

- List and group activities.
- Select time units and estimate the time each activity requires.
- Draw the layout, grids, and enter headings.
- Add activities, lists, and time lines.

Specific Guidelines: Maps

In many technical disciplines, maps orient readers and identify sites mentioned in the communication (Figure 8-13), or they give directions to meeting sites (Figure 8-14).

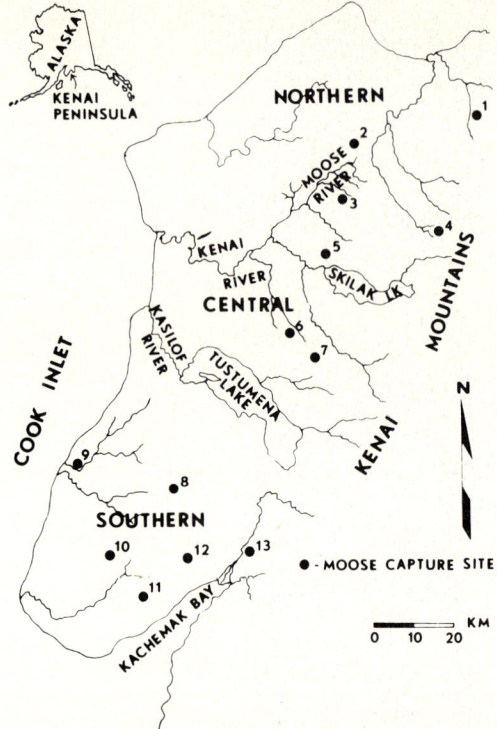

Fig. 1. Kenai Peninsula study area and locations of captured moose (1 = Big Indian Creek, 2 = Moose Research Center, 3 = Moose River Flats, 4 = Mystery Creek Basin, 5 = Skilak Lake, 6 = Funny River, 7 = Benchlands, 8 = Caribou Hills, 9 = Ninilchik River, 10 = Bald Mountain, 11 = Headwater Hills, 12 = Eagle Lake, 13 = Fox River).

FIGURE 8-13. Map.

A general map showing the location of a study area in Alaska. *(Source: T. N. Bailey, A. W. Franzman, P. D. Arneson, and J. L. Davis, An Evaluation of Visual Location Data from Neck-Collared Moose, Journal of Wildlife Management 47 (1), 1983. Copyright © The Wildlife Society. Reprinted by permission.)*

Tips for Maps. For general orientation maps:

- Include one map of the general area in question and a second map that shows an enlargement of the specific area in question.
- Simplify the map; include only key features—roads, rivers, lakes, and so forth—that help users understand the location.
- Orient the map with north to the top; position it with an arrow pointing to north at the page top.
- Provide a distance scale.
- Add a legend identifying all symbols, if needed.

```
Hewlett-Packard Demonstrations
To Highlight
March Meeting

     Members who attend the March
meeting, hosted by HP in Fort
Collins, will have a chance for
some "hands-on" experience with
the text processing system HP
documentation groups use.

     The STC Chapter business
meeting will begin at 7 p.m. and
the demonstration of the text
processing system will follow.

Directions to Hewlett-Packard

     Take I-25 to the Timnath exit
(265).  Go west on Harmony Road
about two miles.  At the top of
the hill you will see the HP
facility on the north side of the
road.

OR
     Take Hwy.287 to Harmony
Road.  Go east on Harmony Road
about three miles to HP.


Use HP's south entrance.  Stop at
the guard house and the guard
will direct you to the correct lobby.
```

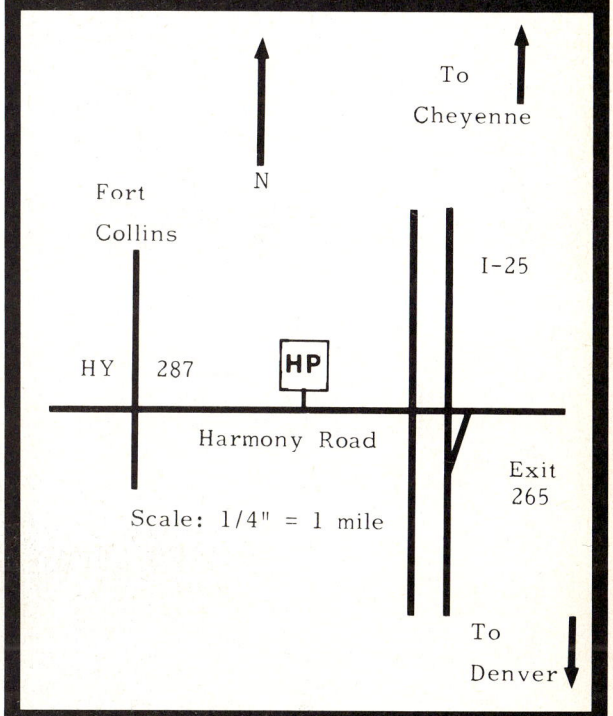

FIGURE 8-14.   Map.

A map showing the location of a meeting of the Rocky Mountain Chapter of the Society for Technical Communication. *(Source: D. Zimmerman, Hewlett-Packard demonstrations to highlight meeting,* Technicalities, *March 1981, p. 4. Used with permission.)*

Maps giving directions to meetings and other gatherings require enough details to guide users to the site on time. For direction maps:

- Assume that users are not familiar with the area.
- Clearly label streets, roads, highways, and key features. Identify key buildings and features of terrain.
- Add an arrow showing north pointing to the page top.
- Include a narrative to help readers understand the routes to the meeting.
- Double-check the map for errors or confusing parts.

Guidelines for Line Art

Line art shows specific details and omits others (Figure 8-15). It minimizes the distraction caused by extraneous details and helps to focus audience attention on the key features. Some line art cartoons lighten technical subjects (Figure 8-16). Most technical and scientific professionals turn to professional artists to produce quality line art.

Tips for Line Art. Unless you have good illustration skills or have access to a sophisticated computer, seek a professional artist to prepare your line art. To develop a good working relationship with professional artists, Carol Wassell (1984), principal graphic artist of Colorado State University's instructional services, makes these suggestions:

- Don't rush to the artist. Think through your illustrations, their purposes, audience, and the message.
- Sketch your ideas, even if they're extremely crude. Provide computer graphics, photographs, or examples of what you want illustrated.
- Meet with the artist, provide your sketches, and explain the art's purpose, audience, and message.
- Explain the illustration's final form—drawing for journal article, overhead transparency, 35mm slide, etc.
- Know the correct width and page size for publications and reports.
- Know what you would delete to simplify art. Many illustrations suffer from clutter that reduces audience comprehension and recall.
- Provide enough lead time—three or more weeks—for the artist to complete your illustrations.
- Remember, artists aren't technical experts, so explain your terms and provide correct spellings.
- Tell artists your budget so they can work within it.
- Check illustrations carefully for possible errors, and request the needed changes.
- Don't send original art to publishers. Send high-quality copies, such as PMTs.
- Package your illustrations carefully before sending them.
- Store originals in a safe place.

Simple drawings communicate best. Put labels on the drawings themselves rather than only in a legend. Referring readers to a legend slows reading and may reduce comprehension. Add color, if possible, to make illustrations easier to understand and to ease identification of key components (Wright 1977).

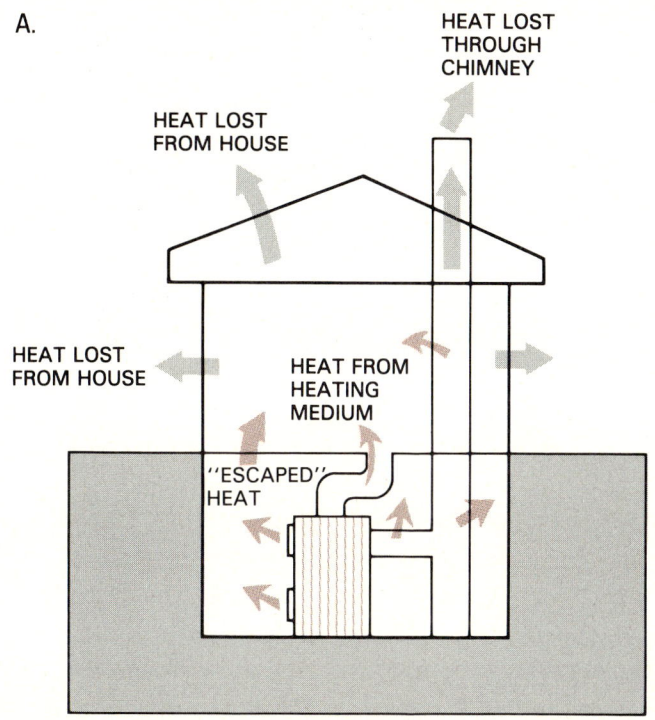
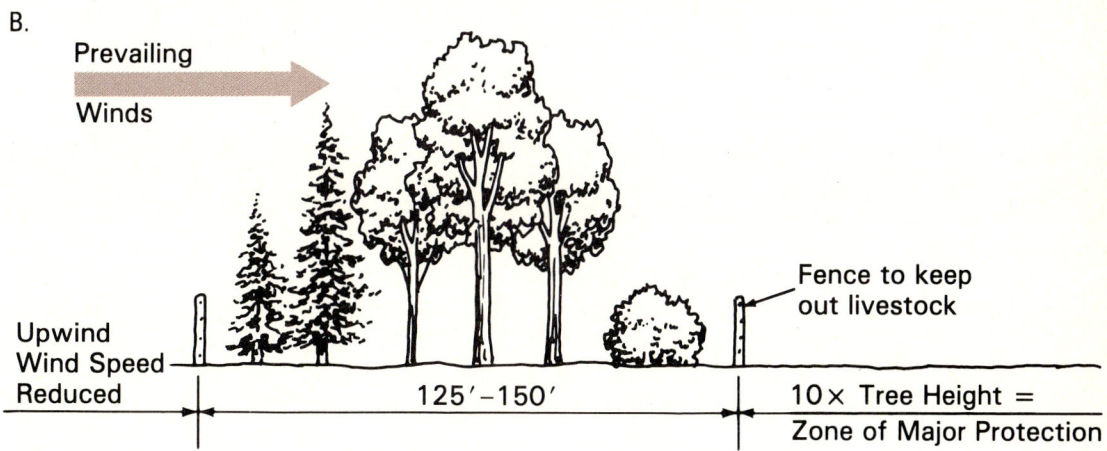

FIGURE 8-15. Line art.

A. Heat generation and loss in a home. B. Tree and wind break for a farmstead.
(Source: A. Small Homes Council, Heating the Home *(Champaign: University of Illinois).*
© 1971 by the Board of Trustees of the University of Illinois. Used by permission from Circular G3.1, Small Homes Council. B. Reproduced with permission from Farmstead Planning Handbook, *MWPS-2, 1st ed., 1974. © Midwest Plan Service, Ames, Ia.)*

SAFETY INSTRUCTIONS

READ YOUR OWNER'S MANUAL AND ALL SUPPLEMENTS (if any enclosed) thoroughly before operating your saw. ①

Operation of a chain saw should be restricted to mature, properly instructed individuals. **DO NOT ATTEMPT OPERATIONS BEYOND YOUR CAPACITY OR EXPERIENCE.**

WEAR CLOSE FITTING AND PROTECTIVE WORK CLOTHING that is made to give protection, such as (A) safety hat, (B) safety shield, safety goggles or safety glasses, (C) safety work shoes, (D) heavy duty work gloves, and (E) good grade ear plugs or sound barriers. ②

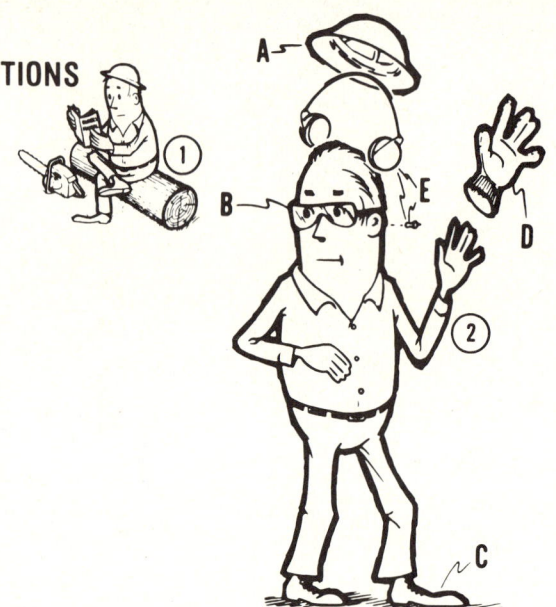

FIGURE 8-16. Cartoons.

Cartoons can lighten technical subjects and help communicate key points. *(Source: Owner's Manual, PRO MAC 610/650 Chain Saws (Los Angeles: McCulloch Corporation). Reprinted with permission.)*

Guidelines for Photographs

Many technical professionals own 35mm cameras and often shoot photographs. With practice, you can refine your skills and use photography to communicate about your profession. Even if you're inexperienced, advances in electronics and photographic processing will enable you to produce technically accurate photographs.

Photographs can help you communicate visually; they add another dimension to your reports and documents. Photographs, either black and white or color, can replace thousands of words and often communicate more effectively than words alone. Photographs can show equipment, objectives, plants, animals, processes, instructions, progress, and much more. They can provide visual records of projects and assignments. For example, if you're an engineer building an electrical power plant, you'll find yourself frequently on site, and you can photograph the project's progress.

In the list that follows and in Figure 8-17, we identify major points that will help you improve the quality of your photographs. In the "For More Help," we list references that provide detailed guidelines. Keep in mind three overall characteristics of good photographs: (1) they have excellent technical quality, (2) they show the content clearly, and (3) they have a pleasing composition.

Tips for Technical Quality. To improve the technical quality of your photographs:

- Know your equipment and how it functions. That means practice.
- Shoot test rolls to learn about new equipment and films.
- Shoot low to medium speed film.
- Use black and white film for black and white prints with little grain.
- Have black and white film processed in a custom laboratory. Request that film be processed in a developer that produces good contrast and little grain.
- Have color transparency films processed in a custom laboratory that produces consistently good results.
- Shoot color slides for slide presentations and publications printed in four colors.
- Avoid making black and white prints from color negatives or prints.
- Use your light meter correctly, and try different exposures.
- Focus carefully when shooting.
- Use a tripod to steady your camera when shooting at slow shutter speeds.

Tips for Improving Content. Before shooting, make a list of ideas. Shoot several shots to illustrate each idea. Be sure to consider the viewer's frame of reference when shooting. If necessary, shoot a series of photographs, starting with the familiar and moving to the unfamiliar. For any technical photograph, add people or familiar content for scale and to add visual interest.

Tips on Improving Composition. Improving composition requires using the camera's and film's technical capabilities and thinking visually to create appealing scenes. Techniques for improving composition include moving in close, framing scenes, varying angles, and using the "rule of thirds" (Figure 8.17). Developed during the nineteenth century, the rule of thirds requires drawing imaginary lines that divide the picture into thirds horizontally, then vertically. To apply the rule, compose your picture so that important subject areas fall at intersections of or along the lines.

To learn more about photography and the principles behind consistently fine photographs, study the texts listed in "For More Help."

172 CREATING DRAFTS

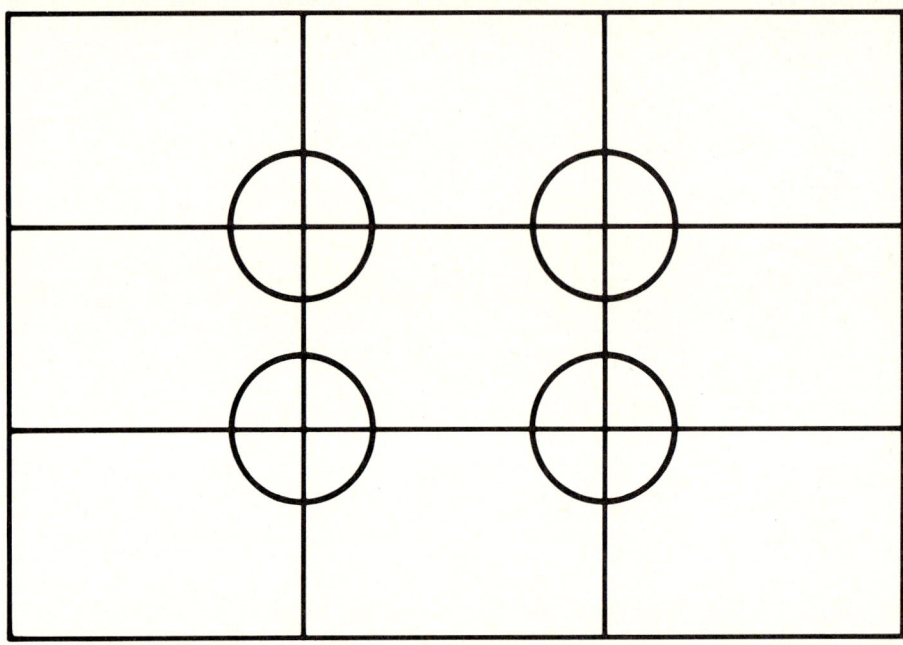

FIGURE 8-17. Techniques for improving composition.
When shooting, use the "rule of thirds." To apply the rule of thirds, imagine drawing two horizontal lines and two vertical lines through the scene, dividing it into thirds. Then place the centers of interest where the lines intersect.

ILLUSTRATIONS: TECHNOLOGY'S UNIVERSAL LANGUAGE 173

Move in close to show the essential information you are illustrating, in this case removing an air filter on a rototiller.

FIGURE 8-17. (continued)   Techniques for improving composition.
Vary your angles to remove distracting background and foreground. Here the first scene of a bulldozer covering riprap has a cluttered background of silos and electrical poles. In the second view, from another angle, the background is less noticeable and readers can center on the bulldozer's work.

Try to frame your subject. The first photograph lacks visual depth and has a cluttered background. By framing the geese with the tree, the second photograph adds visual depth, removes the cluttered background, and suggests a more natural setting.

HIGHLIGHTS

1. Illustrations provide the visuals for scientific and technical fields.
2. Technical illustrations enhance understanding, render abstract concepts, visually summarize and emphasize key points, and simplify and condense information.
3. You can develop visuals at any time, but the earlier you begin thinking about and preparing them, the more effectively they can communicate your ideas.
4. For the most coherent presentation, first develop visuals and then write the narrative.
5. Audience, illustration functions, and topic complexity influence selection of appropriate illustrations.
6. Get illustrations by borrowing, preparing them yourself, using computer graphics, or hiring professional artists.
7. Once you have prepared illustrations, handle, label, and ship them carefully.

PROJECTS

The beer consumption for United States residents 18 years and older, including armed forces abroad, was 24.0 gallons in 1960, 25.5 gallons in 1965, 28.6 gallons in 1970, 32.8 gallons in 1975, and 34.8 gallons in 1980. Per capita consumption of distilled spirits, as measured in U.S. gallons, was 1.87 gallons in 1960, 2.13 gallons in 1965, 2.61 gallons in 1970, 2.49 gallons in 1975, and 2.90 gallons in 1980. Wine consumption in gallons was 1.36 gallons in 1960, 1.44 gallons in 1965, 1.70 gallons in 1970, 2.20 gallons in 1975, and 2.65 gallons in 1980. These data come from the *Statistical Abstract of the United States*, 1982–83, U.S. Department of Commerce, Bureau of the Census.

1. Using the data given above, develop a table reporting the average alcoholic consumption per person in the United States from 1960 through 1980. Assume that your audience consists of nurses attending a training session on alcoholism.
2. Assuming that yours is a lay audience attending a public meeting on alcohol consumption, convert the above data into a bar graph. Would you use one or three bar graphs? Why?
3. Using 1960 as the base year for consumption, convert the above data into a table showing the percentage increase in alcohol consumption for each year specified between 1960 and 1980.
4. Write a narrative stressing the most important change in the data, and refer to the table in Project 3.
5. Select data on either beer, wine, or distilled spirits, and create an isotype illustrating consumption over the two decades.

6. Prepare two circle graphs illustrating your income and your expenditures for the previous academic year. Next, write a succinct narrative on the data.
7. If you're assigned a major paper for your technical communication class, prepare a time chart to help you allocate time for developing your idea, researching the topic, drafting and editing your manuscript, and preparing the final document.
8. Using the guidelines given for developing tables, write a critique of less than 500 words of the table given below. Then prepare a revised table that eliminates the problems and improves the table.

Garden Plot	pH	Organic Matter*	N†	P‡	K§
A	7.6	1.1	49	10	388
B	7.0	.9	56	13	350
C	6.5		38	8	277
D	6.8	1.0	45	11	375
E		2.0	50	15	400
F	7.3	1.3	47	9	423

*Organic matter reported in percent in sample
†N = Nitrates, reported in p.p.m.
‡P = Phosphorus, reported in p.p.m.
§K = Potassium, reported in p.p.m.

TABLE A. The results of soils test reports of five garden plots in Larimer County, Colorado

9. Evaluate the graph below, and then write a critique of not more than 500 words. Prepare a revised graph eliminating the problems that you identified.

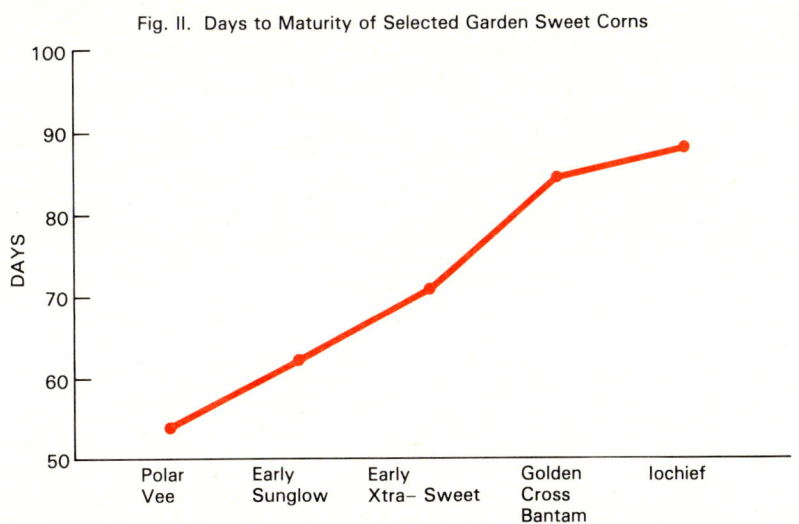

Fig. II. Days to Maturity of Selected Garden Sweet Corns

10. Reconsider the table in Project 8. Think about the potential users and the communication effectiveness. If the table were for a report to high school seniors, how would you improve its communication effectiveness?

FOR MORE HELP

The Chicago manual of style. 1983. 13th ed. Chicago: University of Chicago Press.
The manual provides an excellent, detailed discussion of developing tables, illustrations, and related information.

The communicator's catalog. Current year. Publication S-4. Rochester, N.Y.: Eastman Kodak.
Kodak's detailed listing identifies useful publications on still photography, slide set production, and darkroom techniques.

COUNCIL OF BIOLOGY EDITORS. 1983. *CBE style manual.* 5th ed. Bethesda, Md.: Council of Biology Editors.
The CBE manual discusses the handling of data in equations, tables, and illustrations.

Photographer's handbook. 1970. New York: Time/Life Books.
The handbook succinctly covers many fundamental concepts to help professionals improve their photographs.

TUFTE, E. 1983. *The visual display of quantitative information.* Cheshire, Conn.: Graphic Press.
A book strongly praised for its visual treatment and guidelines.

UPTON, B., AND J. UPTON. 1985. *Photography.* 3rd ed. Boston: Little, Brown, Educational Associates.
The authors present excellent guidelines for improving photographic skills.

PART FOUR

Revising and Rewriting: Strategies for Improving Technical and Scientific Writing

Different writers revise differently. Some, like longtime *New Yorker* writer William Zinsser, must have every paragraph as nearly right as possible before drafting their next, so they revise and polish each paragraph as they go along. Others, such as professional engineer Kenneth Nelson, whose writing habits were described in Chapter 6, revise little. Still others, including academic author Peter Elbow, "dump" their copy first and then revise extensively.

A few researchers have studied how professionals and students revise. They have found that when experienced writers revise, they keep in mind their reason for writing the manuscript, the medium, their familiarity with their audience, their writing task and subject, level of formality, text length, and related factors. Only after the broader issues have been addressed do experienced writers correct grammatical, spelling, and similar mistakes.

The next three chapters provide strategies for revising and rewriting. These activities need to be carefully considered in light of the points we have made about providing solid content, understanding and using communication principles, and presenting a polished appearance. Chapter 9 provides overall guidelines for revising and suggests strategies for checking your content. Chapter 10 provides guidelines for improving communication effectiveness and minimizing clutter. Chapter 11 discusses revising to improve appearances.

How you use the suggestions presented in the following chapters will depend on how you go about writing. We suggest a linear approach, in part because of the linearity of a textbook. You may find that the linear approach works well for you. If not, you can apply the strategies in the order you find most helpful.

9
Revisions: Making Good into Better

OVERVIEW

In Chapter 9, we
- ☐ Present general principles for guiding revisions
- ☐ Suggest ways of evaluating and checking content

GENERAL GUIDELINES FOR REVISING AND REWRITING

Drawing from both research about and the practical approaches of skilled communicators, we recommend the following revision process:

- Allow time between drafts.
- Assume the editor's role when you revise.
- Revise at different levels.
- Make several trips through your drafts.
- Use copyediting symbols.

Allow Time between Successive Drafts

Many skilled writers agree that allowing time between their drafts helps them to put needed distance between them and their work. So put as much time between your drafts as you can afford. Set memos and letters aside for a few minutes or overnight. Put longer documents aside for a day or two, a week, or more. If you don't have more than a few minutes, try to think of something else and take your mind off the draft. When you return to it, you can take a more detached view of its content, effectiveness, and appearance.

Assume the Editor's Role

Once you've drafted a manuscript and let it "cool," you're ready to begin revising. Take a detached look at your writing. Pretend your draft is someone else's. Don't begin admiring how beautifully you've said something. Look for ways to say things better. Keep asking yourself, "How can I improve my draft?"

Revise at Different Levels

We once attended a Society for Technical Communication seminar on improving writing. When the instructor reached the editing segment of the seminar, he distributed a poorly written document. The instructor asked students, most of whom were beginning writers, to edit the document. For a few minutes, the sound of scratching pens filled the room as people shortened sentences, improved punctuation, and corrected spelling and grammar. Then silence. Finally a student muttered, "Hell, it still doesn't make sense."

That was the opening the instructor wanted. It's a waste of time, he pointed out, to correct the fine points until you have the major points in order. When you revise your drafts, take care of the large issues first, and then fine-tune your manuscript.

Make Several Trips through Your Manuscript

When you revise, make several trips through your manuscript. On each trip look for different problems. Few people can check for several different kinds of problems at one time. By breaking your tasks apart and looking for limited problems on each trip, you assure yourself that you're covering the document well. Concentrate on what you're looking for, and correct those weaknesses. If you do spot a problem other than the one you're looking for, correct it or flag it for correction later, and then return to checking what you set out to do.

Use Copyediting Symbols

When you revise a typed manuscript or copy printed from a word processor, copyediting symbols will save you time and effort. Recognized by many typists and word processing operators and by most editors and printers, these symbols provide clear, reliable shortcuts for making editorial changes. (See Chapter 10 for more on copyediting symbols.)

STRATEGIES FOR EDITING CONTENT

Once you are ready to start editing, evaluate the draft's content. Look at the content as if another professional in your field had prepared it and you

REVISIONS: MAKING GOOD INTO BETTER   183

PANEL 9-1. Checklist for Editing Content

When you edit for content, make sure you've clearly considered the following questions:

- What are the objectives, questions, and problems behind the manuscript's content?
- How did I collect the information?
- What information, evidence, or data did I provide?
- What data analysis techniques did I use?
- How did I verify results?
- What conclusions did I draw?

---

were asked to review it. Let your technical background and problem solving experience help you to evaluate content. As you read through the manuscript, ask questions, and make notes in the margins.

What questions should you ask? Consider the points on minimizing content errors presented in Chapter 2 and in the checklist for editing in Panel 9-1. The following discussion elaborates on the points in the panel.

What Are the Goals and Questions behind the Content?

Keep in mind that the questions guiding your problem solving may be posed as objectives, problems, or hypotheses. What you call the question reflects your discipline's approach to problem solving.

When you begin reviewing your draft, look carefully for the question or purpose behind the project you're reporting and the purpose of the draft itself. Does your question come through clearly? Have you provided sufficient discussion of the reasoning and justification for your question?

How Did You Collect the Information?

Will your readers readily understand how you collected the information that you've included? Did you obtain the information by interviewing others? By reviewing literature? By observing? By testing equipment? By analyzing existing information? As you review your draft, consider whether readers can easily and quickly determine the sources of your information. For example:

> The data for my project came from the records of the Mushroom Computer Corporation. The data included sales records—the number and price—of the Model S computer from 1980 through 1985.

Further, you may want to consider whether others will deem your information and data collection methods appropriate for the problem considered. Incorrectly collected data or biases in selection can invalidate the conclusions.

What Information or Evidence Did You Provide?

Ask yourself, "How do you know what you know?" The evidence used in solving technical and scientific problems comes in many forms. Make sure that the evidence clearly reflects the question posed. Thus for a study comparing the endurance of women and men and cross-country runners, you'd need to report how you measured endurance. If you measured endurance as the number of miles runners covered at their fastest pace, then the data should be presented as a table or bar graph with two columns, one for men and another for women. Readers then can easily compare women and men.

Which Data Analysis Techniques Did You Use?

Will your data analysis techniques be clear to readers? Using Mushroom Computer Corporation again, here's how sales growth might be analyzed:

> Using Mushroom's spreadsheet program, I determined the highest percentage increase, the lowest percentage increase, and the average increase. Using low, moderate, and high growth scenarios, I then projected possible sales for the Mushroom Computer Corporation for the next five years.

Suppose that you're comparing and contrasting two personal computers for use in an office. Did you explain how you made the comparison and contrasts? Did you collect manufacturer information on each computer? Did you use a subjective or an objective judgment or an empirical count? Let's say that you're trying to determine which personal computer to buy for your office. You've narrowed the field to two personal computers. You talk with owners of both kinds. Does your draft explain which criteria you used for the comparisons? How you compared and contrasted the owners' responses? Will your readers agree that you compared the two computers appropriately?

Did you shift definitions or base data on different definitions? Suppose that a student were reporting a study in which she tried to determine if recent laws were reducing the number of drunk drivers. The student collected data from 20 states on the number of individuals tested and found drunk on New Year's Eve. Of the 20 states, 10 had passed tough driving while intoxicated laws within the previous five years. The content problem lies in the definition and testing procedures for drunkenness. Do all 20 states define drunkenness in the same way? Do the states use the same test? If not, then the data and any comparisons between them are questionable. The student has to be sure that her bases for comparison are appropriate.

How Did You Verify Your Results?

Will your readers clearly understand the various ways that you verified the information? Returning to the recommendations on purchasing personal computers, did you check the current popular and trade magazines for product reviews and assessments? If so, did the authors of those articles also report maintenance records? Did the authors and records agree?

If your data are based on calculations, do they appear realistic, and do the answers fall within the expected ranges? Let's say that you were calculating the average snowfall for Ames, Iowa, for the last decade. Does your calculation fall within the expected range? If annual snowfalls had been 30, 75, 20, 22, 31, 43, 29, 54, 49, and 50 inches, and your first calculation showed an average of 19 inches, you'd know you'd made a computational error because no one year had less than 20 inches of snowfall. If your calculation showed an average of 70 inches, similarly you would know that you'd erred.

Let your technical education and experience guide your assessment of the data you're presenting. Before you began your data analysis, did you consider the dispersion of the data? Look for technical errors in calculations and data analysis. When reviewing data in tables, check the math. Based on what you know about similar occurrences, are the data realistic? Have others reported projects similar to yours? Can you use others' results to check your results? If so, how do the results compare?

What Conclusions Did You Draw?

After reviewing your data, carefully review your conclusions. Are they stated clearly? Do they fit logically with your objectives and data analysis?

Consider the following example. In a recent survey, buyers of new Chevrolets and visitors to car showrooms were asked whether they had inspected and priced Fords and Dodges. In the report, the writer concluded that Chevrolet buyers and visitors were not really comparison shoppers. But the data show a different possibility (Table 9-1). The conclusion that Chevrolet buyers were not comparison buyers is not valid, because 29

TABLE 9-1. A comparison of car salesroom visiting and buying practices

Salesrooms Visited	Visitors (n=250)*	Buyers (n=175)
Ford	18%†	29%
Chevrolet	82	100
Dodge	28	33

*n = sample size
†Percentages add to more than 100% because of multiple visits.

percent of the Chevrolet buyers reported visiting Ford showrooms, and 33 percent reported visiting Dodge showrooms. Furthermore, 18 percent of the visitors reported visiting Ford showrooms, and 28 percent reported visiting Dodge showrooms. The writer should point out that people visited other showrooms, possibly to comparison shop. The writer does not *know* whether the visiting represented comparison shopping, and so conclusions based on the data need to be presented cautiously.

When you find a content problem, correct it. At times, you may find yourself returning to problem solving strategies or gathering more information to make your case.

HIGHLIGHTS

1. General guidelines for revising manuscripts include: (1) allowing time between writing and revising; (2) taking the editor's role when revising; (3) revising at different levels; (4) making several trips through the draft and looking for different problems on each trip; and (5) using copyediting symbols.
2. Revising and editing content require (1) checking the underlying questions, objectives, or problems; (2) critically reviewing how information was collected; (3) assessing the information and evidence provided; (4) reviewing data analysis techniques; (5) checking to make sure results were verified; and (6) reassessing conclusions.

PROJECTS

1. Before you took the technical or scientific communication course in which you're now enrolled, how did you revise your manuscripts? In a narrative of not more than 500 words, explain the revising process that you used.
2. People in each field may create different errors in reasoning, logic, and content. Investigate the kinds of content errors that occur in your field. Prepare a report of not more than 500 words that explains the common content errors in your field.
3. Using your report from Project 2, prepare a checklist to guide your manuscript editing.
4. If you're writing a major paper for the semester, use your checklist as you revise. Attach a copy of your checklist to your draft, with problem areas noted in the margins. Turn both in to your instructor for review and critique.
5. What errors in logic, statistics, or reasoning have the authors committed, according to your technical or scientific background? Consider each scenario:

a. A report on the chances of flooding noted that floods greater than 20,000 cubic feet per second occur once in 100 years. In that the last flood was in 1897, the author reasoned that another flood would occur by 1997.
b. On March 29, 1985, the *Rocky Mountain News* (Fuentes 1985) reported that Joe Granville, a maverick of the stock market, said stocks were headed for a crash reminiscent of 1929. His supporting evidence included the failures of savings and loan institutions in Ohio, the defeat of the farm bill, the agricultural recession, and the fact that the words used by President Hoover prior to the crash of 1929 were exactly the same as those President Reagan was using to describe the 1985 economy.
c. In 1983, the number of volunteers for military service jumped substantially. Some individuals pointed out that youth had become more patriotic in recent years; other individuals pointed out that the military's recruitment program was producing superior results.
d. In an advertisement, a home gardener reported she had used "Great Grow" fertilizer on her zucchini plants the previous year and that they had produced some 200 pounds of squash. She then went on to strongly endorse "Great Grow" for boosting all vegetable production in home gardens.
e. A UPI news story in the *Rocky Mountain News* on August 2, 1985, reported that President Harry Truman and others had said that dropping the atomic bomb on Hiroshima and Nagasaki saved between 500,000 and 1 million lives that would have been lost had the United States tried to invade Japan. The story then cited a report by Dr. Arjun Makhijani and John Kelly, "Target: Japan—the Decision to Bomb Hiroshima and Nagasaki," in which World War II documents projected American losses to be as low as 40,000 and 153,000 and a proposed invasion to be "relatively inexpensive" in terms of casualties (UPI 1985).

FOR MORE HELP

CAMPBELL, S. 1974. *Flaws and fallacies in statistical thinking*. Englewood Cliffs, N.J.: Prentice-Hall.
    The author provides an easy-to-read book that is useful for catching common errors in presenting statistics.

FISCHER, D. H. 1970. *Historians' fallacies*. New York: Harper & Row.
    Fischer provides an excellent treatment of errors in reasoning as applied to history and many other disciplines.

RADNER, D., AND M. RADNER. 1982. *Science and unreason*. Belmont, Calif.: Wadsworth.
    The authors discuss pseudoscience and faulty reasoning in scientific fields.

RUNYON, R. P. 1981. *How numbers lie*. Lexington, Mass.: Lewis Publishing.
   The book provides a light treatment and numerous real-life examples of how statistics are frequently misused.

WILLIAMS, F. 1968. *Reasoning with statistics*. New York: Holt, Rinehart and Winston.
   Williams provides a succinct overview of statistics without a long discussion of theory.

10
Effective Editing

OVERVIEW

In Chapter 10, we discuss
- ☐ Recasting to eliminate gender bias
- ☐ Reworking to clarify organization
- ☐ Revising to enhance paragraph unity
- ☐ Editing to improve sentences

As we have pointed out, technical writing requires a clear, concise style that produces effective communication with one interpretation. To meet the goal, reconsider the communication principles of Chapter 1 as you revise your drafts.

Be especially careful of any passage that may create problems for the receiver, especially noise—interference—of any sort. What kind of noise might create problems for an audience receiving or acting on your message? We have already spoken about errors that bias the selection of content and those that bias the reporting of results. Noise can also come from the gender bias that can slip into your writing. To eliminate gender bias from your drafts, follow the suggestions in Panel 10-1.

Poor organization, weak paragraph structure, and awkward, wordy writing also create noise problems. So once you've checked your draft for content and gender bias problems, you're ready to concentrate on improving the details that produce clear, concise organization, paragraphs, sentences, clauses, phrases, and words (Panel 10-2).

When you edit your drafts, it is advisable to use the standard copyediting symbols, as explained in Panel 10-3, pp. 204–205.

REVIEWING AND EDITING OVERALL ORGANIZATION

When you set out to check the overall organization of your manuscript, you need to consider the content and the logical order in which to present it for the audience you're trying to reach. What is correct for one audience

PANEL 10-1. Eliminating Gender Bias in Language

To avoid sexist language, pay careful attention to details. The following article from *Simply Stated,* the monthly newsletter of the Document Design Center, provides clear guidelines for eliminating sexist language from your drafts.

Clear writing experts recognize that part of writing understandable documents is understanding and responding to the needs of the intended audience. This might mean limiting the use of jargon or technical terms for a general audience, or using a larger type size for an elderly audience. It is the writer's job to maintain the audience's willingness to go on reading the document. Readers who are continually stumped by big words or offended by a pompous tone are likely to stop reading and miss the intended message.

Today, part of striking the right tone is handling gender-linked terms sensitively. Use of gender terms is a controversial issue. Some writers use the generic masculine exclusively, but this offends many readers, because it seems to be based on the presumption that all people are male unless proven female. Other writers are experimenting with ways to make English more neutral. Some have even tried to create a gender-neutral pronoun for the third person singular, but this has not caught on.

Avoiding gender bias in writing involves two kinds of sensitivity: (1) being aware of potential bias in the kinds of observations and characterizations that it is appropriate to make about women and men, and (2) being aware of certain biases that are inherent in the language and how you can avoid them. The second category includes using gender-specific nouns and pronouns appropriately, and this article presents guidelines for handling these problems. The best approach right now is to write around the problem when you can, to avoid offending your audience. Here are a few ways to do that:

1. Use a gender-neutral term when speaking generically of your fellow creatures. For example:

man	the human race
mankind	humankind, people
manpower	work force, personnel
man on the street	average person

2. Avoid clearly gender-marked titles. Use neutral terms when good ones are available. For example:

chairman	chair
spokesman	speaker, representative
policeman	police officer
stewardess	flight attendant

    When you can, avoid awkward coinages like "spokesperson."

3. If you are speaking of the holder of a position and you know the gender of the person who currently occupies it, use the appropriate gender pronoun. For example, suppose the "head nurse" is a man:

    Before: The head nurse must file her report by September 1, 1982.

---

may be inappropriate for another. Below we highlight some common organizational problems.

Is Content in Appropriate Places? Consider the basic report organization given in Panel 7-4. Remember to put points where they belong. A methods section should not include a preliminary discussion of the results nor points that belong in the introduction. New materials or points should not appear in concluding sections. Conclusions should present a brief summary of the key findings, not launch into a further discussion of points not adequately covered in the discussion section.

PANEL 10-1. (continued)

    After: The head nurse must file his report by September 1, 1982.

4. Rewrite sentences to avoid using gender pronouns. Use the appropriate title instead.
   Before: You should see your doctor first, and he should call the pharmacist directly.
   After: You should see your doctor first, and the doctor should call the pharmacist directly.
5. Recast your statement in the plural, thus avoiding the third person singular pronoun.
   Before: Each student should bring his text to class.
   After: All students should bring their texts to class.
6. Address your readers directly in the second person if you can appropriately do so.
   Before: The student must send in his application by the final deadline date.
   After: Send in your application by the final deadline date.
7. Replace third person singular possessives with articles.
   Before: Every branch chief should draft his preliminary schedule by Friday.
   After: Every branch chief should draft a preliminary schedule by Friday.
8. Write your way out of the problem by using the passive voice.
   Before: Each department head should do his own projections.
   After: Projections should be done by each department head.

We recommend this solution only if nothing else works, since the active voice is generally clearer and more effective than the passive.

9. Use a third person singular pronoun to refer to a third person singular antecedent, unless you are sure your audience is as willing as you are to break the rules of English grammar.
   Before: Every child should brush his teeth after meals.
   After: All children should brush their teeth after meals.
   or: Every child should brush his or her teeth after meals.
   but not: Every child should brush their teeth after meals.
   Using "their" is becoming more acceptable in speech. But if you use it in writing, you are likely to distract your reader.
10. Avoid "s/he," "he/she," and "his/her." They look awkward, but even worse, they interfere when someone is trying to read a text aloud. If none of the other guidelines has been helpful, use the slightly less awkward forms "he or she," and "his or hers."
11. Remember, the goal is to avoid constructions that will offend your readers enough, because of their views on grammar or politics, to distract them from your text

(*Source: Simply Stated, no. 28, August 1982. Reprinted with permission from Simply Stated, the monthly newsletter of the Document Design Center, American Institutes for Research.)*

PANEL 10-2. Checklist for Tightening and Editing Drafts

Check the overall organization of the draft.

- Have you sequenced points correctly?
- Have your included all needed points?

Check paragraphs.

- Do paragraphs keep to one central idea?
- Do paragraphs reflect a continuity of logic?

Develop clear, strong sentences.

- Are you using vigorous verbs?
- Do you use simple constructions?
- Do you speak plainly?

Are Points in Correct Chronological Order? Check to see that communications report sequences of events as they happened. Suppose that you're reporting the sequence leading up the collapse of a highway overpass. Narrate events step by step as they happened. Subsequent inquiries, lawsuits, and other legal actions may hinge on how you report the order.

Have You Documented the Process Properly? Let's say that you're reporting how you cultured bacteria in the laboratory. Make sure that your report follows the steps that you followed. If you varied from a standard laboratory procedure, make clear why and how you varied your approach. Remember that others may later use your record to replicate (repeat) your study and to verify your findings.

Are Instructions Complete? When giving instructions, make sure that you have included all steps in the proper order. Poorly organized instructions often result in frustrated users, even injuries. Lawsuits based on poorly prepared instructions are not uncommon (a point we cover more fully in Chapter 18).

Living Up to Promises? If you declare that you'll discuss five points or list 16 items, make sure you discuss five, not three, points and list 16, not 17, items. Your declared intentions help readers prepare their minds for the subsequent points.

REVIEWING AND EDITING PARAGRAPHS

Once you've reviewed and reconsidered the larger organization, you're ready to reconsider individual paragraphs and how you might improve them. First make sure that your paragraphs have internal unity, that each paragraph sticks to one main idea. Also make sure that each paragraph has continuity, that the logic and structure of each sentence within the paragraph lead clearly to the next. Frederick Crews and Sandra Schor (1985) caution you to be especially watchful that you don't contradict yourself in a paragraph. If you make one point, don't reverse yourself in the next sentence or present a conflicting point.

Make sure your paragraphs are the appropriate length for the style of the document or publication. Watch for especially long paragraphs; readers shy away from them. Paragraphs that go on and on often create comprehension problems for the reader. Break up or shorten especially long paragraphs. To break a paragraph in two, first decide if you've presented more than one central idea. If so, break the paragraph at the point where you shift from one idea to the next. You may need to divide a paragraph into two, three, or more paragraphs. Here's an example:

> The "Systems Approach" to agricultural development requires an interdisciplinary team consisting of farmers, technicians, researchers, and policy makers. Farmers serve as the central building block while the other team members provide the needed support. Everyone becomes involved in the specific agricultural development project. The key activities in agricultural development include training the team members, building organizations, conducting valuations, and carrying out monitoring and evaluation activities. Training . . .

From here the paragraph continues and provides succinct definitions of each key activity. The paragraph shifts ideas with the sentence beginning, "The key activities . . ." That sentence is a realistic place for creating a second paragraph. The first paragraph focuses on the idea of the interdisciplinary team. The second focuses on main activities.

In breaking up some paragraphs, you may find that you need to develop certain paragraphs more fully. If so, consider which additional points you'll need to make. These points should expand or clarify the original main point.

To shorten paragraphs, look for sentences that do not add to the central meaning or do not adequately amplify points. Edit individual sentences to shorten paragraphs. Techniques for shortening individual sentences follow.

GUIDELINES FOR CLEAR, CONCISE SENTENCES

In most cases, drafts benefit from a harsh trimming. Cut 10, 20, 30 percent or more from your drafts, and you'll almost certainly have a better document. Say what you want to say, but say it simply. Researchers and practitioners alike support the techniques we suggest:

- Use strong verbs.
- Use simple constructions.
- Speak plainly.

Strategies for Strengthening Verbs

You'll find improving your verb usage one of the most powerful techniques for tightening your writing. When you revise your drafts:

1. Use active voice.
2. Use action verbs.
3. Eliminate "There is . . ." and similar constructions.

Use Active Voice.   In doing our research, we read four dozen books and articles offering advice on how to write better. All but two agreed on one point: avoid the passive voice. You hear the active voice in verb forms like "I take," "you see," "we appreciate." You hear the passive voice in verb forms like "it was taken by me," "it was seen by us," and "the book was appreciated." In a sentence that uses the active voice, the subject (or agent of the action) does the action of the verb. Passive constructions turn the sentence around; the agent becomes the object of the action. Some passive constructions drop the agent entirely. Consider the following:

Example 1. Active Voice: Sally completed her experiment.
Example 2. Passive Voice: The experiment was completed by Sally.
Example 3. Passive Voice: The experiment was completed.

In *Example 1* the agent, Sally, did the action; readers understand who did what. *Example 1* is also shorter than the passive form in *Example 2*. *Example 3* is the same length as *Example 1*, but readers don't know who did the action.

Using the active voice leads to

- Stronger verbs
- Shorter sentences
- Clear specification of who or what carried out the action
- Vigorous writing
- Clear, simple sentences

Skilled writers argue against the passive voice because it usually lengthens sentences. It often hides the agent (who is doing the action). "Results were deemed inconclusive," the report says. By whom, the reader asks? By you? By your adviser? By your laboratory assistant? In such cases, the passive becomes neither scientific nor logical, but a refuge for the timid. Passives encourage vagueness, the enemy of scientific precision. Avoid the passive, except when logic tells you to use it.

To distinguish active from passive constructions,

- Look for the verb "to be"—*is, was, were, are,* and other "small verbs." They often signal passive voice.
- Check for the preposition "by." It usually signals the passive voice.
- Ask "Who did what to whom?" Is the subject of the sentence the agent of the action? If so, the sentence probably is in the active voice.

In active constructions, the subject can take many forms:

Example 4. Person: Karen completed her report.
Example 5. Pronoun: She completed her report.
Example 6. Organization: The National Cancer Institute evaluates health communications.

Example 7. Thing: The computer crackled before its screen went blank.
Example 8. Plant: Our trees produced 50 bushels of apples last year.
Example 9. Animal: Garter snakes eat insects.

Sometimes people confuse verb *tense* with verb (active and passive) *voice*. Verb tense indicates when an action is carried out: today he writes, yesterday he wrote, tomorrow it will be written. You can use active and passive constructions in any verb tense.

Although experienced writers strive to use the active voice, they use passive constructions when they:

- Do not need to identify the agent
- Do not know who the agent is
- Want to focus attention on the subject
- Want to avoid possible legal problems
- Want to build a climax at the sentence's end
- Want to provide variety

Use Action Verbs. Not only the active voice, but action verbs and avoiding "trapped" verbs will improve your writing.

1. *Eliminate Weak Verbs.* Whenever you read *am, is, are, was, were, have, had, would,* or *should,* see if you can replace them with stronger verbs. Compare *Examples 1* and *2*.

Example 1: We noted that his work *is* a bit sloppy.
Example 2: We noted his sloppy work.

When you eliminate the small verbs, your sentences grow more forceful and brisk. In *Example 2,* we eliminated the hedge words "a bit" and made "sloppy" modify "work."

2. *Eliminate Trapped Verbs.* Check for small verbs and "trapped" verbs—sometimes called *nominalizations*. The verb form is "trapped" inside a noun. Released from the noun, the verb gains force. You also can strengthen your sentences by replacing the small verb with the nominalization's verb form. In the paired sentences below, study the italicized nominalization, and compare it to the verb in the second sentence:

Example 3: The *assessment* of the experiment's success was made by the scientist.
The scientist *assessed* the experiment's success.
Example 4: The *detainment* of the tourists will be made by the country's military.
The country's military will *detain* the tourists.
Example 5: The technician made an *examination* of the cracked beam.
The technician *examined* the cracked beam.

The first sentences in *Examples 3* and *4* contain passive voice. Releasing the trapped verbs makes each sentence more powerful and shortens the sentences as well.

To identify nominalizations one source recommends these steps:

- Look for words ending in *tion, al, ance, ence, ment,* or *ure.*
- Look for nouns embedded in the phrase, "the ——— of."
- Check nouns at the beginning of sentences.
- Check sentences with no mention of who does what. (Felker et al. 1982, 36)

To which we would add:

- Check sentences with small verbs such as *is, was, were, have, had, has, would,* or *should.*
- Check sentences with passive voice.

If you are in doubt about how to convert nominalizations to their verbs, check your dictionary.

3. Avoid Padded Constructions. Sentences that have padded, wordy constructions, such as "there," "it," "this," "these," and similar words combined with forms of the verb "to be," can be strengthened. "There are" and similar constructions aren't grammatically wrong; an occasional "there is" doesn't present a problem. But avoid using these fillers regularly. Consider:

Example 1: There is a television show on the whooping crane on Channel 6 tomorrow at 7 P.M.

Channel 6 will air a show on whooping cranes at 7 P.M. tomorrow.

Example 2: It is required that you report to the president's office by 3:30 P.M. today.

Report to the president's office by 3:30 P.M. today.

Strategies for Simplifying Constructions

Improving verb usage improves writing. Simplifying constructions improves it even further. When reviewing your drafts:

- Use simple sentences.
- Edit unnecessary clauses.
- Revise excessive prepositional phrases.
- Break up "stacked" modifiers.

Use Simple Sentences. The simple sentence—subject-verb-object—does not guarantee brevity, but it does usually restrict you to one or two ideas.

Thus you usually avoid the long, drawn out sentences that lose readers. When editing your drafts, look for long sentences. If appropriate, break them up. In searching for sentences to simplify, look for:

- Complex sentences that may confuse readers
- Long sentences in which you change thoughts
- Sentences that run on and on
- Run-on sentences in which two independent clauses lack conjunction or punctuation
- Sentences joined by series of conjunctions
- Sentences of independent clauses joined by semicolons or commas

Consider:

Example 1: I would disagree and I think the researcher is in an unfair position and I suggest the researcher return next week to explain his data.
Example 2: It is one that we have generally been aware of in the computer field; it is so new we just don't have a large body of people concerned about it.

The sentence in *Example 1* is a run-on. To be grammatically correct, it must be broken down into two sentences and the independent clause punctuated. The sentence in *Example 2* is grammatically correct, if wordy.

Edit Unnecessary Clauses. Occasionally, you can tighten sentences with unnecessary clauses by:

- Eliminating clauses
- Converting them to appositions
- Replacing them with adjectives
- Replacing them with prepositional phrases

1. Eliminating Unnecessary Clauses. Look for clauses that contain no new information or repeat information. Omit them:

Example 1: Our lawn mower, *which cuts the grass,* needs a new spark plug.
Example 2: Our lawn mower needs a new spark plug.

2. Converting Clauses to Appositions. If you want to add information, shorten clauses, and place them in apposition:

Example 3: The Apple computer, *which was introduced more than 10 years ago,* remains popular.
Example 4: The Apple computer, *introduced more than 10 years ago,* remains popular.

3. Replacing Clauses with Adjectives. If a clause contains only one new idea, recast it as an adjective:

> *Example 5:* Bill Rudolph, *who is an engineer,* designed and evaluated the spring testing machine.
> *Example 6: Engineer* Bill Rudolph designed and evaluated the spring testing machine.

4. Creating Prepositional Phrases. You'll find it easy to create prepositional phrases from clauses:

> *Example 7:* Destroy specimens *that have no tags.*
>   becomes
> Destroy specimens *without tags.*

Revise Excessive Prepositional Phrases. Recasting prepositional phrases can help you add information and make sentences more clear and interesting. A problem arises, however, when you use too many phrases. Then your writing becomes wordy. Consider the italicized prepositional phrases:

> *Example 1:* Ed Carpenter, the geologist *from the University of Montana,* discovered high concentrations *of cadmium in the water of home owners of the county of Larimer.*

Compare *Example 1* with *Example 2:*

> *Example 2*: University of Montana geologist Ed Carpenter discovered high cadmium concentrations in Larimer county home owners' drinking water.

Study *Examples 3* and *4:*

> *Example 3:* The speech *of the chief of marketing of XYZ Computers* predicted a downturn *in the sales of personal computers of all kinds.*
> *Example 4:* The marketing chief of XYZ Computers predicted a downturn in personal computer sales.

Tips for working with prepositional phrases include:

- Converting prepositional phrases to adjectives.
- Converting prepositional phrases to adverbs.
- Converting prepositional phrases to possessives. (Burnett, Powers, and Ross 1973, 74–5)

Let us look at each of these.

1. Converting Prepositional Phrases to Adjectives. When you find prepositional phrases modifying nouns, convert the prepositional phrases to adjectives. Be careful: some nouns do not have adjective forms; check your dictionary.

2. Converting Prepositional Phrases to Adverbs. When prepositional phrases modify verbs and verb forms, convert them to adverbs:

> *Example 5:* Gardeners who handle herbicides, plant killers, *in a careless manner* may kill their vegetables.
> *Example 6:* Gardeners who carelessly handle herbicides, plant killers, may kill their vegetables.

3. Converting Prepositional Phrases to Possessives. Some prepositional phrases show ownership; some show association. Convert them to possessives:

> *Example 7:* The slide presentation *of the engineer* explained how his company would remove the nuclear wastes.
> *Example 8:* The engineer's slide presentation explained how his company would remove the nuclear wastes.

4. Eliminating Unnecessary Prepositional Phrases. Look for phrases that add nothing or repeat information unnecessarily. When sentences end with a prepositional phrase, delete the phrase if it adds nothing:

> *Example 9:* The investigators found no significant change in the water quality between January 1980 and June 1985 *in this study*.

Break Up "Stacked" Modifiers. Stacked modifiers crop up often in technical writing:

> product assessment experiment program
> agricultural conservation development project leader
> agency predator reduction technique test
> cancer detection test kit evaluation project

Stacked modifiers make writing difficult to understand. Some readers have difficulty following stacked modifiers, because the connectives have been dropped—prepositions and possessives that make relationships clear.

Recasting the stacked modifiers produces phrases that are easier to understand:

> program of experiments in product assessment
> project leader in charge of development program for agricultural conservation
> test of the agency's predator reduction technique
> project to evaluate the cancer detection test kit

Break up stacked modifiers by:

- Creating prepositional phrases by using such words as *of, for, about,* and *in*
- Creating possessives
- Using commas
- Recasting the constructions

Breaking up stacked modifiers usually creates longer constructions.

Strategies for Speaking Plainly

To improve your writing further, avoid jargon, terms with many definitions, complex words, and unnecessary modifiers and qualifiers. To improve clarity and conciseness, we suggest that you:

- Use jargon sparingly
- Define your terms
- Use short words
- Use specific, concrete, and definite words
- Drop unnecessary modifiers and qualifiers
- Avoid clichés
- Eliminate wordiness

Use Jargon Sparingly. A term that seems precise to you, the writer, may be jargon to your reader. To communicate effectively, develop a feel for which technical terms your readers will understand and which ones they won't. The word *jargon* comes from a Greek root that means "sacred language, known only to the initiated." Just because you know jargon, you don't need to use it all the time. Most audiences don't know jargon.

But when only the precise technical term will carry the appropriate message, use it. If you cull all unnecessary jargon, your readers will permit such lapses.

Here's a list of random jargon words:

baud	=	the transmission rate in bits per second of a computer system
byte	=	a string of binary digits, usually 8
gigabyte	=	one billion bytes
in vitro	=	in test tubes, beakers, or containers
in vivo	=	in a living organism
lactase	=	an enzyme that breaks down milk sugar (lactose)

Define Your Terms. Sometimes words have different meanings in different fields. For example, "dwell," "lake," and "barn" are common enough words. But to the mechanical engineer, "dwell" means a period of time; to the chemical engineer, "lake" is a dye compound; and to the atomic physicist, a "barn" is a measurement of an atom (Vinci 1975).

Keep in mind that you can define terms not only with words, but with mathematical symbols, equations, and illustrations.

Use Short Words. Whenever possible, use a short word instead of a long word, an English word instead of a foreign word. Write to communicate, not to impress. To help you start, here's a list of long, unfamiliar, and foreign words with their shorter, crisper counterparts:

excised = cut out
euthanasia = mercy killing
phenomenal = large, great (use a specific amount)
prioritize = rank
dialogue = talk

utilize = use
employ = use
compel = force
caveat emptor = let the buyer beware
en masse = in a mass, body, or group

Use Specific, Concrete, and Definite Words. General, vague, and abstract words do not tell your readers what you really mean and garble your message. Replace general, vague, and abstract words with specific, concrete, and definite words. John O'Hayre, in *Gobbledygook Has Gotta Go* (1978), says that general and abstract words represent broad categories, unlimited numbers, and immeasurable concepts. These vagaries evoke different meanings for different readers.

By using specific, concrete, and definite words, you will

- Cut down on possible misinterpretations of your message
- Increase readers' interest

Below are lists of general words that are better as specific words; abstract words that are better as concrete words; and vague words that are better as definite words. (The list of general words is from O'Hayre 1978, 19.) Undoubtedly, you can think of other examples.

General Words	*Specific Words*
people	Pete, Sally, and Joe
structures	outhouse, office building
animals	aardvarks, pelicans, horned toads
devices	cider press, bombs
clothing	parka, mukluks, moccasins
mountains	Rocky Mountains, Mt. Rainier, Long's Peak

Abstract Words	Concrete Words
effect	cured 80 percent of the herpes cases
expensive	$10 million
color	slate gray
large size	size 48 belt
shallow depth	20 fathoms

Vague Words	Definite Words
some	one dozen, 46, half a cup
many	5,000 to 10,000
large	300-pound rock
small	3 grams
big	6 feet tall and 245 pounds

In each example listed below, compare the paired sentences:

Example 1. General: Some of the researchers became ill earlier in the year.
   Specific: Jones, Smith, and Carpenter contracted bubonic plague in March.
Example 2. Vague: Add a little bit of nitric acid to the solution.
   Definite: Add two milliliters of nitric acid to distillate A.
Example 3. Abstract: He has a good business.
   Concrete: His business grossed $2.5 million last year.
      or
   He has 2,000 satisfied customers.

When you must use abstract or general words, do so carefully.

Drop Unnecessary Modifiers and Qualifiers.   Words like "rather, very, little, pretty are leeches," E. B. White said, "that infest the pond of prose, sucking the blood of words" (Strunk and White 1979, 73). Unnecessary modifiers—adjectives and adverbs—creep in to pep up language, but they add no pep at all. Find better ways of expressing your thoughts.

Some modifiers and qualifiers are hedge words. William Zinsser (1980, 104–5) identifies the hedging qualifiers as:

a bit	a little
sort of	kind of
rather	quite
very	too
pretty much	in a very real sense

Hedge words suggest that you're not sure of your content. Use hedge words, and readers may question your competence. When editing, eliminate the unnecessary modifiers and qualifiers that have crept into your writing.

Hedged: It is extremely important to disconnect the power supply before removing the cover; otherwise serious injury may result.
Direct: You may kill yourself if you don't unplug the computer before taking off its cover.

When you find yourself trying to intensify an expression by using adjectives or adverbs, try rephrasing it.

Avoid Clichés. Cincinnati writing consultant Phyllis Martin compiled *The Word Watcher's Handbook* (1977), which she described as a "deletionary." In it she listed clichés to avoid. *Denver Post* columnist Jack Kisling favorably reviewed the book. His tongue-in-cheek review is packed with clichés:

> In a nutshell this book is jam-packed with helpful hints for eager beavers who don't know beans about English, and unless they're blind as bats, it should help them throw off the yoke of stuffed-shirt phrases that rub people the wrong way. . . .
> Remember, most clichés are as old as the hills. . . . There's no time like the present to give tired phrases the good letting alone they richly deserve. (Kisling 1980, 21)

The rule for avoiding clichés is easy to know but tough to follow: distrust the phrase that comes quickest to mind.

Eliminate Wordiness. If you use more words than you need, you clutter your sentences and confuse readers. You can eliminate clutter by:

- Cutting needless words
- Replacing several words with one or two words
- Eliminating repetition and redundancy

When you revise your drafts, study each word and ask, "Do I need this word?" Consider how the phrases in the left column below can be written more simply and clearly:

in order to	=	to
The fact that we conducted	=	We conducted
the major problem is, to say the least,	=	the major problem is
My personal experience	=	My experience
Now more than ever	=	Now
Their own observations	=	Their observations

You probably will see clutter as you read your drafts. By eliminating wordiness, you tighten your writing.

> *Example 1. Wordy:* This research project report of mine focuses on four case studies done in the state of South Carolina.
> *Succinct:* My report focuses on four case studies completed in South Carolina.

PANEL 10-3. Copyediting

Copyediting symbols are marks used for editing the typewritten page. Such symbols are a shorthand or code for indicating the changes you want made in retyping or typesetting. The symbols save you time, energy, and potentially money, because you need not retype copy that needs only minor changes. Editors, word processing operators, and most secretaries know these symbols.

In typing copy, double space, and leave adequate margins for copyediting. For most copy that is to be typeset or retyped, copyedit with a number two pencil. Then, if you change your mind, you can erase easily. But check with the publisher, editor, secretary, or word processing operator you're working with. Some prefer brightly colored copyediting symbols and request that you use red, felt-tipped pens.

Too many corrections create "dirty copy," a page with so many marks it's hard to read and understand. Retype dirty copy.

Copyediting Symbols

Symbol	Meaning
㉖	Spell out
⟨one hundred⟩	Abbreviate
⁄Technical	Use lower case
english	Capitalize
Techn⋏cal	Insert letter
misspe‸lled ~~missipelede~~	Mark out grossly misspelled words and print the correct above

> *Example 2. Wordy:* In recent years, there has been considerable interest in research in curing concrete.
> *Succinct:* Since 1975 researchers have completed 200 tests on curing concrete.

These editing techniques will help improve your writing, so use them. When you review your drafts, think of them as someone else's. Look for the wordy and awkward constructions and recast them. Cut the clutter. With practice, you'll probably cut your drafts up to 50 percent and develop a clear, concise personal style that will serve you well.

PANEL 10-3. (continued)

Copyediting Symbols

Symbol	Meaning
cour/se	Delete and close-up
gather ~~retrieve~~ materials	Delete and join to remaining text
John completed the task⊙	Insert period
The ⌊elements⌉trace⌋	Transpose words
ta͡nce	Transpose letters
. . . dogs. ⌊New practices	Start new paragraph
dog's fleas	Insert apostrophe
runon#words	Insert space between words
to ~~provide~~ insight (stet)	Retain the original
in the shallows. ⌒ ⌒In most rivers . . .	Close up or join
What did you want ?	Insert question mark
feet, inches∧and pounds	Insert comma
feet⁄	Delete letter

HIGHLIGHTS

1. When editing for communication effectiveness, check for potential noise.
2. When editing for organizational problems, determine if you (1) present your content in the appropriate places, (2) have points in chronological order, (3) document properly, (4) include all points, and (5) live up to your promises.

3. When editing paragraphs, check to make sure that they (1) reflect unity and continuity, (2) avoid contradictions, and (3) are short.
4. Strategies for developing clear, concise writing include using stronger verbs, using simple constructions, and speaking plainly.
5. Strategies for strengthening verbs include using the active voice, strong verbs, and recasting padded constructions, such as "there is."
6. Strategies for simplifying constructions include (1) using simple sentences, (2) recasting unnecessary clauses, (3) revising prepositional phrases, and (4) breaking up stacked modifiers.
7. Strategies for speaking plainly include (1) using jargon sparingly; (2) defining terms; (3) using short words; (4) using specific, concrete, and definite words; (5) dropping modifiers and qualifiers; (6) avoiding clichés; and (7) eliminating wordiness.

PROJECTS

Section A

1. Select a paragraph in need of revision from one of your drafts, and make a photocopy. Determine the weak points. Revise the paragraph. Then write a short explanation of why you revised the paragraph. Submit to your instructor the photocopy of the original paragraph, your explanation of why the paragraph needed revising, and the revised paragraph.
2. Bring a photocopy of one of your drafts to class. Exchange photocopies with a classmate. Find a weak paragraph in your classmate's draft and revise it. Then write a short explanation of why you revised the paragraph. Submit the original draft, your revised paragraph, and your explanation to your instructor.

Section B

Review each paragraph below, and revise or edit as needed.

> The rock squirrel (*Spermophilus variegatus*) is found in rocky areas of evergreen and deciduous treelands, scrublands, and savannas of both uplands and wetlands and the species may feed in agricultural row crops. Rock squirrels eat flowers, seeds, and snails (Young 1979). They have been observed eating green vegetation, cypress cones, Texas persimmons, plantago (*Plantago* spp.), and hackberry flower buds (Johnson 1979), as well as cedar berries, walnuts, pecans, acorns, hackberries, wild grapes, soft insects, and occasionally turkey poults (Ramsey 1956) and cultivated fruits and vegetables (Ramsey 1956; Davis 1964). They are good climbers and feed in trees as well as on the ground (Steiner 1975).
> —Adapted from *Terrestrial Habitat Evaluation Criteria Handbook*, 1980, p. 5.6–1.

Thus, the Farming Systems Research and Development leaders should adapt the available training materials to local conditions. The training staff must provide trainees with adequate field experience and make the best use of qualified trainers. The size of the groups to be trained depends on the capabilities of the trainers, the needs of the program and the ability of trainers, and the number of qualified candidates.

Section C

For each item, identify the weak construction. Recast it succinctly.

1. Since the Space Age's beginning, there has been much research done in ceramics.
2. To install a solar hot water system, the panels are bolted to your house's roof.
3. The report by me will be based on five experiments.
4. The output of the generation system is some 110 volts.
5. Tables 1 and 2 contain the summarization of the results of the experiment.
6. It has long been known by dog trainers that dominance over a pet can be created with proper training.
7. The take-home examination should be completed not later than next Friday.
8. The bottom line remains that we must increase our productivity in the soils testing section.
9. The justification of the change in procedures will be made by the division chief.
10. The military is proposing new devices that officers claim will end the need for hostile confrontations between nations.
11. After a lengthy debate, the two technicians are in agreement on the procedure to follow in solving the problem.
12. No further reports of the diseases have been reported so far this month.
13. The anti-tobacco chewing program brochure budget will be submitted by Keep Illegal Substances from Students.
14. The Pacific Northwest Forest Experiment Station was established by the Forest Service more than 25 years ago.
15. The Department of Defense proposes a salary incentive program to improve the morale of enlistees of the Branches of the Army, Navy, Marines, and Air Force.
16. The shop foreman reported the mechanic was a little bit ill and it contributed to very poor work habits.

17. There are many different ways that you can use to plan your reports including brainstorming, pattern notes, tree diagrams, traditional outlines.
18. The letter of resignation was drafted in a quick manner by Sam.
19. The Friday seminars will be planned by Jane Smith, who is the new assistant professor.
20. The field experiments on large concentrating systems have been very positive although quite limited.
21. The objective of this study was the determination of whether timing irrigation or reducing the irrigation of corn can save water and maintain yields.
22. It was learned by the State of New York that to achieve a proper asphalt/aggregate mix, 70% recycled materials and 30% virgin materials needed to be used.
23. The project, which will test recycled asphalt, will be a very costly one in excess of $1.25 million.
24. The report of the investigation team has been submitted to the chief of operations of the unit.
25. Home owners who handle paints in a careless manner may harm themselves, their families, their homes, and their pets.
26. The process of the summarization of the literature on ergonomics and computers is one that will begin next week by the staff.
27. The new agency will have the responsibility for the coordination of the mapping of satellite orbits.
28. The inventory is the summarization of genetic data collected by the Federal government.
29. The data are shown in Table 1. Clearly, more students who completed a technical communication course scored better on the tests than students who have not completed a course.
30. The purpose of the project of the juniors of State University is the determination of the quality of instruction by professors above the associate rank.
31. The paradigm described below outlines the scope and general direction for the company.
32. The goal of the Administration and Operations Division in the overall function in the Agency is to facilitate the accomplishment of approved agency tasks and activities.
33. There was no problem of waterlogging and salinity in the field of the farmer.
34. It is estimated that the farm's productivity of wheat will exceed its yields from last season.

35. The booklet identifies the common plants around the state. It also describes their habitats and provides their scientific names.
36. John Jones discusses the project which is changing how some home gardeners start their seeds.
37. One should know how to distinguish between different kinds of poisonous snakes in order to protect oneself from the chance of being bitten.
38. There was no effect on the inflation rate by controlling the money supply.
39. Ideally, each technical report would be judged by its reviewers as to its verbosity and wordy constructions.
40. She said we sort of had a problem in trying to get the report completed by the first workday of next week.

Section D

Correct any flaws in the organization of the following accident report:

Both drivers and golfer had exchanged insurance information. Car window of J. D. Roberts was broken on arrival at scene. Roberts complained he had noticed the windshield break while driving on north side of City Park, along Shields Ave. Golfer at scene admitted driving from tee on fourth hole and slicing severely. When Roberts noticed window break, he put on brakes, causing car to stop suddenly. Resulting crash broke rear window as well and dented back of his car. Golfer pointed out that players on licensed golf courses are not legally responsible for broken windows in passing cars. Driver of second car was not injured. Neither was Roberts. Driver of second car was Cecil Ray. Golfer was Dan Shook. Ray said he couldn't stop in time.

FOR MORE HELP

BERNSTEIN, T. 1971. *Miss Thistlebottom's hobgoblins*. New York: Farrar, Straus and Giroux.
   Bernstein provides a succinct, realistic review of language problems for writers.
———. 1965. *The careful writer*. New York: Atheneum.
   Bernstein clearly answers many questions on common usage.
DEGEORGE, J., G. A. OLSON, AND R. RAY. 1984. *Style and readability in technical writing*. New York: Random House.
   The text provides a series of useful exercises to help writers improve their personal style.
WILLIAMS, J. M. 1981. *Ten lessons in clarity and grace*. Glenview, Ill.: Scott, Foresman.
   Williams suggests many ways for writers to enhance the quality of their work through tight editing.

11
Revising to Improve Appearance

OVERVIEW

In Chapter 11, we
- ☐ Suggest ways for improving a document's appearance
- ☐ Discuss the value of peer reviews

APPEARANCES: HOW DOES YOUR DRAFT LOOK?

When others review your memos, letters, reports, and articles, their first impressions provide the mental framework for how they'll view your work. Many first judge your work on its overall appearance. Consider the points in Panel 11-1 as a good checklist for improving appearance. Your work's appearance depends upon how well you conform to a standard style.

To resolve any disputes about matters of style and to ensure consistency, many organizations and professional associations either present their own style in a style manual or adopt the widely accepted style of a standard style manual. Such manuals contain the unique rules and guidelines for producing communications. Panel 11-2 lists a selection of style manuals.

Style manuals cover many issues. Some give rules for capitalization, spelling, punctuation, abbreviation, preparing illustrations, grammar, and writing mechanics. Other style guides explain how to prepare typed and printed documents that will have a consistent, unified appearance. Other style guides explain the review process a document must follow prior to publication. Some style manuals identify other style manuals that writers and editors should use when preparing manuscripts.

Organizations and professional fields insist that employees, writers, and others concerned follow style guides for three reasons. First, conforming to the style shows professionalism. It tells readers that you know about the communication style of your profession and take pride in following it.

Second, conforming to the style makes communications consistent, and consistency builds a good image. Conforming to a specified style suggests care and attention to details. Furthermore, when everyone in an organization uses a consistent style, they create a united front; everyone goes about work in the same way. Third, conforming to the style saves your employer time and money. People do not have to spend endless hours correcting points covered in the style guide.

Occasionally you will miss style points. That's why all writers need editors and reviewers: they catch errors. But do strive to follow style. If your drafts look like you've disregarded your employer's editorial style, people may question your sincerity and interest in your job.

If your employer or the publication for which you're writing does not have its own style manual, your supervisor or the editors often can specify an acceptable manual, possibly one of those listed in Panel 11-2.

To further explore style, we can identify four kinds: editorial, typographic, idiomatic, and personal style (Hill and Cochran 1977). In the following discussion, we elaborate on each kind of style.

PANEL 11-1. Checklist for Manuscript Appearance

When you evaluate a manuscript for appearance, look for errors, omissions, or other problems. Ask the following questions:

- Did I use the appropriate style?
- Did I use standard and acceptable spellings?
- Did I use standard punctuation and grammar?
- Did I use appropriate headings, legends, and typographical style?

PANEL 11-2. Selected Style Manuals

The Associated Press Stylebook. 1977. New York: Associated Press.

CBE Style Manual Committee. 1983. *CBE style manual: A guide for authors, editors, and publishers in the biological sciences.* 5th ed. Bethesda, Md.: Council of Biology Editors.

The Chicago Manual of Style. 1982. 13th ed. Chicago: University of Chicago Press.

Government Printing Office Style Manual. 1973. Washington, D. C.: U.S. Government Printing Office.

Handbook for Authors of Papers in American Chemical Society Publications. 1978. Washington, D.C.: American Chemical Society.

Publication Manual of the American Psychological Association. 1983. 3rd ed. Washington, D.C.: American Psychological Association.

Webster's Standard American Style Manual. 1985. Springfield, Mass.: Merriam-Webster.

EDITORIAL (STYLEBOOK) STYLE

Here we discuss style as used in editorial offices and stylebooks. You often must look closely to see the style differences among organizations. For instance, consider the serial comma—the comma before conjunctions:

    dogs, cats, and fishes
    dogs, cats and fishes

The first example conforms to *The Chicago Manual of Style* (1982) and the *Government Printing Office Style Manual* (1973), followed by most scholarly journals and government publications. The second conforms to the *Associated Press Style Manual* (1980), followed by newspapers.

When it comes to literature citations and bibliographic styles, you'll find wide differences. One style calls for titles to be underlined, another for them to be surrounded by quotation marks.

If punctuation presents a problem for you, and you don't know of the appropriate style manual for punctuation, use the guidelines in Panel 11-3. You also can use guidelines on punctuation, mechanics, signs and symbols, and other standards in a standard collegiate dictionary. Appendix A lists selected collegiate dictionaries.

Resolving Editorial Style Issues

As you write, you'll encounter style questions that you cannot answer with a specific style manual. You can resolve such questions in any of four ways.

First, you can rewrite around the problem to eliminate it. Rewriting often proves to be the most cost-efficient technique.

Second, you can develop a sequence in which you use style manuals. Start with the style manual specifically for your organization, and then move to a more general manual. Suppose that you were writing an article for an agronomy publication and came upon a style point not covered in the *Publications Handbook and Style Manual* of the American Society of Agronomy (Buxton et al. 1984). You might turn to the *CBE Style Manual* (1983). If you couldn't find the answer there, you might turn to *The Chicago Manual of Style* (1982). If you're trying to resolve a grammar question, you might check handbooks on grammar, such as the *Harbrace College Handbook* (Hodges and Whitten 1986) and *The Borzoi Handbook for Writers* (Crews and Schor 1985).

Third, look through back issues of a publication in your field to see how the editors handled a similar problem. This approach may take time, and it is not always successful.

Fourth, call or write to the editor, and ask for guidance.

REVISING TO IMPROVE APPEARANCE

PANEL 11-3. How to Punctuate

How to punctuate

By Russell Baker

International Paper asked Russell Baker, winner of the Pulitzer Prize for his book, Growing Up, and for his essays in The New York Times (the latest collection in book form is called The Rescue of Miss Yaskell and Other Pipe Dreams), to help you make better use of punctuation, one of the printed word's most valuable tools.

When you write, you make a sound in the reader's head. It can be a dull mumble—that's why so much government prose makes you sleepy—or it can be a joyful noise, a sly whisper, a throb of passion.

Listen to a voice trembling in a haunted room:

"And the silken, sad, uncertain rustling of each purple curtain thrilled me—filled me with fantastic terrors never felt before..."

That's Edgar Allan Poe, a master. Few of us can make paper speak as vividly as Poe could, but even beginners will write better once they start listening to the sound their writing makes.

One of the most important tools for making paper speak in your own voice is punctuation.

When speaking aloud, you punctuate constantly—with body language. Your listener hears commas, dashes, question marks, exclamation points, quotation marks as you shout, whisper, pause, wave your arms, roll your eyes, wrinkle your brow.

In writing, punctuation plays

"My tools of the trade should be your tools, too. Good use of punctuation can help you build a more solid, more readable sentence."

the role of body language. It helps readers hear you the way you want to be heard.

"Gee, Dad, have I got to learn all them rules?"

Don't let the rules scare you. For they aren't hard and fast. Think of them as guidelines.

Am I saying, "Go ahead and punctuate as you please"? Absolutely not. Use your own common sense, remembering that you can't expect readers to work to decipher what you're trying to say.

There are two basic systems of punctuation:

1. The loose or open system, which tries to capture the way body language punctuates talk.

2. The tight, closed structural system, which hews closely to the sentence's grammatical structure.

Most writers use a little of both. In any case, we use much less punctuation than they used 200 or even 50 years ago. (Glance into Edward Gibbon's "Decline and Fall of the Roman Empire," first published in 1776, for an example of the tight structural system at its most elegant.) No matter which

system you prefer, be warned: punctuation marks cannot save a sentence that is badly put together. If you have to struggle over commas, semicolons and dashes, you've probably built a sentence that's never going to fly, no matter how you tinker with it. Throw it away and build a new one to a simpler design. The better your sentence, the easier it is to punctuate.

Choosing the right tool

There are 30 main punctuation marks, but you'll need fewer than a dozen for most writing.

I can't show you in this small space how they all work, so I'll stick to the ten most important—and even then can only hit highlights. For more details, check your dictionary or a good grammar.

Comma [,]

This is the most widely used mark of all. It's also the toughest and most controversial. I've seen aging editors almost come to blows over the comma. If you can handle it without sweating, the others will be easy. Here's my policy:

1. Use a comma after a long introductory phrase or clause: *After stealing the crown jewels from the Tower of London, I went home for tea.*

2. If the introductory material is short, forget the comma: *After the theft I went home for tea.*

3. But use it if the sentence would be confusing without it, like this: *The day before I'd robbed the Bank of England.*

4. Use a comma to separate elements in a series: *I robbed the*

(Source: Reprinted by permission of International Paper Company.)

(continued)

PANEL 11-3. (continued)

Denver Mint, the Bank of England, the Tower of London and my piggy bank.

Notice there is no comma before *and* in the series. This is common style nowadays, but some publishers use a comma there, too.

5. Use a comma to separate independent clauses that are joined by a conjunction like *and, but, for, or, nor, because* or *so*: *I shall return the crown jewels, for they are too heavy to wear.*

6. Use a comma to set off a mildly parenthetical word grouping that isn't essential to the sentence: *Girls, who have always interested me, usually differ from boys.*

Do not use commas if the word grouping *is* essential to the sentence's meaning: *Girls who interest me know how to tango.*

7. Use a comma in direct address: *Your majesty, please hand over the crown.*

8. And between proper names and titles: *Montague Sneed, Director of Scotland Yard, was assigned the case.*

9. And to separate elements of geographical address: *Director Sneed comes from Chicago, Illinois, and now lives in London, England.*

Generally speaking, use a comma where you'd pause briefly in speech. For a long pause or completion of thought, use a period.

If you confuse the comma with the period, you'll get a run-on sentence: *The Bank of England is located in London, I rushed right over to rob it.*

Semicolon [;]

A more sophisticated mark than the comma, the semicolon separates two main clauses, but it keeps those two thoughts more tightly linked than a period can: *I steal crown jewels; she steals hearts.*

Dash [—] and Parentheses [( )]

Warning! Use sparingly. The dash SHOUTS. Parentheses whisper. Shout too often, people stop listening; whisper too much, people become suspicious of you. The dash creates a dramatic pause to prepare for an expression needing strong emphasis: *I'll marry you—if you'll rob Topkapi with me.*

Parentheses help you pause quietly to drop in some chatty information not vital to your story: *Despite Betty's daring spirit ("I love robbing your piggy bank," she often said), she was a terrible dancer.*

"Punctuation puts body language on the printed page. Show bewilderment with a question mark, a whisper with parentheses, emphasis with an exclamation point."

Quotation marks [" "]

These tell the reader you're reciting the exact words someone said or wrote: *Betty said, "I can't tango."* Or: *"I can't tango," Betty said.*

Notice the comma comes before the quote marks in the first example, but comes inside them in the second. Not logical? Never mind. Do it that way anyhow.

Colon [:]

A colon is a tip-off to get ready for what's next: a list, a long quotation or an explanation. This article is riddled with colons. Too many, maybe, but the message is: "Stay on your toes; it's coming at you."

Apostrophe [']

The big headache is with possessive nouns. If the noun is singular, add *'s*: *I hated Betty's tango.*

If the noun is plural, simply add an apostrophe after the *s*: *Those are the girls' coats.*

The same applies for singular nouns ending in *s*, like Dickens: *This is Dickens's best book.*

And in plural: *This is the Dickenses' cottage.*

The possessive pronouns *hers* and *its* have no apostrophe.

If you write *it's*, you are saying *it is*.

Keep cool

You know about ending a sentence with a period (.) or a question mark (?). Do it. Sure, you can also end with an exclamation point (!), but must you? Usually it just makes you sound breathless and silly. Make your writing generate its own excitement. Filling the paper with !!!! won't make up for what your writing has failed to do.

Too many exclamation points make me think the writer is talking about the panic in his own head.

Don't sound panicky. End with a period. I am serious. A period. Understand?

Well . . . sometimes a question mark is okay.

Russell Baker

Today, the printed word is more vital than ever. Now there is more need than ever for all of us to *read* better, *write* better and *communicate* better.

International Paper offers this series in the hope that, even in a small way, we can help.

If you'd like additional reprints of this article or an 11"x 17" copy suitable for bulletin board posting or framing, please write: "Power of the Printed Word," International Paper Company, Dept. 13, P.O. Box 954, Madison Square Station, New York, NY 10010. © 1984 International Paper Company

INTERNATIONAL PAPER COMPANY
We believe in the power of the printed word.

Printed in U.S. on International Paper Company's Springhill® Offset, basis 60 lb.

Developing a Reference Library

If you develop a personal reference library, you'll be able to answer most of your style questions. To begin your communications library, obtain a good collegiate dictionary, a basic style manual, a standard grammar reference, and, possibly, a thesaurus. Appendix A, a bibliography of relevant publications to technical communication, lists useful standard references for many technical and scientific fields.

TYPOGRAPHIC STYLE

Typographic style constitutes typeface, type size, column width, page design, white space, layout, and other design features. If your technical reports are set in type and your organization has editors and graphic artists, you may not need to concern yourself with the details of typographic style. But when you work with editors and graphic designers, you'll find helpful a working knowledge of printing and typesetting terminology, as discussed in Chapter 14.

You'll also find yourself concerned with typographic style when you have papers typed or printed on word processing equipment. The way you style headings, for example—all capitals, underlined, centered, or flush left—will depend upon your organization's typographic style. Secretaries and word processing operators often know about typographic style.

IDIOMATIC STYLE

Idiomatic style constitutes the acceptable grammar, syntax, word use, and spellings that you learned in English class. As a writer, you need to concern yourself with such conventions, because they form the basis for consistency in language. "For More Help" and Appendix A identify references that can answer questions on idiomatic style. If you don't have these references, you'll find that many dictionaries include guidelines on punctuation, mechanics, and related conventions as well as on word meanings.

When checking your spelling, the rule is: if in doubt, look it up. If you have problems with spelling, review the fundamentals covered in Panel 11-4; make a habit of questioning the accuracy of spellings in your drafts. Most style manuals specify the dictionary authors are to follow. Appendix A lists the more commonly used dictionaries for technical and scientific fields.

Computer programs are available for checking matters of manuscript appearance. These programs can check spelling, grammar, syntax, and matters of style. But keep one caution in mind: *Although such programs*

PANEL 11-4. How to Spell

How to spell

By John Irving

International Paper asked John Irving, author of "The World According to Garp," "The Hotel New Hampshire," and "Setting Free the Bears," among other novels—and once a hopelessly bad speller himself—to teach you how to improve your spelling.

Let's begin with the bad news.

If you're a bad speller, you probably think you always will be. There are exceptions to every spelling rule, and the rules themselves are easy to forget. George Bernard Shaw demonstrated how ridiculous some spelling rules are. By following the rules, he said, we could spell <u>fish</u> this way: <u>ghoti</u>. The "f" as it sounds in enou<u>gh</u>, the "i" as it sounds in w<u>o</u>men, and the "sh" as it sounds in fic<u>ti</u>on.

With such rules to follow, no one should feel stupid for being a bad speller. But there are ways to improve. Start by acknowledging the mess that English spelling is in—but have sympathy: English spelling changed with foreign influences. Chaucer wrote "gesse," but "guess," imported earlier by the Norman invaders, finally replaced it. Most early printers in England came from Holland; they brought "ghost" and "gherkin" with them.

If you'd like to intimidate yourself—and remain a bad speller forever—just try to remember the 13 different ways the sound "sh" can be written:

<u>sh</u>oe   suspi<u>ci</u>on
<u>s</u>ugar   nau<u>se</u>ous
o<u>ce</u>an   con<u>sci</u>ous
i<u>ss</u>ue   <u>ch</u>aperone
na<u>ti</u>on   man<u>si</u>on
<u>sch</u>ist   fu<u>ch</u>sia
p<u>sh</u>aw

Now the good news

The good news is that 90 percent of all writing consists of 1,000 basic words. There is, also, a method to most English spelling and a great number of how-to-spell books. Remarkably, all these books propose learning the same rules! Not surprisingly, most of these books are humorless.

Just keep this in mind: If you're familiar with the words you use, you'll probably spell them correctly—and you shouldn't be writing words you're unfamiliar with anyway. USE a word—out loud, and more than once—before you try writing it, and make sure (with a new word) that you know what it means before you use it. This means you'll have to look it up in a dictionary, where you'll not only learn what it means, but you'll see how it's spelled. Choose a dictionary you enjoy browsing in, and guard it as you would a diary. You wouldn't lend a diary, would you?

A tip on looking it up

Beside every word I look up in my dictionary, I make a mark.

"Love your dictionary."

Beside every word I look up more than once, I write a note to myself—about WHY I looked it up. I have looked up "strictly" 14 times since 1964. I prefer to spell it with a <u>k</u>—as in "stric<u>kt</u>ly." I have looked up "ubiquitous" a dozen times. I can't remember what it means.

Another good way to use your dictionary: When you have to look up a word, for any reason, learn—and learn to *spell*—a *new* word at the same time. It can be any useful word on the same page as the word you looked up. Put the date beside this new word and see how quickly, or in what way, you forget it. Eventually, you'll learn it.

Almost as important as knowing what a word means (in order to spell it) is knowing how it's pronounced. It's go<u>vern</u>ment, not goverment. It's Fe<u>bru</u>ary, not Febuary. And if you know that <u>anti</u>- means against, you should know how to spell <u>anti</u>dote and <u>anti</u>biotic and <u>anti</u>freeze. If you know that <u>ante</u>- means before, you shouldn't have trouble spelling <u>ante</u>chamber or <u>ante</u>cedent.

Some rules, exceptions, and two tricks

I don't have room to touch on <u>all</u> the rules here. It would take a book to do that. But I can share a few that help me most:

What about -<u>ary</u> or -<u>ery</u>? When a word has a primary accent on the first syllable and a secondary accent on the next-to-last syllable (sec′re-tar′y), it usually ends in -<u>ary</u>. Only six important words like this end in -<u>ery</u>:

(Source: Reprinted by permission of International Paper Company.)

PANEL 11-4. (continued)

cemetery monastery
millinery confectionery
distillery stationery
(as in pap*er*)

Here's another easy rule. Only four words end in -*efy*. Most people misspell them—with -*ify*, which is usually correct. Just memorize these, too, and use -*ify* for all the rest.

stupefy putrefy
liquefy rarefy

As a former bad speller, I have learned a few valuable tricks. Any good how-to-spell book will teach you more than these two, but these two are my favorites. Of the 800,000 words in the English language, the most frequently misspelled is <u>alright</u>; just remember that <u>alright</u> is <u>all</u> <u>wrong</u>. You wouldn't write <u>alwrong</u>, would you? That's how you know you should write <u>all right</u>.

The other trick is for the truly *worst* spellers. I mean those of you who spell so badly that you can't get close enough to the right way to spell a word in order to even FIND it in the dictionary. The word you're looking for is there, of course, but you won't find it the way you're trying to spell it. What to do is look up a synonym—another word that means the same thing. Chances are good that you'll find the word you're looking for under the definition of the synonym.

Demon words and bugbears

Everyone has a few demon words—they never look right, even when they're spelled correctly. Three of my demons are <u>medieval</u>, <u>ecstasy</u>, and <u>rhythm</u>. I have learned to hate these words, but I have not learned to spell them; I have to look them up every time.

And everyone has a spelling rule that's a bugbear—it's either too difficult to learn or it's impossible to remember. My personal bugbear among the rules is the one governing whether you add -<u>able</u> or -<u>ible</u>. I can teach it to you, but I can't

remember it myself.
You add -<u>able</u> to a full word: adapt, adaptable; work, workable. You add -<u>able</u> to words that end in <u>e</u>—just remember to drop the final <u>e</u>: love, lovable. But if the word ends in two <u>e</u>'s, like agree, you keep them both: agreeable.

You add -<u>ible</u> if the base is not a full word that can stand on its own: credible, tangible, horrible, terrible. You add -<u>ible</u> if the root word ends in -<u>ns</u>: responsible. You add -<u>ible</u> if the root word ends in -<u>miss</u>: permissible. You add -<u>ible</u> if the root word ends in a soft <u>c</u>

"*This is one of the longest English words in common use. But don't let the length of a word frighten you. There's a rule for how to spell this one, and you can learn it.*"

(but remember to drop the final <u>e</u>!): force, forcible.

Got that? I don't have it, and I was introduced to that rule in prep school; with that rule, I still learn one word at a time.

Poor President Jackson

You must remember that it is permis<u>ible</u> for spelling to drive you crazy. Spelling had this effect on Andrew Jackson, who once blew his stack while trying to write a Presidential paper. "It's a damn poor mind that can think of only one way to spell a word!" the President cried.

When you have trouble, think of poor Andrew Jackson and know that you're not alone.

What's really important

And remember what's really important about good writing is not good spelling. If you spell badly but write well, you should hold your head up. As the poet T.S. Eliot recommended, "Write for as large and miscellaneous an audience as possible"—and don't be overly concerned if you can't spell "miscellaneous."

Also remember that you can spell correctly and write well and still be misunderstood. Hold your head up about that, too.

As good old G.C. Lichtenberg said, "A book is a mirror: if an ass peers into it, you can't expect an apostle to look out"—whether you spell "apostle" correctly or not.

Today, the printed word is more vital than ever. Now there is more need than ever for all of us to *read* better, *write* better, and *communicate* better.

International Paper offers this series in the hope that, even in a small way, we can help.

If you'd like additional reprints of this article or an 11″ x 17″ copy suitable for bulletin board posting or framing, please write: "Power of the Printed Word," International Paper Company, Dept. 12, P.O. Box 954, Madison Square Station, **New York, NY 10010.** © 1983, INTERNATIONAL PAPER COMPANY

INTERNATIONAL PAPER COMPANY
We believe in the power of the printed word.

Printed in U.S. on International Paper Company's Springhill® Offset, basis 60 lb.

can help, they do not replace the need for a strong working knowledge of style and a thorough check of hard copy.

PERSONAL STYLE

Personal style constitutes how you think and write. It's *your* way of expressing yourself—your choices of words, your pacing, your way of expressing thoughts, and your way of organizing and structuring your ideas.

Consider the personal style of anthropologist and naturalist Loren Eiseley, who began a chapter with this simple, intriguing sentence: "If there is magic on this planet, it is contained in water" (1959, 15). Not everyone appreciates Eiseley's personal style. A dog-eared copy of that book on our office bookshelf contains a previous owner's scribbled comment: "He is *very poetic*. Can't imagine a 'scientist' being ecstatic *about this*." Yet Eiseley's book is now a standard, and many people who are scientists today owe him a debt for introducing them, in his poetic and charming way, to science.

Eiseley is not the only scientist to make technical subjects meaningful to wide audiences. Dr. Lewis Thomas, in *Lives of a Cell* (1974), astronomer Carl Sagan, in *Cosmos* (1985), and many others have made science clear to millions of readers. Naturalist Aldo Leopold helped the modern environmental movement with his reflective *Sand County Almanac:*

> A March morning is only as drab as he who walks in it without a glance skyward, ear cocked for geese. I once knew a Phi Beta Kappa, who told me that she had never heard or seen the geese that twice a year proclaim the revolving seasons to her well insulated roof. Is education possibly a process of trading awareness for things of lesser worth? A goose who trades his is soon a pile of feathers. (Leopold 1968, 18)

In those four short sentences, and with few adjectives, Leopold speaks to us all. Personal style almost singlehandedly produces the major impact of writing and grabs readers' attention—or loses it.

PEER REVIEWS

Whenever possible, have other professionals review your work before it goes to your intended audience. Some company and agency policies require that peers review drafts. Sometimes drafts go to your boss and then up the chain of command to headquarters. Of course, the extent and details of reviews depend upon organization policy and structure.

When you can select your reviewers, select people who know the subject area, the environment in which you work, and the intended audience. The more they know about your audience, the better their advice on tailoring the document to that audience. And the more they know about your subject, the more suggestions and constructive criticism they can give you.

When you get your peers' reviews, read them with care. Try not to become defensive. Use the reviews to help you restructure and improve your draft. Some reviews will be helpful, and some won't. Glowing reviews may be less helpful than constructive criticism.

If you get highly critical reviews, carefully consider their comments. If you have solid information and you've approached the problem carefully, maintain your perspective and argue your case. You may have a better grasp of your problem and your audience than your reviewers do. But always keep your mind open, and realize that you might be in error. Reviewers can help you to avoid serious errors and save you from embarrassment.

HIGHLIGHTS

1. Different fields use different style manuals. You need to learn which style your field uses and begin the transition to it in your professional writing. A panel listing selected style guides is included.
2. Following a style reflects professionalism, makes communication consistent, and saves time and money.
3. Four kinds of style are: editorial, typographic, idiomatic, and personal.
4. Peer reviews of documents can be helpful in providing feedback.

PROJECTS

The following projects will familiarize you with a style manual or guide.

1. Which style manual do professionals in your field follow in preparing reports and articles? To find out the appropriate publication, check the guidelines for authors in research journals covering your field, or ask a professor from your major department. If your field does not have its own style manual, refer to either *The Chicago Manual of Style* or *Webster's Standard American Style Manual* to answer the following questions. Prepare a succinct report on the following points: (1) the name of the style manual, (2) publisher, (3) field or fields covered, (4) publication date, (5) address of publisher, (6) the major topics covered, and (7) publication cost.

To familiarize yourself with the style manual for your field, succinctly answer the questions asked below.

1. Which subjects does the style manual cover?
2. Does the style manual recommend a specific citation style? If so, which one? Provide an example of the citation style for within the narrative and for notes that appear at the bottom of pages or at the end of a document.
3. Does the style manual provide guidelines for illustrations? If so, how do they differ from those given in Chapter 8? Consider specifically the points on tables and figures. From the guidelines in the style manual, prepare a checklist for editing illustrations in your field.
4. Which dictionary does the style manual recommend to authors in your field? Provide a complete citation for that dictionary.
5. Does the style manual for your field include a section on punctuation? If so, what guidelines does it provide for: using a colon to introduce a list, appropriately punctuating the end of a list, and placing commas with words in a series?
6. Does the style manual discuss word usage and clarity? If so, give five such guidelines.
7. Does the style manual discuss handling numbers? If so, when do you spell out and when do you use Arabic numerals? Provide examples and explanations of handling numerals.
8. Does the style manual provide guidance on abbreviations? If so, give five examples, and briefly discuss the manual's approach to abbreviations.

FOR MORE HELP

CREWS, F., AND S. SCHOR. 1985. *The Borzoi handbook for writers*. New York: Knopf.
This book is packed with guidelines on standard grammar and mechanics.

HODGES, J. C., AND M. E. WHITTEN. 1986. *Harbrace college handbook*. 10th ed. New York: Harcourt Brace Jovanovich.
Hodges and Whitten provide details and exercises on standard grammar and mechanics.

STRUNK, W., JR., AND E. B. WHITE. 1979. *The elements of style*. New York: Macmillian.
A succinct classic with fundamentals that apply to technical and scientific writing.

PART FIVE

Preparing Professional Communication Products

In Part V, we focus on techniques for producing specific communication products. We provide general strategies by which you can tailor communication products to your individual needs. The chapters rest on the assumption that you've grasped the concepts presented in earlier parts of the book. We assume that now you can prepare products with solid content and polished appearance and that communicate effectively.

Chapter 12 suggests techniques for preparing definitions, descriptions, and instructions. Chapter 13 discusses technical and scientific proposals, progress reports, and technical reports and their components. Chapter 14 provides an overview of word processing, discusses the link between word processing and printing, and explains the printing process. Chapter 15 presents refined methods for giving oral presentations. Chapter 16 explains the fundamentals of correspondence and introduces certain common types of letters. Finally, Chapter 17 outlines job searching from the communication perspective.

12
Definitions, Descriptions, Instructions

OVERVIEW

In Chapter 12, we
- ☐ Show the crucial role definitions, descriptions, and instructions play in technical communication
- ☐ Describe how to create precise definitions
- ☐ Discuss major categories of descriptions
- ☐ Explain how to write accurate descriptions
- ☐ Explain how to write clear instructions
- ☐ Identify problems common to instructions and ways of overcoming them

THE IMPORTANCE OF DEFINITIONS

"When *I* use a word," Humpty Dumpty told Alice rather scornfully, "it means just what I choose it to mean—neither more nor less." With that, Lewis Carroll, in *Through the Looking-Glass,* referred to a debate on meaning that was raging among the philosophers of his day. What made the statement funny then, as now, was Humpty Dumpty's easy assumption that he alone controlled the meaning of the terms he used.

Too often we use technical and scientific terms without considering whether our audiences will use them in exactly the same way. If they do not, we lose ground in the struggle to communicate.

A definition problem created confusion for many first-time personal computer users in the early 1980s. When IBM introduced its "PC," which rapidly became the industry standard, dozens of less costly clones emerged as "IBM compatible." Some were and some weren't. Some clones would not run software or read diskettes programmed for the IBM. Diskettes prepared on other computers would not run on IBM PCs. As a result, some buyers who bought equipment touted as "IBM compatible" were frustrated and disappointed. Gradually, as writer David Salisbury explained in the *Christian Science Monitor,* four clearer definitions for "IBM compatible" emerged:

- Operationally compatible. These machines run nearly all IBM PC software right out of the box. They also have keyboards and display screens quite similar to the PC's. They may be capable of using PC add-on boards, which provide such functions as clock/calendars, telephone communication devices, and additional memory.
- Functionally compatible. These computers, while very similar to the PC, have enough differences that they cannot run every one of the important IBM programs. But the companies that make them carry enough clout to have the top programs rewritten for them. While the machines can read PC diskettes, they cannot use PC boards. They have screens and keyboards that are different from the PC's.
- Data compatible. These machines can't run all the major software but can exchange information with PCs by reading and writing to PC diskettes.
- MSDOS compatible. MSDOS (or PCDOS) is the PC's operating system, the program that does many of a computer system's basic housekeeping chores. Machines using the same operating system, but unable to read PC diskettes or run PC-labeled software, fall into this category. (Salisbury 1984, 26)*

Unwary customers thus learned to their dismay that imprecise definitions could lead to frustration and financial loss.

Similarly, terms with precise meanings in one field can change meaning when transferred to another field. Ecologists use "indicator species" to mean plants and animals sensitive to changes in the environment. But to the United States Forest Service, the term applies to species for which it manages an area. To minimize definition problems, people in many fields have developed special dictionaries and glossaries.

If you want to see the pervasiveness of definition problems, make two photocopies of Panel 12-1, complete one, and ask a friend to complete the second. Then calculate how closely the two of you agree. If you're like most people, you'll not agree closely. Because people work from different frames of reference, you must produce precise, easily understood definitions. By carefully defining your terms, you increase your audience's understanding of your points.

TOWARD PRECISE DEFINITIONS

A good definition creates new meaning from known meaning. It starts with a succinct definition, and expands it, perhaps by using examples. For

---

* Reprinted by permission from *The Christian Science Monitor.* © 1984 The Christian Science Publishing Society. All rights reserved.

DEFINITIONS, DESCRIPTIONS, INSTRUCTIONS 225

PANEL 12-1. Comparing Concepts

Draw a line through the items that you would not include under the concepts named. Compare responses with a friend. Divide the number of times you agree on each list by the total number of elements on that list. The ratio gives a coefficient of comparison; it represents the level of agreement between responses.

FRUIT	VEGETABLE
Peach	Tomato
Cucumber	Potato
Sugarcane	Mushroom
Jicama	Rice
Olive	Soup
Egg	Seaweed
Pumpkin	Flour
Salad	Alfalfa
Almond	Vitamins
Avocado	Avocado

(Source: Susan Dressel, Physical Science Laboratory, New Mexico State University, Las Cruces, N.M. Reprinted by permission of Susan Dressel.)

LANGUAGE
Chinese
Morse code
Poetry
Lozi
Mathematics
Signaling or signing
Music
FORTRAN
Hieroglyphics

ANIMAL
Zebra
Fur
Egg
Woman
Robin
Whale
Sponge
Unicorn
Protein
Sea cucumber

DISEASE
Bleeding
Measles
Tonsillectomy
Medication
Low blood pressure
Alcoholism
Inflation
Psychosis
Pain

GAME
Tennis
Crossword puzzle
Stock market
Olympics
Toy
Backgammon
Politics
Computer
Swimming
Cocktail party

---

example, 4,000 years ago, when writing first developed, the Sumerians used a simple drawing of a mound to represent the word for mountain. When everyone agreed that the mound represented the word, written language was on its way. Because they raided nearby mountain tribes for slaves, the Sumerians combined the symbol for mountain with the one for woman. The result was "mountain-woman," or slave. For that term to make sense, the audience had to know the meaning of the components.

You'll rarely find it necessary to create new terms. But you'll often need to declare the special sense in which you use existing terms. Begin by considering what your audience knows about the term, what you know about it, and then draft a definition that clarifies what you mean. Avoid making circular definitions. Do not, for example, say that "a definition is something that defines meaning." Using that to tell someone the meaning of "definition" makes no sense.

Real objects are more easily definable than abstractions. Your senses of touch, sight, and smell can help you to understand and define tangible things. The Sumerians had a difficult time defining "thought." Cartoonists, who also deal with pictorial language, often draw a light bulb turning on to suggest an idea occurring. People create ways of defining abstractions in words, symbols, and drawings. For example, in statistics the idea of calculating the mean or average is represented by $\frac{\Sigma X_i}{N}$, where

X_i = the observed values
N = the number of observed values
Σ = the summation (addition) of the observed values
—— = division

Common techniques for preparing definitions include examples, words, explications, comparisons, visuals, symbols, and word histories. Let us turn to these now.

Defining by Example

Often you can give specific examples of what you're trying to define. Reproduce a newspaper's front page, and you've given a workable definition of "newspaper." But if you're in charge of granting special postage rates to publishers of newspapers, you must worry about things like frequency of publication. Is something that looks like a newspaper, but is published only once, *really* a newspaper? Defining by example, though extremely useful, tends to blur differences between the example and the general class of things being defined.

Defining with Words

Word-based definitions prove useful for defining abstractions. In *Sundials to Atomic Clocks* (1977), James Jespersen and Jane Fitz-Randolph define "time" as "a physical quantity that can be observed and measured with a clock of mechanical, electrical, or other physical nature" (5). The authors provide definitions from two dictionaries. Then they expand their definition:

> At least part of the trouble in agreeing on what time is lies in the use of the single word *time* to denote two distinct concepts. The first is the *date* or *when* an event happens. The other is *time interval,* or the "length" of time between two events. This distinction is important, and is basic to the problems involved in measuring time (5).

Their definition clearly takes their general, nontechnical audience into account.

Defining by Explication

Another way to construct a precise definition with words is to review how others have used a given concept. Specialists often define a single term in different ways or use similar definitions for different terms. Social scientists handle the problem by "explication." The process entails reviewing how others have defined a term. Then the author states explicitly how the term will be used in the article or research. In one Stanford University study, Dr. Richard Carter sought to explicate "understanding":

> Just what "understanding" is, however, has not been clearly spelled out. In some communities, understanding is inferred from acquiescence to the policy of educational leaders. Elsewhere, it may be the hoped-for result of an information program, but still an unknown entity. The first step, then, must be to define what is meant when we talk about understanding. We have defined understanding as *a common perception among a group of people of the existing situation*. They may differ widely in their ideas as to what should be done in the situation, but in a state of understanding, they at least agree what the situation is. Whether there be agreement or disagreement, conflict or acquiescence, there is a basis—understanding—for development, for progress. (Carter 1962, 93)

Defining by Comparison

In defining terms by comparison, introduce a new term, and then relate it to what the audience knows. For example, in an ornithology class, an instructor might define a shoveler by saying, "It's like a mallard." The instructor might then note the shoveler's shovel-shaped bill and compare it to the mallard's bill. Comparisons often work best when a drawing or photograph is provided.

Defining by Visuals

Illustrations—photographs, line art, and drawings—provide some of the easiest and simplest ways to define and describe objects, as Figure 12-1 illustrates. Good illustrations can cut a lengthy narrative and provide details a narrative doesn't.

Defining by Symbols

Mathematics, physics, chemistry, computer science, finance, accounting, and other fields rely heavily on symbols and equations to define concepts. For example, Andrea Lubov uses an equation as a definition in *Issuing Municipal Bonds* (1979, 3):

$$\text{Average maturity} = \frac{\text{Total bond years}}{\text{Total bonds}}$$

A chemistry publication would define carbon dioxide as

$$CO_2$$

Such definitions rely on the audience's knowledge of the terms.

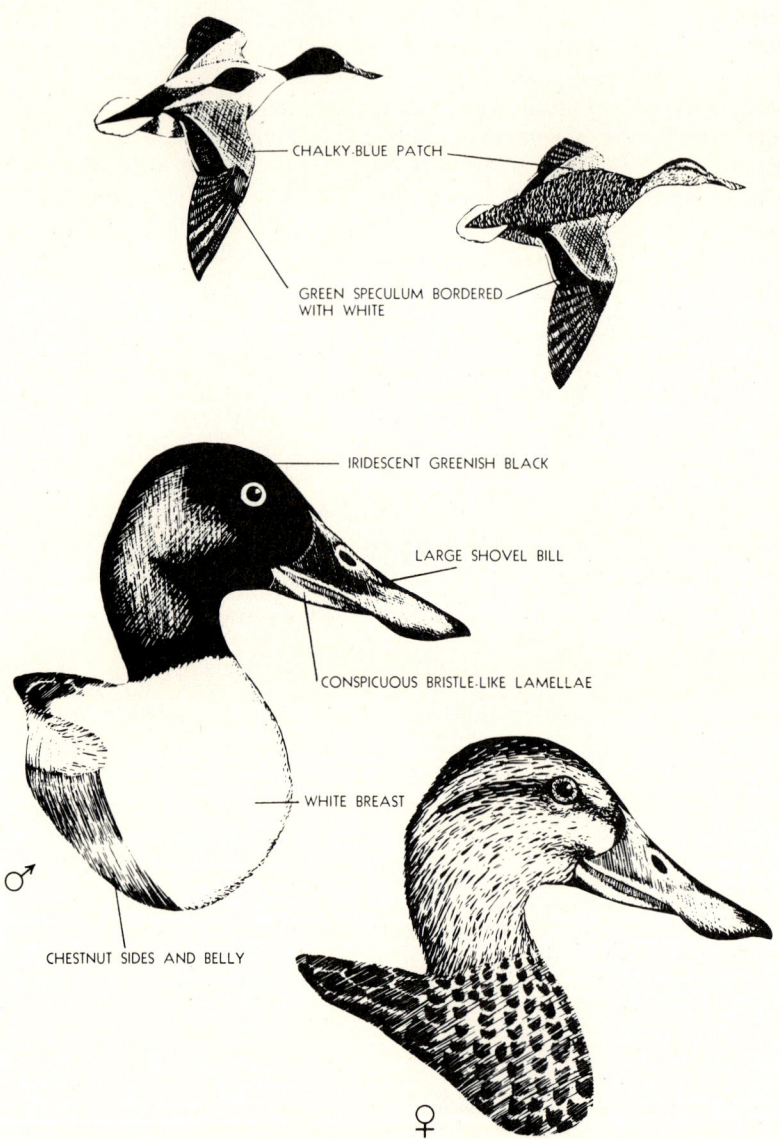

FIGURE 12-1. Here a line drawing provides a succinct way of defining a shoveler. *(Source:* Kansas Waterfowl Identification Guide *(Pratt, Kans.: Central Flyway Council, 1964), p. 29. Reprinted by permission of the Central Flyway Council.)*

Defining by Word History

Frequently, the history, or etymology, of a word helps audiences understand a word. For example, in computer science, the term "bug" means a problem in a computer program that prevents the program from running correctly. In the early days of computer development, Grace Hopper, a programming expert, coined the term after finding a program wouldn't run because a moth had shorted a circuit (Keerdoja, Vercammen, and Lord 1983).

Standard college-level dictionaries, like *The Random House College Dictionary*, carry etymologies for many words, but not all. Dictionaries usually give a word's etymology by using symbols and abbreviations explained elsewhere in the dictionary. For example, the etymology of "define" is that it came from the Middle English word *def(f)ine(n)*; that came from the Anglo-French and Old French *define(r)*, meaning "(to) put an end"; and that came from the Latin *definire,* which meant "to limit or explain."

PLACING DEFINITIONS

You can place definitions (1) in the narrative where you use the term, (2) in footnotes at the bottom of the page, and (3) in a glossary.

Placing Definitions in the Narrative

Two common approaches to defining terms in technical and scientific fields are to define terms (1) in apposition, or immediately after a word or symbol and (2) in sentences immediately after introducing the terms. Defining terms in apposition works well for short, parenthetical definitions:

> Roundup®, *isopropylamine salt of glyphosate,* proves to be an extremely effective weed killer.

Immediately after the brand name, the chemical composition is defined. After the apposition, the sentence further defines the compound as a weed killer.

The second way to define terms places the definition in a sentence immediately after the introduction of the term:

> [I]nformation stored on a disk is stored as a file. DOS automatically keeps and updates a list of every file you save on every disk you use. This list is called the *directory.* (Wolverton 1985, 26)

By placing definitions near the word, you force the reader to review the definitions—a distinct advantage.

Placing Definitions in Footnotes

Footnotes, logic suggests, interrupt reading. Readers must stop, look at the bottom of the page, read the definition, and then return to their reading. Many readers skip footnotes. Footnotes also increase typing and printing costs because of the special effort required. Avoid footnotes whenever possible.

Placing Definitions in a Glossary

A glossary includes a list of terms and their definitions. If the glossary consists of terms readers are at least vaguely familiar with, place it at the end of a document. But, if you have several terms that you know will be new to the readers, place them at the document's beginning. When you have a short glossary—a dozen terms or so—set them within a boxed border. The big advantage of a glossary is that people who know the terms need not refer to it, and those who do not know them can find it in a convenient place.

When to Define?

Deciding which terms to define depends on your assessment of your audiences. Newspaper reporters constantly define terms, reasoning that people who already know the definition will not be insulted, whereas people who do not know must be given explanations. Scientists sometimes become defensive about their technical language, asserting that it's the reader's job to look up terms and learn them. But even if they write only for peers, that argument is faulty, because they probably use some terms in special ways. Moreover, scientists often read works from other fields. In those cases, definitions become extremely important.

Always define when you use a term in a special way. Beyond that, define when you feel reasonably certain that some members of the audience will not know the meaning of the terms you use. It's better to err on the side of caution.

DESCRIPTIONS

In technical and scientific fields, definitions by themselves do not adequately communicate information. Spend any time browsing through technical and scientific literature or attending professional meetings, and you'll see the magnitude and diversity of descriptions. Descriptions play an important role in popular publications, too.

In a piece about Alaska (Jaynes 1984, 9), *Time* magazine reported that in the winter of 1899, Ed Jesson, a gold prospector, looked for transportation from Nome to the Klondike and found only a bicycle. So Jesson pedaled from Nome to the Klondike. An Indian who had never seen a bicycle described what he saw: "White man, he set down, walk like hell." That was excellent description.

Descriptions abound in technical and scientific fields. Medical doctors and nurses describe case histories; biologists describe environmental conditions; soils technicians describe their analyses. Engineers describe feasibility; police describe traffic accidents. Computer programmers describe new languages and programs.

Scientific articles describe how scientists develop ideas and carry out investigations. Methods sections describe details of investigations. Findings sections describe study results.

All such descriptions are *functional;* they emphasize how things work or fit together. Functional description makes up a major part of technical and scientific communication.

PREPARING DESCRIPTIONS

To create effective functional descriptions, you need a strategy. Your approach to preparing descriptions should follow the basic eight-step technical communication process. In this section, we emphasize points that merit special attention.

Consider Your Audience

Before preparing your description, assess your audience's frame of reference. What does your audience know about similar mechanisms, topics, concepts, or processes? Ask:

- Who is my audience?
- What does my audience know about what I'm about to describe?
- Does the audience know of something similar? If so, can I refer to it? How safe are my assumptions?
- How will my audience use the description?

When Clyde M. Christensen, a professor of plant pathology and physiology, prepared the University of Minnesota's Agricultural Extension Service Bulletin *Edible Wild Mushrooms* (1985), he had to consider carefully who would read the bulletin and why. He knew that readers were not likely to have had experience gathering, preparing, and eating wild food. Christensen had to assume that some readers would eat wild mushrooms. He had to assume that some lacked experience in identifying safe mushrooms. There lay the critical communication problem: inexperienced mushroom pickers seldom live to tell about their mistake. One nibble of the wrong kind of mushroom can kill. Christensen began his bulletin with a warning:

> Caution: Do not eat any wild mushroom, gathered either by yourself or by someone else, unless you can identify it with 100 percent certainty and know that it is safe for eating. Should you become ill from eating wild mushrooms, contact a physician or the nearest hospital. If necessary, they will phone the Poison Treatment Center nearest you for treatment suggestions. These centers are throughout the United States, usually in hospitals in large cities. (Christensen 1985, 2)

Then Christensen provided a line drawing and named the parts of a mushroom. The bulletin also presented color photographs and a narrative describing nine common edible mushrooms.

Consider Your Purpose

Why are you preparing the description? What do you want the audience to know or to do after reading, hearing, or listening to your description? Keep in mind the communication functions suggested in Chapter 1—to inform, instruct, persuade, and document.

Often a description functions clearly to document. The medical records in your doctor's office document your health history, for example. Reviewing your medical records before examining you helps your physician build a frame of reference within which to consider your current condition.

Organize Your Descriptions

To organize your thoughts, list the points you want your audience to know. Then ask:

- Does the audience need an introduction to the topic? If so, how much detail should that introduction provide?
- Does the audience need any background information?
- Does the audience need an orientation to the components of the concept being described? If so, how detailed must the orientation be?
- Should I divide the description into major and minor parts? If so, what are they?
- Which specifics must I include to help the reader comprehend my description?

Your answers will help you decide on the points you need to make. If you are describing a physical object, let your senses help you. Ask:

- What does the object look like?
- How does it smell?
- How does it taste?
- What does it feel like?

Whenever possible, let its content help dictate how you organize your description. Knowing the major parts of what you're describing will help you to structure your descriptions.

In *Design for Lively Learning* (1971), a manual on teaching techniques for volunteer leaders and consultants, David J. Miller uses six categories to structure descriptions of teaching technique:

- What it is
- What the teacher can do
- Strong points of the technique
- Weak points of the technique
- Physical requirements
- Suggested arrangement

Miller let content help him organize his descriptions of 17 teaching techniques.

In developing dozens of popular booklets on technical subjects, Shell Oil Company began each booklet with a 150-word introduction and then used a question and answer format (Panel 12-2). The questions posed determined the booklets' organizations. In sum, if you have a choice, let the subject's content, the points relevant to the audience, and the points you want to make dictate your organization.

Consider Your Accuracy

In reading technical and scientific descriptions, your audience is interested in the facts about the topic, not value judgments or opinions. To test your ability to distinguish between observation, what you see happen, and inference, or what you reason must have happened, read and complete Panel 12-3. Then compare your answers with those in Appendix 12-A at the chapter's end.

In *Communication and Organizational Behavior* (1985), William Haney notes that statements of observation

- Can be made only after or during observation
- Must not go beyond what was observed
- Can be made only by the observer

PANEL 12-2. Format Helps Communicate

Here is a page from the Shell Answer Book #14, which uses a question and answer format. Note the use of boldface type, the box for additional descriptions, and illustrations to break up the narrative.

Skillful Driver Pledge

I will enjoy driving more and maintain a good mental attitude by following these steps:
1. I will be courteous to other drivers and cooperate with them by driving friendly.
2. I will consider safety above all else.
3. I will feel responsible for all who ride with me.
4. I will remain calm and vigilant at the wheel.
5. I will keep an especially watchful eye for pedestrians.
6. I will try to help other drivers when they are in difficulty.
7. I will remember that the roads belong to two-wheeled vehicles as well as cars.
8. I will see that regular maintenance and safety inspections are performed on my car.
9. I will not drive if under the influence of drugs or alcohol.
10. I will obey all traffic laws.

Take the Skillful Driver Pledge. *Will you make a commitment to become a better driver? Self-awareness is the beginning. We'll give you a free Skillful Driver Pledge you can stick inside your car so you'll be reminded of your commitment every time you drive. Just stop by a participating Shell dealer and pick one up. If you like, get more than one.*

Bob Bondurant told me that keeping your hands at the 9 o'clock and 3 o'clock positions gives you the maximum amount of steering wheel maneuverability.

Q. What's the correct position for your hands on the steering wheel?
A. Many people are taught that the 10 o'clock and 2 o'clock positions are best. But Bob Bondurant says the 9 o'clock and 3 o'clock positions are better. You have more control and your arms will be less likely to tire.

Q. The speed limit is always a safe speed. True or false?
A. False. Speed limits are set for the best driving conditions. Slow down in heavy traffic, darkness and bad weather. But remember, if you're on a freeway, don't go too slowly. That can be very dangerous. Concentrate on staying within the speed limit and "going with the flow."

Q. Are there any tricks to keeping your mind alert while driving?
A. One way is to keep your body alert. Sit up straight in the seat with your buttocks deep in the seat. Your car will tell you how it's responding. But you have to listen with your body as well as your ears. When you turn a corner too swiftly, you'll feel it first in your bottom and your legs.

(Source: C. Nau, The Driving Skills Book, *Shell Answer Book 14* (1978). © 1978 Shell Oil Company. Reprinted by permission of Shell Oil Company.)

> ### PANEL 12-3. Observation or Inference?
>
> Read the story and mark the questions either true, false, or don't know. Appendix 12-A at the end of this chapter contains the correct answers.
>
> A businessman had just turned off the lights in the store when a man appeared and demanded money. The owner opened a cash register. The contents of the cash register were scooped up, and the man sped away. A member of the police force was notified promptly.
>
> _____ 1. A man appeared after the owner had turned off his store lights.
> _____ 2. The robber was a man.
> _____ 3. The man who appeared did not demand money.
> _____ 4. The man who opened the cash register was the owner.
> _____ 5. The store owner scooped up the contents of the cash register and ran away.
> _____ 6. Someone opened the cash register.
> _____ 7. After the man scooped up the contents of the cash register, he ran away.
> _____ 8. The cash register contained money, but the story does not state how much.
> _____ 9. The robber demanded money of the owner.
>
> (Source: Adapted from William V. Haney, Communication and Interpersonal Relations: Text and Cases, 5th ed. (1985: Homewood, Ill.: Irwin), p. 214. © 1985 Richard D. Irwin. Reprinted by permission of Richard D. Irwin.)

In contrast, statements of inference

- Can be made at any time
- Can go beyond observation—well beyond, to the limits of imagination
- Can be made by anyone

When preparing descriptions, distinguish between what's fact and what's your opinion by following Haney's checklist:

- Did I personally observe what I am talking or writing about?
- Do my statements stay with, and not go beyond, my observations?
- When I deal with inferences, do I assess their probabilities?
- When I communicate with others, do I label my inferences as such and get them to label theirs?

Consider Words and Illustrations

Too often when we think of communication, we think only of words. Yet much technical and scientific information comes through visuals that supplement, reinforce, and replace words.

In *Edible Wild Mushrooms*, Christensen recognizes the role of visuals. He illustrated morel mushrooms with a color photograph and a succinct narrative:

> The accompanying picture describes the morel better than words. There are several kinds of morels, but all are similar in appearance and in edibility.

They are 4 to 8 inches high, the caps are light brown, the stems white, and both cap and stem are hollow and brittle. The morels appear in spring after abundant rainfall. Wooded areas are their favorite haunts, but they also come up in grassy pastures and even in lawns, commonly around or near recently dead trees or stumps. (Christensen 1985, 5)

Consider Your Format

Format—the physical arrangement of descriptions on a page—can help readers focus on key elements. For example, headings point readers to key topics. Question headings serve as clues to what's coming and help readers understand your points.

For short descriptions, narrative formats work well. For a series of questions or key points, you can use "bullets"—small dots—as this text does, or a similar typographical device. Bullets signal readers that points are important.

Varying typefaces also shows readers where important information lies. The Shell Answer Guides, for example, use bold print for all questions (Panel 12-2).

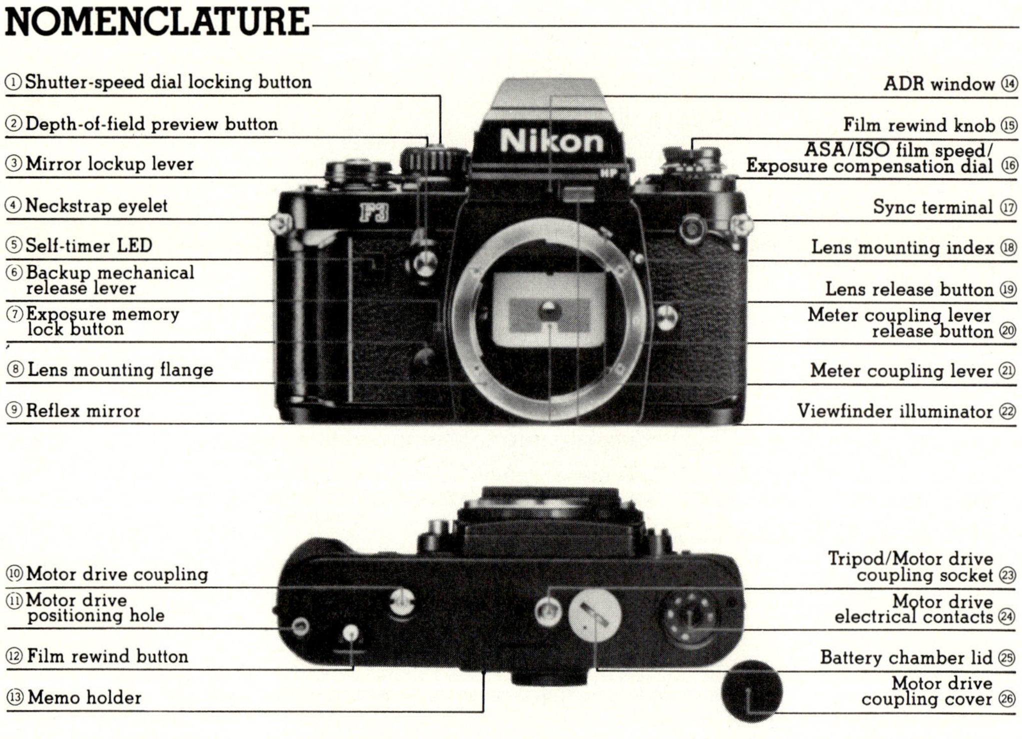

NOMENCLATURE

① Shutter-speed dial locking button
② Depth-of-field preview button
③ Mirror lockup lever
④ Neckstrap eyelet
⑤ Self-timer LED
⑥ Backup mechanical release lever
⑦ Exposure memory lock button
⑧ Lens mounting flange
⑨ Reflex mirror

ADR window ⑭
Film rewind knob ⑮
ASA/ISO film speed/Exposure compensation dial ⑯
Sync terminal ⑰
Lens mounting index ⑱
Lens release button ⑲
Meter coupling lever release button ⑳
Meter coupling lever ㉑
Viewfinder illuminator ㉒

⑩ Motor drive coupling
⑪ Motor drive positioning hole
⑫ Film rewind button
⑬ Memo holder

Tripod/Motor drive coupling socket ㉓
Motor drive electrical contacts ㉔
Battery chamber lid ㉕
Motor drive coupling cover ㉖

FIGURE 12-2. Visual description.

An illustration describing the parts of a Nikon Camera. *(Source:* Nikon F3 High-eyepoint Instruction Manual, *pp. 2, 3. Used with permission of Nikon Nippon Kogaku K.K.)*

Consider Your Narrative

When writing descriptions, apply the principles covered in the chapters on writing. Keep in mind that you're trying to communicate specific information carefully and accurately. Succinct, tightly written sentences are the most effective for your descriptions.

COMMON DESCRIPTIONS

Five common types of functional descriptions are those devoted to (1) hardware and conditions, (2) spatial relationships, (3) events and case studies, (4) classifications, and (5) comparisons and contrasts.

Describing Hardware and Conditions

Most equipment manuals begin with a parts illustration and identifying key. (See Figure 12-2.)

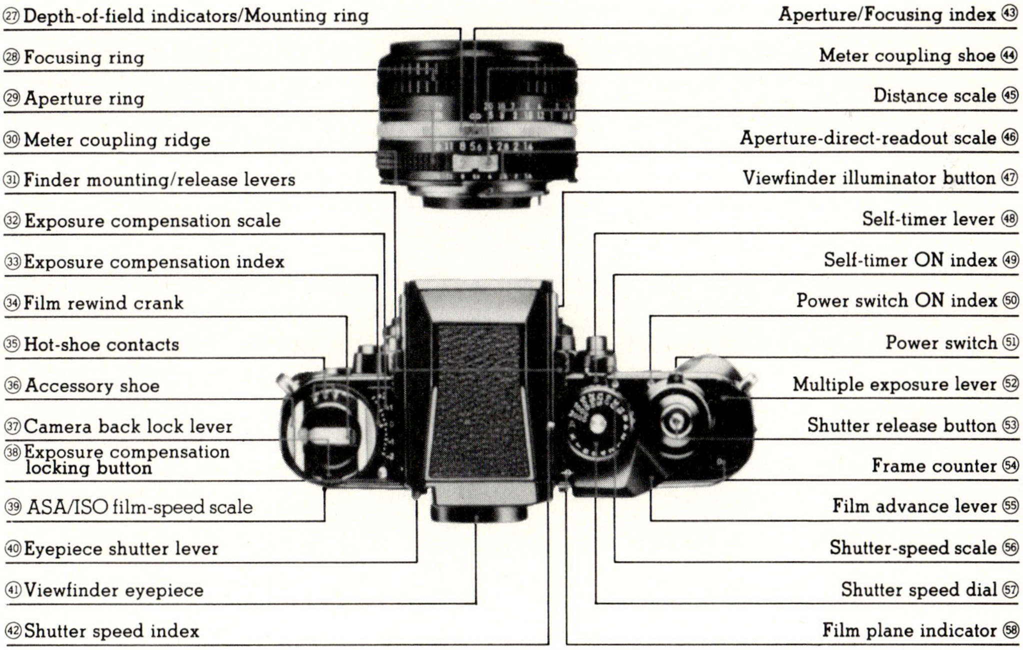
㉗ Depth-of-field indicators/Mounting ring
㉘ Focusing ring
㉙ Aperture ring
㉚ Meter coupling ridge
㉛ Finder mounting/release levers
㉜ Exposure compensation scale
㉝ Exposure compensation index
㉞ Film rewind crank
㉟ Hot-shoe contacts
㊱ Accessory shoe
㊲ Camera back lock lever
㊳ Exposure compensation locking button
㊴ ASA/ISO film-speed scale
㊵ Eyepiece shutter lever
㊶ Viewfinder eyepiece
㊷ Shutter speed index
Aperture/Focusing index ㊸
Meter coupling shoe ㊹
Distance scale ㊺
Aperture-direct-readout scale ㊻
Viewfinder illuminator button ㊼
Self-timer lever ㊽
Self-timer ON index ㊾
Power switch ON index ㊿
Power switch (51)
Multiple exposure lever (52)
Shutter release button (53)
Frame counter (54)
Film advance lever (55)
Shutter-speed scale (56)
Shutter speed dial (57)
Film plane indicator (58)

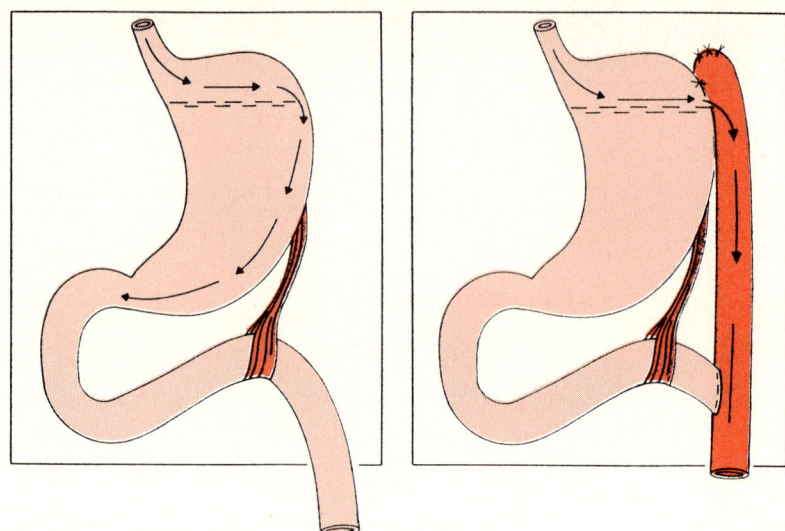

FIGURE 12-3. Gastric bypass and gastroplasty surgery.
(Source: George Lechner, Gastric Bypass and Gastroplasty: Treatment for Morbid Obesity, Reprinted with permission from the January issue of Nursing 81. Copyright © 1981 Springhouse Corporation. All rights reserved.)

Many descriptions begin with a succinct introduction and an expanded narrative. For example, for nurses reading *Nursing81*, surgeon George Lechner described two stomach operations to treat obesity:

> If a morbidly obese patient on your unit is scheduled for weight-reduction surgery, chances are he'll undergo gastric bypass or gastric partitioning. . . . Let's compare how the two operations produce weight loss: in the *gastric bypass operation,* the lower 90 percent to 95 percent of the stomach is stapled off so that only the upper 5 percent to 10 percent receives food. This upper stomach, which now will hold only 2 to 4 tablespoons of food at a time . . . [Figure 12-3]. *Gastric partitioning* is similar to gastric bypass. The stomach is divided with staples into a large lower section (90 percent to 95 percent) and a small upper section. However, the 1 cm opening is left in the staple line, preferably at the edge of the stomach. This allows food to pass slowly from the upper section to the lower section. With either surgery, a patient can eat only small quantities of food at one time. (Lechner 1981, 59)

Here's how an Ozark Mountain farm could be described using a bulleted list:

DESCRIPTION

- 160-acre Ozark farm accessible by hard-surfaced road
- Three bedroom, solar heated, native stone ranch-style home with utilities

- 1000-square-foot barn and five smaller buildings
- 600-square-foot solar- and wood-heated greenhouse
- One-acre organic garden with berry patches
- 10-acre orchard of bearing fruit and nut trees
- Five ponds stocked with bass, sunfish, and catfish
- 40 acres of pastures and 20 acres of croplands
- 80 acres mixed hardwoods
- 10 year-round springs

The list format quickly gives readers the key physical characteristics of the farm.

Describing Spatial Relationships

Natural history, environmental studies, construction, health, agriculture, and other fields often use site descriptions. In an issue of *The Journal of Wildlife Management,* wildlife biologists describe a study area in southwestern New Mexico:

> The Fort Bayard Administrative Site, located 10 miles east of Silver City, New Mexico, contains approximately 13,500 acres between the elevations of 6,000 and 7,000 feet. Since 1870, annual rainfall has averaged 16.7 inches with a range of 5.7 to 31.2 inches. The topography of the area ranges from open, level plateaus to steep, precipitous slopes cut by numerous small water courses. In general, vegetation consists of pinyon-juniper (*Pinus* spp.–*Juniperus* ssp.) overstory with an understory of shrub and herbaceous plants. At lower elevations, some grassland sites are devoid of overstory. At higher elevations, dense stands of pinyon-juniper and shrubs are characteristic. A few small homesteads that were formerly cultivated are scattered along the main drainages. (Boeker et al. 1972, 56–57)*

The ideal spatial description gives sufficient detail so that readers can picture the area. Information often includes distances from known points, elevations, terrain features, vegetation, and maps, as in Figure 12-4.

Describing Events

Literally hundreds of different descriptions fall into this classification: laboratory reports, case studies, patient histories, employee evaluations, and many others. Newspaper articles, although not always technical, provide useful examples:

---

* E. L. Boeker, V. E. Scott, H. G. Reynolds, and B. A. Donaldson, Seasonal Food Habits of Mule Deer in Southwestern New Mexico, *Journal of Wildlife Management* 36: 56–57. Copyright © The Wildlife Society. Reprinted by permission.

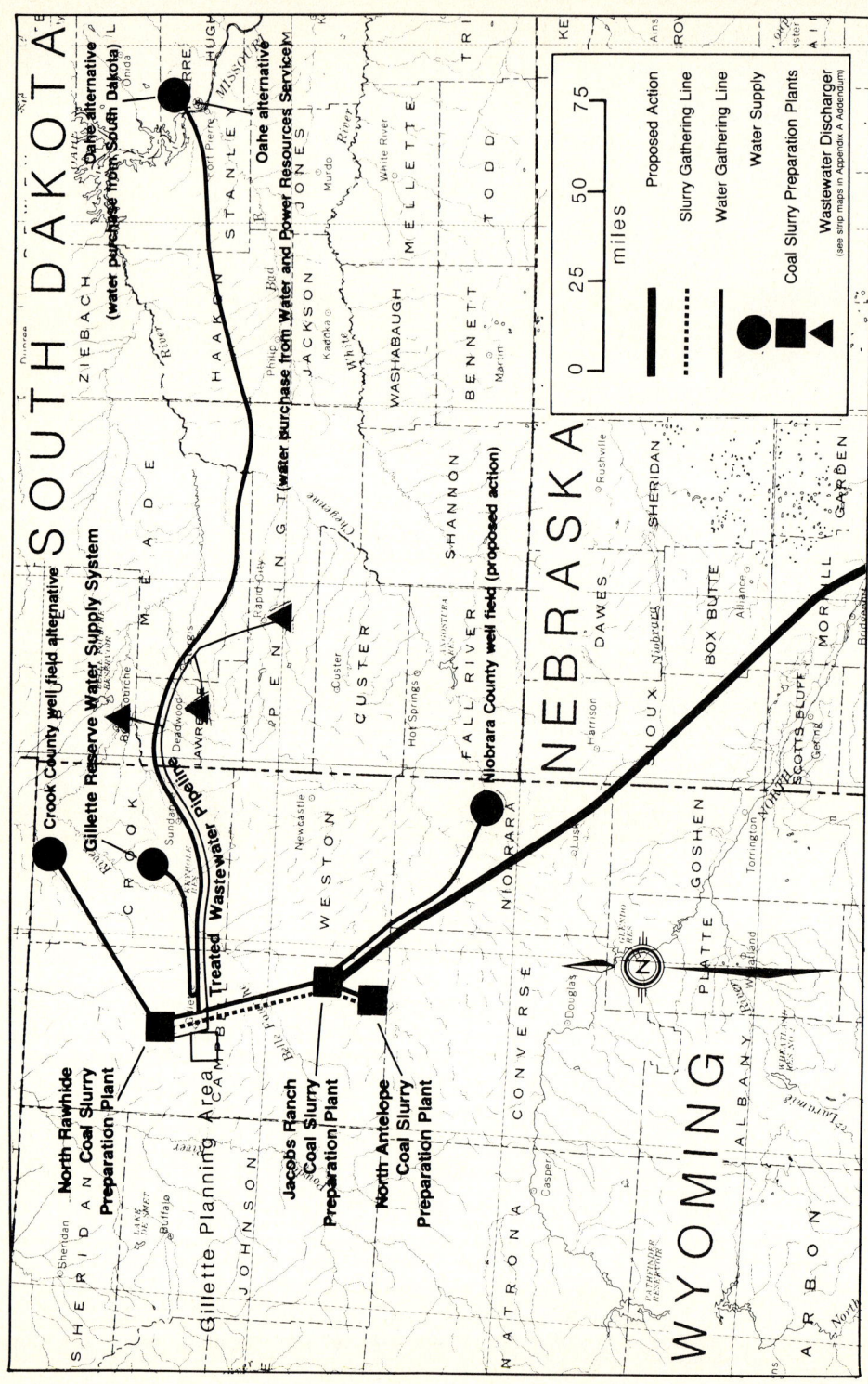

FIGURE 12-4. Map of proposed water supply systems.
(Source: Bureau of Land Management, Final Environmental Impact Statement on the Energy Transportation Systems, Inc., Coal Slurry Pipeline Transportation Project (Denver, Colo.: U.S. Department of the Interior, 1981), map 1–2.)

> ST. PAUL ISLAND, Alaska (UPI)—Aleut islanders clubbed 500 fur seals Monday in the start of a two-century-old slaughter that sparked a major debate over native rights and animal-protection issues.
> About 35 men left the village of St. Paul in a thick fog to begin the annual year's northern fur seal kill, the last full-scale harvest. St. Paul, in the Pribilof Islands 800 miles west of Anchorage, has the largest Aleut community in the world with 550 native residents.
> The work went quickly, with the take of 2- to 3-year-old males completed within 4½ hours. Opponents, led by the Humane Society of the United States, say the kill threatens a declining seal population and is economically unsound since seal skins are hard to sell. Islanders say the society's efforts to stop the kill ignores a history of repression and the cultural needs of the Pribilof natives. (United Press International 1984, 24)*

The UPI piece not only describes the harvest, but carefully separates fact from opinion by quoting both sides on the controversial harvest. The reporter carefully kept to the facts—the observations—and neither reported speculation nor made inferences.

In an article in *NursingLife*, Judith S. Kieffer provided a succinct case history:

> You wouldn't question the time a doctor takes to make a medical diagnosis; quality medical care depends on it. The same is true of a nursing diagnosis, but many nurses don't take this to heart. And, as a result, their patient care can sometimes be haphazard. That was the case with Barney Greene.
> Mr. Greene, a 52-year-old night watchman, developed diabetes mellitus a year ago. During his hospitalization, nurses taught him how to test his urine, follow his 1,200-calorie American Diabetic Association diet, and to take his medication. Mr. Greene followed those instructions carefully, but a year after his discharge he had to be readmitted for a below-the-knee-amputation. His problems had begun on the job. As a night watchman, he had to spend long hours on his feet wearing stiff regulation shoes. When a blister developed on his foot about 3 months ago, he treated himself with hot water soaks in Epsom salts.
> Why hadn't this well-motivated patient used proper foot care or sought proper foot care sooner? (Kieffer 1984, 18)†

The article moves from the case history to answer the question. Nurses reading the article and most other readers, for that matter, can understand the case history. Furthermore, the case history presents key information and no unnecessary details.

Describing Classifications

Classification systems provide ways of systematically describing similarities and differences among things, animals, plants, devices, and equipment.

---

* Reprinted with permission of United Press International, Inc. and *The Rocky Mountain News*.
† Reprinted with permission from the May/June issue of *NursingLife*. Copyright © 1984 Springhouse Corporation. All rights reserved.

Widely used classification systems are the periodic table of chemical elements and the binomial nomenclature system for plants and animals.

To prepare a classification, start with the universal or common characteristics for all items to be classified. For example, a classification of mammals begins with the most fundamental and common characteristics: warm-bloodedness, four-chambered heart, and complete double circulatory system. Once you establish common characteristics, select a characteristic on which members *differ,* and divide them into subgroups. If possible, keep to one characteristic for the division. Then select additional characteristics, and break the subgroup into increasingly smaller units. For animals, the divisions are, in descending order: Phylum, Subphylum, Superclass, Class, Subclass, Infraclass, Cohort, Superorder, Superfamily, Family, Subfamily, Tribe, Subtribe, Genus, Subgenus, Species and Subspecies.

In *Keys to Oregon Freshwater Fishes* (1973), Carl E. Bond begins by providing diagrams of spiny-rayed fish and of cycloid and ctenoid scales (Panel 12-4). Then a dichotomous key helps readers to identify fish (p. 244).

Another approach to classification is to establish major categories and put observed items into the categories. To develop such a classification system, select the organizing criteria. Then define and describe each major category. Thus you might classify people into smokers and nonsmokers. First, you might define and describe what you mean by smoker and nonsmoker. Then you might classify smokers by what they smoke—cigarettes, pipe, cigars. Then you might classify by material smoked—tobacco, marijuana, corn silk, and so forth. Frequency might be another category.

An approach akin to classification is partitioning. To partition, begin by describing the components of a large unit, and then provide detailed descriptions of the components.

The *KayPro 2 User's Guide* (1983) uses a partitioning description to familiarize computer users with the machine. The description includes a narrative and illustrations for each partition: "Keys to the Left of the Keyboard," "Keys to the Right of the Keyboard," "The Keypad," and 34 additional topics. Partitions consist of subdivisions. Thus "Care of Diskettes" includes subdivisions on handling diskettes with care, keeping diskettes covered, and eight others. Each subdivision then contains a succinct discussion of key points.

Descriptions that Compare and Contrast

You may be assigned the task of investigating two or three different brands of a product, such as typewriters, computers, chain saws, trucks, cars, or telephones, and then recommending which unit your organization should buy. A description that compares and contrasts can make your task easier.

First, decide on what you are to compare and contrast. Then develop

PANEL 12-4. Keys to Oregon Freshwater Fishes

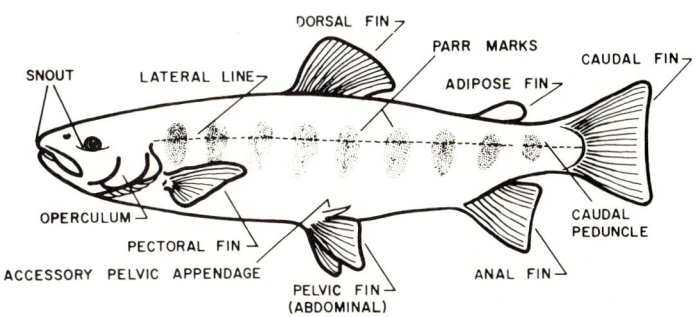

Figure 1   Soft-rayed fish showing external features.

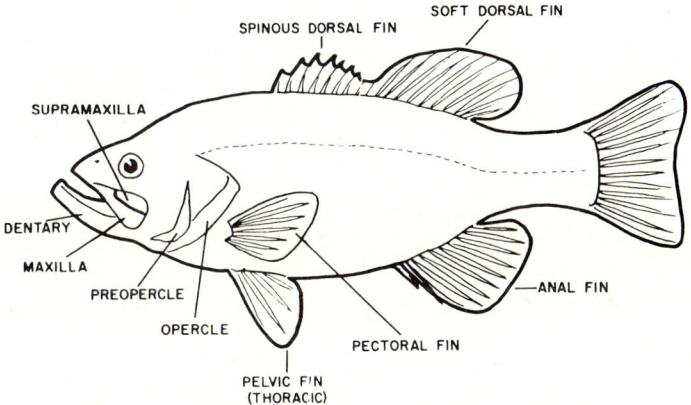

Figure 2.   Spiny-rayed fish showing external features.

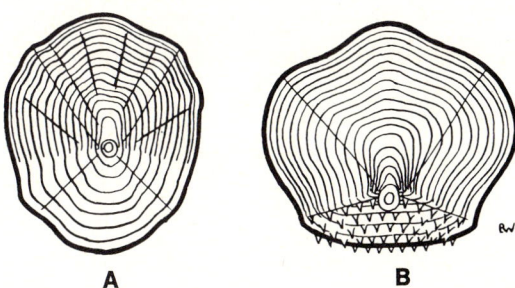

Figure 3.   (A) Cycloid scale showing several radii. (B) Ctenoid scale showing "teeth" or ctenii.

(Source: Carl Bond, Keys to Oregon Freshwater Fishes, Technical Bulletin 58 rev. (Corvallis, Ore.: Oregon State University Agricultural Experiment Station, 1973), pp. 2–3. Reprinted by permission of Carl Bond.)

PANEL 12-4. (continued)

Section 1. Key to Families

1a. No paired fins, jaws, or scales; 7 round gill openings on each side of pharyngeal region.
   Family PETROMYZONTIDAE—lampreys, page 9.

1b. Paired fins and jaws present; scales usually present; 1 gill opening on each side. .... see 2

2a. Both eyes on same side of head; body very flat.
   Family PLEURONECTIDAE—flounders, mostly marine, page 39.

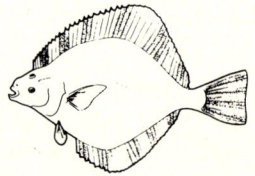

2b. One eye on each side of head. .... see 3

3a. Tail heterocercal; body with bony scutes; snout produced into rostrum; 4 barbels in advance of inferior mouth.
   Family ACIPENSERIDAE—sturgeons, page 11.

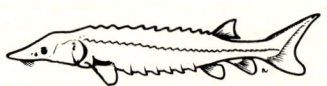

3b. Tail not heterocercal, both lobes of caudal fin nearly the same length; if scutes are present, the mouth is terminal. .... see 4

4a. An adipose fin present. .... see 5

4b. No adipose fin. .... see 8

3

TABLE 12-1. Typewriter comparison

Criteria	EM-100	EX55
Cost	$895	$995
Office test of one week	No	Yes
Pitch	6, 10, 12, 15, Proportional	10, 12, 15, Proportional
Typing speed	20 cps	20 cps
Line spacing	1, 1½, 2, 3	1, 1½, 2
Type ribbons	Single-strike Film Multi-Strike Film Fabric	Single-strike Film Multi-Strike Film Fabric
Memory	500 Character Auto-correction	2 Line Auto-correction
Boldface typing	Yes	No
Auto-centering	Yes	Yes

the criteria upon which to base your comparisons. A table is helpful. Use a vertical column for each item to be compared; list criteria for comparison along the left-hand side. Enter the essential information for the units to be compared.

Let's say that you've been assigned to determine which electronic typewriter your office should purchase for less than $1,000. Only two models of office electronic typewriters for less than $1,000 exist, the Brother EM-100 and the Silver-Reed EX55. Using the literature, you develop Table 12-1. You could continue building the table by adding criteria, but it illustrates the basic approach. Your narrative would refer to the table and point out the key differences between the two machines.

Preparing Process Descriptions

Procedures or process descriptions, like recipes, must be complete to be useful. As the CBE Style Manual explains:

> Describe subjects, materials, and methods used, including experimental design, in sufficient detail to enable other scientists to evaluate your work or to duplicate your research procedure. The usual sequence for experimental studies is design of the experiment, subjects (plant, animal, human), materials, procedures, and methods for observation and interpretation. Avoid unnecessary details. If you used well known methods without modification, simply name the methods or, at most, cite the papers in which they are described. If you used modifications of previously described methods, describe the modifications.

Give details of unusual experimental designs or statistical methods, and precisely describe animals used. (*CBE Style Manual* 1983, 21)*

Organize a process description chronologically. Start at the beginning, and proceed through to the end. Be sure to keep points in sequence. When describing a piece of equipment and a process, describe the equipment first and then the process. Separate the descriptions into distinct types.

Try to use a first person approach: "First, I examined," when explaining how you carried out a task. Using the first person helps you to avoid passive voice and dangling modifiers. Furthermore, people usually write more clearly and concisely when they use the first person. If you're describing a process carried out by someone else, use the person's name: "John welded the cross member to the I-beam." If the process has been completed, use the past tense: "Broom and Meiller conducted their initial survey of Lodi, Wisconsin, in June 1975." If the process is under way, keep to the present tense: "Some 34 million Americans raise gardens annually."

INSTRUCTIONS

In preparing instructions, keep in mind how and when people read and use them. For example, we've queried our classes about reading instructions before operating a new piece of equipment, such as a stereo, camera, or bicycle. About 75 percent tell us that they turn to the instructions only after trying the equipment and having it fail. Some people do study instructions thoroughly before trying to assemble or use new equipment, especially if the equipment is strange, complicated, and expensive.

Nevertheless, for almost any set of instructions, some users will be quite familiar with the product, some entirely unfamiliar with it, and some slightly familiar. Furthermore, people who use instructions have different abilities and frames of reference. Some people have more mechanical ability and knowledge than others. Some people can look at a piece of equipment or a drawing and understand it immediately, while others remain forever puzzled. When people do read instructions, they often use them piecemeal and look up the points they want to know.

Preparing Instructions

Tips on Picking and Using Strawberries (1980), a bulletin from the University of Illinois Cooperative Extension Service, first provides a general orientation and then explains nutrition, how to buy strawberries, and how the plant develops. The instructions and our critique follow.

---

* CBE Style Manual Committee, *CBE Style Manual: A Guide for Authors, Editors, and Publishers in the Biological Sciences*, 5th ed. Reprinted by permission of the Council of Biology Editors.

HOW TO PICK STRAWBERRIES

Strawberries look better and keep longer when they are picked and handled correctly. Because they are a very tender fruit, they will bruise and discolor any time they are squeezed. *Handle them carefully at all times,* whether you are picking them, or placing them in the container, or handling filled containers.

Some strawberry varieties are much easier to pick than others. For example, when they are mature, Surecrop berries usually snap off readily with a portion of the stem attached. Sparkle, on the other hand, will bruise unless you pinch the stem off. The surest way to pick fruit with a minimum of bruising is:

1. Grasp the stem just above the berry between the forefinger and the thumbnail, and pull with a slight twisting motion [Figure 12.5].
2. With the stem broken about a ½ inch from the berry, allow it to roll into the palm of your hand.
3. Repeat these operations using both hands until each holds three or four berries.
4. *Carefully place*—don't throw—the fruit into the containers. Repeat the process, picking with both hands.
5. Don't overfill your containers or try to pack the berries down.*

* *Tips on Picking and Using Strawberries* (Urbana: University of Illinois Cooperative Extension Service). Reprinted by permission of the University of Illinois.

Title informs readers of topic.

The introduction gives an overview of picking strawberries and explains why they must be handled carefully.

Note use of direct address to reader as "you." Note italics. Note cautions to reader about handling requirements of different varieties.

Note numbered sequence and use of illustrations.

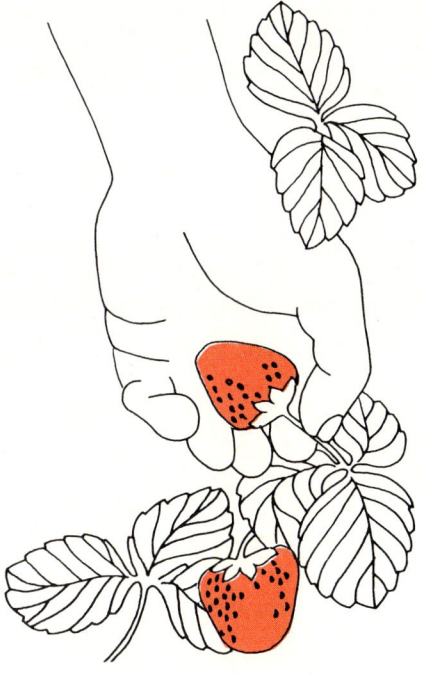

FIGURE 12-5. How to grasp strawberries.

Instructions often begin with an overview and a list of equipment or tools that will be needed. Then instructions explain each step in carrying out the activity in question. Panel 12–5 (pp. 250–251) provides simple instructions for installing carpeting on stairs. Note the use of illustrations, sequencing of activities, and narrative for each step.

To prepare instructions that communicate, you'll need to carry out three major activities: (1) consider your audience, (2) plan and produce the instructions, and (3) evaluate the instructions.

Considering Your Audience

As you develop instructions, keep in mind your reader's orientation. What does your reader know about the process on which you're about to give instructions? Why is the person using the instructions? What misinformation might the reader have about the process? What assumptions can you safely make about the group of people who will comprise your readers?

Next, consider your purpose. By following the instructions, what should a reader be able to do or know? In most cases, you want behavioral changes; a reader should be able to do something new after reading the instructions. But knowing what the instructions say does not guarantee that a reader will be able to follow them. Missing parts and tools, lack of understanding, and other factors may hamper a reader's ability to complete the process described in a set of instructions.

If a person could be injured or killed carrying out the process, provide adequate warnings (Panel 12-6, p. 252). In Chapter 18, we discuss what can happen when you do not give users adequate warning of danger should they or others misuse a product or ignore instructions.

As you consider your audience, give it clear reasons to read your instructions. Try to sort the information that readers must have from information you simply want them to have.

Planning and Producing Instructions

When you prepare instructions, you will improve your readers' understanding by:

1. Orienting readers properly
2. Using an easy-to-follow format
3. Using visuals
4. Writing in a personal and direct style

Orienting Your Readers. Your orientation helps tie together your and your readers' frames of reference. Keep your readers' perspectives in mind. Consider the following points:

- Don't assume much knowledge or experience on the part of readers.
- Start at the beginning of the process, and organize your points sequentially, first to last.
- Keep instructions reader-oriented. Tell readers how to stand, which direction is to be right, left, top, bottom, front and back.
- Provide a list of supplies, equipment, and materials which readers will need if they are to carry out your instructions.
- Provide visuals of equipment, and clearly label parts.

Using an Easy-to-Follow Format. By designing a format that makes your points obvious, you make your instructions communicate clearly:

- Keep lists to between five and nine items. Short lists facilitate short-term memory.
- Use numbers, bullets, dots, dashes, or other typographical devices to mark items in lists.
- Number steps that must be carried out in sequence.
- Use topic headings to break up instructions.
- Use clear, distinct labels to help users refer back to key points.

Using Effective Illustrations. Illustrations help users follow points and understand narrative. In fact, illustrations by themselves can provide effective instructions, as Figure 12–6 (p. 253) shows. Illustrations can reduce a narrative's length. Consider the following when preparing instructions:

- Provide illustrations whenever possible and appropriate. Research suggests that illustrations reduce user errors. Narrative alone slows users and increases errors (Stone and Glock 1981; Stone et al. 1981; Bieger 1982).
- Orient illustrations to the reader's point of view. If there's *any* possibility of confusion, identify top, bottom, right, and left. If you change orientation, keep the reader informed.
- Sequence your visuals properly.
- Use several illustrations to show steps in a process. By doing so, you break the instructions into a series of easy-to-follow activities.
- When considering page layout, allow room for the illustrations to go near the copy that explains them. If it is physically impossible for an illustration to be on the same two-page spread as its reference, consider adding a page number to the reference to help readers find the art.

PANEL 12-5. Combining Visuals and Narrative

Step Softly

INSTALLING CARPET ON STAIRS TO MAKE THEM SAFER AND QUIETER

Gleaming surfaces of polished wood look beautiful on dining room or living room floors, but bare wood on stairways looks like an accident waiting to happen. A covering of carpet adds traction to the stairs, making them safer. Carpet makes stairs quieter, too, by muffling creaky boards and heavy footsteps.

Badly-installed carpet, on the other hand, can look worse and be more hazardous than no covering at all. In this installment of Fix It Right, we'll show you how to carpet a stairway quickly and well.

Your first step is choosing the right carpet. ("Carpet Savvy" in the October 1985 NEW SHELTER outlines the many colors, textures, and fibers that are available. If you would like a copy of that issue, send $2 to: Carpet Savvy, NEW SHELTER, 33 E. Minor St., Emmaus, PA 18049.) In general, a carpeting with a fairly short nap will provide the safest, firmest footing and longest wear. You'll need just enough nap to hide the staples.

Padding can make your carpet feel thicker and increase its longevity. You can buy carpeting with padding already attached, or buy the padding in a separate roll. We suggest buying the two separately, since you'll need padding only on the treads and not on the risers. Attached padding will also make the carpet harder to bend, and if you want to turn under the edges you'll have to trim off two inches of padding.

Professional installers work with a long roll of carpet and do six or seven steps at a time. This saves cutting time as well as carpet. Novices, however, will reduce their chances of ending up with crooked carpet if they cut the rug one step at a time.

Tools & Supplies Needed

Measuring tape	⅜- or ½-inch staples
Chalkline	Hammer
Ruler or straight edge	⅝-inch carpet tacks
Utility knife	Carpet
Large scissors	Padding
Stapler	

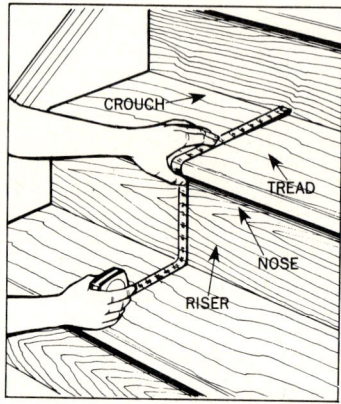

1 Measure the steps. Start at the crouch (the place where the back of the tread meets the bottom of the riser), pull the measuring tape around the nose (the round part of the tread that extends over the top of the riser), and down the riser to the next crouch. Add one inch to this measurement just to be safe. Add up the length measurements for all the stairs, then measure the widths.

2 Take your measurements to a carpet store. You may be able to save money by buying a remnant, if you can find one you like. For most sets of stairs, you'll need two 12-foot lengths. Mark the grain direction on each piece so the two will match after installation.

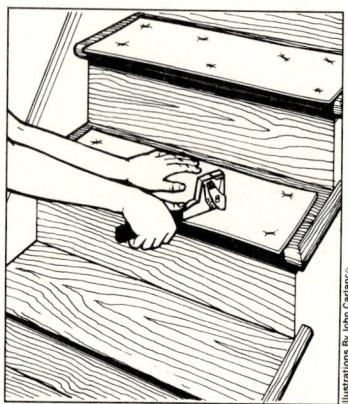

3 Cut padding for the stair treads. If the carpet won't be covering the stairs' entire width, cut the padding an inch short on each side. This will leave space for turning the carpet edges under for a finished look. Lay the padding an inch from the crouch and staple it along the back and sides. The pad should hang an inch over the nose of the step so it can be pulled around and fastened later.

4 Mark each piece of carpet and cut it with a utility knife. If you're planning to turn under the edges, add an extra inch. Butt the back edge of the carpet to the crouch and staple it to the tread. Make sure the staples will be embedded between the fibers of the nap.

(Source: Rodale's New Shelter 7(3): 80–81. Reprinted by permission of Rodale's New Shelter magazine.)

PANEL 12-5. (continued)

5 Pull the carpet taut and staple it to the underside of the nose. Put your first staple near the middle of the step and work toward the edges to keep the carpet smooth and tight. If you're turning under the edges, be sure to make the fold before you staple too close to the ends. You may have to use ⅝-inch carpet tacks or small nails to fasten through the doubled edges.

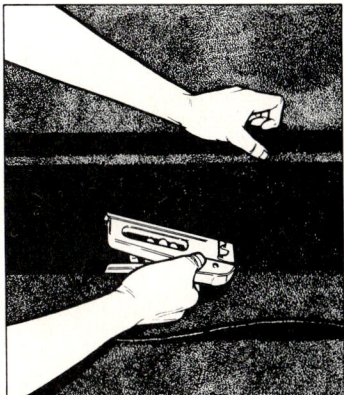

7 Staple the carpet to the bottom of the riser. These staples are often the most visible since they're facing you as you climb the stairs. Try to work them deep into the nap of the carpet so they won't show. But if the fibers on your carpet are ¼-inch or shorter, hiding the staples from view probably won't be possible. Instead, try using brads.

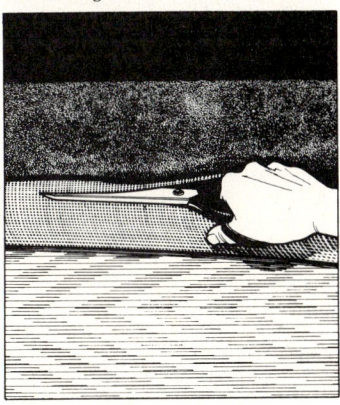

6 Smooth the carpet over the riser and trim it at the crouch with your utility knife or scissors.

In next month's Fix It Right, we'll show you how to install soffit and ridge vents in your attic.

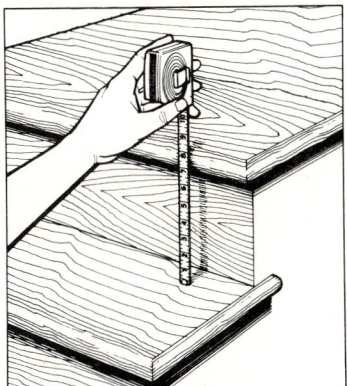

8 Repeat steps four through seven until you've reached the bottom. Then cut a piece of carpet to fit the top riser from the crouch to the bottom of the nose. Don't run the carpet over the nose unless the carpet will continue down the hall.—*Fred Matlack*

Copyright © 1986 Rodale Press, Inc. All rights reserved.

PANEL 12-6. Instructions Often Provide Critical Warnings

Westinghouse —
Plain English
Through Labeling

This is the second in a series of reports on Plain English projects from specific companies.

During the 1970's, at least five different industry standards existed concerning product safety labels. The goal of each was the same — to help manufacturers provide adequate warnings of potential hazards. But each standard had different requirements and recommendations. Most dealt with the label's format but provided little guidance on how to identify or classify a hazard or how to write a label clearly to warn of potential hazards. Engineers and product designers were confused. Westinghouse realized that they needed a better system.

Designing a Product Safety Label Program

How could a corporation that produces 300,000 varieties of some 8,000 basic products ensure that all labels would be consistent and readable? The diversity of products made uniform labels impractical, and the size and organizational structure of the corporation made centralized label design impossible. But a coherent product safety label program was essential to protect the user from hazards and the company from product liability suits. Also, Westinghouse needed to coordinate labels so products from different divisions could be combined into unified systems.

A committee of six engineers and an attorney was formed to develop a product safety label program. The result is a handbook that covers all aspects of developing a label from analyzing the product and audience to writing, designing, and specifying the label. The handbook gives working engineers and product designers professional guidance on typography, language, readability, layout, location, and materials.

Anatomy of a Label

The handbook identifies seven label elements. Using and arranging these elements consistently reinforces the label's safety messages and improves the effectiveness of the whole labeling system.

An example of how a label could be written is illustrated in the next column.

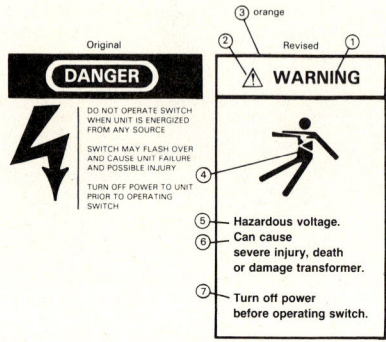

Here are the seven elements in the revised label:

1. Signal word. (DANGER, WARNING, CAUTION, NOTICE) A label must always contain a signal word at the top. This is the main attention-getter. One of the four words listed above must be used.

2. Hazard Alert Signal. ⚠ This symbol always appears with the signal words danger, warning, and caution. It is an international standard that means, Look Out!

3. Color. Each signal word has a corresponding color: danger (red), warning (orange), caution (yellow), and notice (blue).

4. Symbols and pictographs. These reinforce the verbal message and provide nonverbal communication for illiterate or non-English-speaking readers.

5. Identification of the hazard. This is the first verbal message following the signal word. Always in bold type.

6. Result of ignoring the warning. This describes what will happen if the warning is ignored.

7. Avoiding the hazard. This gives instructions on how to avoid injury.

For more information about Westinghouse's step-by-step approach to producing product safety labels, write to:

John F. Gormley, Director
Product Safety
Environmental Affairs
Westinghouse Building
Gateway Center
Pittsburgh, PA 15222

(Source: Simply Stated in Business, no. 3, January 1984, p. 2. Reprinted with permission from Simply Stated in Business, the bimonthly insert of the Document Design Center, American Institutes for Research.)

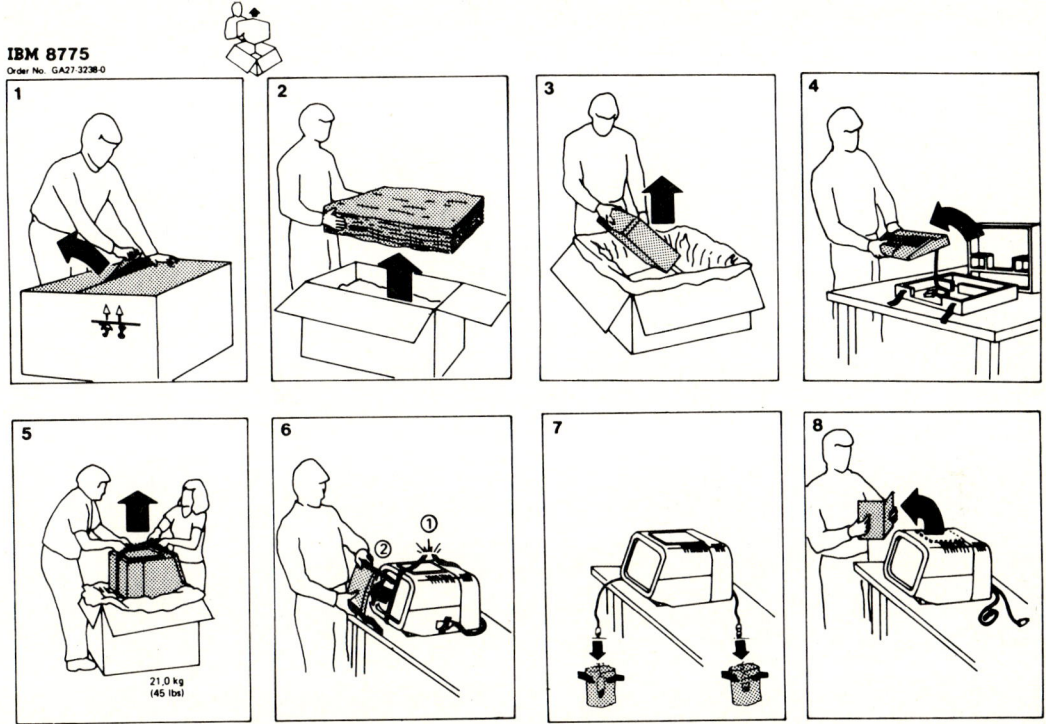

FIGURE 12-6. Instructions without words.
(Source: H. E. Vogt, Wordless Instructions—Say It with Pictures, Proceedings of the 31st International Technical Communication Conference *(Seattle: Society for Technical Communication, 1984),* p. VC-26. Copyright © 1984 Society for Technical Communication. Reprinted by permission of the Society for Technical Communication and Herbert E. Vogt.)

Writing in a Personal and Direct Style.   The techniques for good writing covered in earlier chapters apply equally to writing instructions. In addition, the following points help ensure a more understandable set of instructions:

- Talk directly to your users in the second person. (Note that most of this section is written in a "you" orientation.)
- Write in the active voice. Make your sentences short and precise. Users then will know exactly what they are to do.
- Remember readers' frames of reference. Complex, unfamiliar terms, even in short sentences, may confuse them.
- Draft your instructions, and set them aside for a few hours or, better yet, a day or two. Return to them, and edit for clarity.
- Check carefully to make sure your content is correct and your points are in sequence.

Evaluating Your Instructions. Always check the quality and content of your instructions. Make sure that you've clearly and carefully explained each step in a process. And when you do your evaluation, be sure to consider the following points:

- Read through your instructions several times to check for organizational errors, missing points, or points out of sequence.
- Have someone familiar with the process read through your instructions to see if you've left anything out or presented points out of sequence.
- Test your instructions on someone unfamiliar with the process.
- Revise as needed to overcome the problems.
- Keep in mind that misinterpretations are always possible, but strive to minimize them.

Most problems with instructions occur when a writer fails to follow one or more of the points just mentioned. If you follow those points, you'll find and correct any problems before they reach your audience and someone calls or writes to tell you about them.

HIGHLIGHTS

1. Definitions are critically important for technical and scientific communication because they align the frame of reference of communicator and audience.
2. Terms may be defined by examples, word definitions, explication, comparison, symbols, visuals, and word histories.
3. For readers' convenience, place definitions in narrative or glossary. Footnotes are a distant last choice.
4. Process descriptions tell how someone carried out a particular task; instructions tell readers how they can carry out the task.
5. When preparing process descriptions, describe the subject, materials, and methods you used. Use a chronological order and avoid unnecessary details. But do give readers sufficient details to carry out the process.
6. To prepare instructions that communicate, consider your audience, plan and produce the instructions carefully, and then evaluate the instructions to be sure they are clear and complete.
7. When preparing instructions, orient your readers, use an easy-to-follow format, use visuals, and write in a personal style.
8. To evaluate your instructions, carefully read through them, have individuals familiar with the process check your instructions, test your instructions on novices, and revise as needed.

PROJECTS

1. Study Panel 12-1. Prepare a similar panel by selecting four concepts from your major field. For each concept, list 10 examples. Make six copies of your lists. Then select six people: two classmates in your major, two professors in your major, and two classmates in other majors. Have each person cross out items that are not examples of the concepts. Then calculate the agreement between the classmates in your major, the professors, and classmates in other majors. Write a report of not more than two typed double-spaced pages discussing your panel and your analysis of agreements among the groups.
2. Select two terms commonly used in your major. Define them according to at least three different techniques discussed in this chapter. Consider other students in your major as the audience.
3. Using the same terms as in Project 2, use two different techniques to define each of the terms for a lay audience.
4. Select a piece of equipment that you will use in your career, and describe it for a technical audience and then for a nontechnical audience.
5. Write a three-page or less case history of your experience:
    a. Using a word processor or personal computer for the first time.
    b. Using a mathematical or statistical program on a personal computer for the first time.
    c. Shooting your first roll of film with a single lens reflex camera.
    d. Conducting an experiment in a laboratory.
    e. Applying for telephone service.
    f. Taking part in an activity of your choice.
6. Select two different brands of product from the list below, and contrast them. Keep your description to three pages or less.
    a. Personal computers
    b. Hand calculators
    c. Telephones
    d. Backpacks
    e. Beer
    f. Ice cream
    g. Typewriters
    h. Stereo systems
7. Develop a classification system for one of the following topics, and then apply it to five members of the set:
    a. Dogs
    b. Pizza
    c. Junk food junkies
    d. Beer drinkers
    e. Health enthusiasts
    f. Joggers
8. Prepare a spatial description, three-page maximum, of one of the following locations:
    a. Your hometown
    b. Your favorite vacation spot

c. Your school
   d. A prominent site near your school
   e. The student center at your school

9. With the partitioning technique, describe one of the items below. Keep your description to no more than three pages.
   a. A dog
   b. A computer
   c. Telephone
   d. Personal calculator
   e. The human body
   f. A computer software program

10. Observe another student carrying out out a laboratory experiment. Then prepare a description of the process you observed. Keep the narrative to less than 800 words.

11. Prepare a set of instructions for operating a technical or scientific piece of equipment. Limit your instructions to five pages.

12. Find a poorly done set of instructions of no more than three pages. Photocopy it. Assess its quality in no more than three pages.

13. Revise the poorly done instructions chosen for Project 13. Improve the instructions. Keep your revision to fewer than five pages.

14. Prepare a set of instructions made up entirely of captions and visuals. Keep your instructions to fewer than five pages.

15. Sketch a set of instructions made up entirely of visuals.

FOR MORE HELP

BROCKMANN, R. J. 1986. *Writing better computer user documentation: From paper to screen.* New York: Wiley.
Brockmann covers preparing both manuals and on-line documentation.

GRIMM, S. 1982. *How to write computer manuals for users.* New York: Van Nostrand Reinhold.
A text recommended by both instructors and students that gives useful advice.

PRICE, J. 1985. *How to write computer manuals: A handbook for writing software manuals.* Menlo Park, Calif.: Benjamin-Cummings.
The author provides excellent advice for writing computer manuals.

WEISS, E. H. 1985. *How to write a usable user's manual.* Philadelphia: ISI Press.
Weiss includes useful tips for writing manuals.

ZEMKE, R., AND T. KRAMLINGER. 1982. *Figuring out things—a trainer's guide to needs and task analysis.*
The authors provide excellent guidelines for analyzing tasks for preparing instructions.

APPENDIX 12-A. Answers to the Story

Carefully consider the answers and the points made; reread the story to note the details of observation and inference.

(DK = Don't Know.)

1. DK—Do you know that the "businessman" and the "owner" are one and the same?
2. DK—Was it necessarily a robbery involved here? Perhaps the man was the rent collector—or the owner's son.
3. False—An easy one, to boost the test taker's morale.
4. DK—Was the owner a man?
5. DK—May seem unlikely but the story does not preclude it.
6. True—Story says the owner opened the cash register.
7. DK—We don't know who scooped up the contents of the cash register or that the person necessarily *ran* away.
8. DK—The dependent clause is doubtful; the cash register may or may not have contained money.
9. DK—Again, a robber?

(Source: William V. Haney, Communication and Interpersonal Relations: Text and Cases, *5th ed.* (1985: Homewood, Ill.: Irwin), p. 222. © 1985 Richard D. Irwin. Reprinted by permission of Richard D. Irwin.)

13
Document Format

OVERVIEW

In Chapter 13, we
- ☐ Discuss the role of format in presenting technical material economically and in easily retrievable form
- ☐ Describe common forms of proposals, reports, abstracts, and other written communications
- ☐ Derive basic principles for following required or "standard" formats and for developing an effective format when none is specified

DOCUMENT FORMAT: A KEY TO EFFECTIVE COMMUNICATION

What some people consider disorganized writing in reality is poor formatting. By format we mean the framework of a document or message. Follow all the precepts of good writing, and you may still produce repetitive and confusing writing, unless your document possesses sound format.

Two examples illustrate. During the 1970s boom in citizens' band radio sales, purchasers had to apply for a CB license from the Federal Communications Commission (FCC). The FCC issued with each radio a booklet that answered specific questions. The FCC had five employees who spent their days answering phone calls from frustrated CB owners who could not find necessary answers in the booklet. The booklet's usage and grammar were impeccable, but people complained about confusing and unclear writing.

With help from the Document Design Project of the American Institutes for Research, the FCC gave its booklet a new format (Battison and Landesman 1981a). Using a problem solving approach, writers devised a set of the most common problems faced by a new CB owner who was trying to apply for a license. Then the writers phrased the problems in question form and followed these with clear, brief answers. The new format saved time and phone calls, and the FCC employees soon found room in their days to do other work.

The Document Design Project solved another format problem for the federal government. By redesigning the form used by applicants for college student financial aid, the Project reduced errors by 7 percent (Landesman 1981). Multiplied by the hundreds of thousands of students and their families using the form, that figure meant huge economies in time and money.

Both examples reveal professional communicators at work.

At its core, format's role is to get important information concisely to those who need it. As a professional, you will use traditional formats, and you will devise your own.

SHIFTING FROM "IN SCHOOL" TO "ON THE JOB" WRITING

Suppose that your teacher asks you to prepare a "compare and contrast" assignment in which you choose from your field three publications. You're asked to pick a research journal that publishes academic, experimental studies, a trade journal featuring articles on industrial applications, and a general interest magazine containing popularized accounts of developments and their implications. You are to compare and contrast the three publications' complexity of language, frequency of use of scientific terms, assumptions about readers' backgrounds, article length, and other items.

A school version of your completed assignment might consist of an essay, which took each publication in turn, addressed each topic in turn, and concluded with a summary. You would pay careful attention to your composition, and you might easily produce a paper running four or more pages.

But an on the job version of the same assignment should reflect your awareness that time—yours and your readers'—is money. The right format will help. For instance, by setting up a simple table, with one column each for the research journal, the trade journal, and the popular magazine, and a horizontal row for each of the topics you wish to compare, you can present the results of your study concisely. Panel 13-1 provides an example. The written part of your report, emphasizing the important differences or similarities, might take a page or less. (Look again at Panel 5-2 in Chapter 5, which contains a similar "compare and contrast" effort.)

Once you understand the major principle, that format and writing together reveal meaning, you have begun the transition from writing for writing's sake to writing professionally. Moreover, you will be able to use format—whether it's a traditional one imposed by your field or one you invent for a special case—as an aid to communication, just as you use charts, graphs, photos, and well turned sentences.

GOOD FORMAT ANTICIPATES READERS' NEEDS

Analyze any easy to use document, and you will find format playing a key role. It provides necessary information at just the right places. Essential

PANEL 13-1. Comparison of Research Journal, Trade Journal, and Popular Magazine

	Journal of Latest Research	Trade Journal News	Pop Science Weekly
Average sentence	22 words	18 words	14 words
Readability	College-plus	Some college	9th grade
Technical terms	Rarely explained	Usually defined	Translated
Average article	4,000 words	1,200 words	250 words
Readers' backgrounds	Ph.D, research interests	25,000 managers selected by job function	1.5 million paying subscribers

to effective formatting is an understanding of how readers use a document. Readers operate differently with different types of documents. Fiction readers start at the beginning of a story and go straight through (unless the story gets so exciting that they have to sneak a peek at the ending). Creative writers therefore pace their stories, set scenes, build character, and create revelations and climaxes. Nonfiction magazine writers similarly know that interested readers go from start to finish. They format their articles almost like fiction. Newspaper journalists use a different format. They rank facts in order of decreasing importance. Thus newspapers help busy people stay informed with the least investment of time.

Not a lot of research has been done on how readers of scientific and technical material proceed. In fact, the major studies so far have produced results that at first appear contradictory. More than two decades ago, James Souther (1962), now a professor at the University of Washington, carried out detailed interviews of 70 engineering managers at the Pittsburgh offices of Westinghouse Electric. What he found destroyed forever the idea that technical readers start at the beginning and read straight through. Souther's managers told him that the parts of a report they most often read were the abstract or summary. They used the summary to discover if a report had something of value to them. Those who read further went to the introduction, the second most frequently read part. The conclusion section was the third most frequently read part; the main body was fourth. The least frequently read was the appendix.

However, a more recent study questions some of Souther's results. It was done at the Langley Research Center of the National Aeronautics and Space Administration by Thomas E. Pinelli, Virginia E. Cordle, Myron Glassman, and Raymond F. Vondran, Jr. (1984a, 1984b). They sent questionnaires to 1,100 engineers and scientists, half employed at Langley and half not, asking questions about how they used Langley-produced documents. About one quarter of the group were managers. Chief among the findings was that managers and nonmanagers used technical documents in about the same way. By about a two-to-one margin, they preferred both a brief abstract and a longer, detailed summary. Unlike the managers in Souther's study, the Pinelli group managers declared that they read about as much of the technical documents as the nonmanagers.

Well aware that their study cast some doubt on what they called the "folklore" of how people along the administrative line read technical documents, the Pinelli group looked for explanations. An important one, they theorized, was that administrators in the Langley-NASA group were mostly former "bench scientists" who retained habits they learned while lab workers. A second explanation lay in the well known tendency of survey respondents to give answers that flatter themselves. Perhaps, the Pinelli group suggested, respondents should be observed at their reading instead of describing how much they read.

Other researchers (Flower, Hayes, and Swarts 1983; Huckin 1983) have found that readers of technical material pull out information and incorporate it into *schemata*—blends of experience and knowledge that explain how things work. Readers seek to place new information into the existing framework. If they see that something new works like something they are familiar with, they understand the new. Readers evaluate whether new information is relevant and either explore further or go on to something else. These findings confirm that people read technical information differently from other material. They skip around, taking what they need and passing over what they don't need. They rely on schemata to decide what to spend time on.

What helps the technical professional reader collect information in this highly selective fashion? A document hierarchically organized around the topic, question, or problem (Huckin 1983). A hierarchical presentation is clearly organized into major and minor divisions. At one level, *hierarchy* refers to division, paragraph, and even sentence organization. At a different level, *hierarchy* refers to an external framework that is either imposed on or adopted by a writer. In the latter sense, a hierarchy is a document format.

The most important factor in these studies is one they all agree on. Readers of technical documents require and even take for granted a format that clearly denotes the documents' different parts and the specific functions of each part.

KEY PARTS OF TECHNICAL DOCUMENTS

Not all technical communications contain all the parts described below. If you understand how to plan these, you will be able to follow other formats and derive new ones for any type of document. (The proposal and the progress report are not really parts of a technical document, although they belong in the overall process.)

Below we consider the technical report, which deals with a field project or laboratory experiment. The report tends to contain more separate parts than any other technical document. Major divisions of a technical report are as follows:

- title
- proposal
- problem statement
- justification/literature review
- methods
- timetable and budget
- bibliography/reference list
- results
- discussion/conclusion
- recommendations
- abstract
- citations
- progress reports

In actuality, sometimes you'll merge these divisions. In certain fields, a division called "introduction" includes the problem statement and justification/literature review. In some fields, citations (footnotes, endnotes, references) are combined with the bibliography. These variations in format will not stump you if you understand the purpose of each part.

Bear in mind another important point. Although you may prepare your report in one sequence, you present it in another sequence. Thus you will probably do the literature review before you state the problem in final form. You accumulate most of the bibliography before you examine it for your literature review. The abstract, which you usually do at the end of your project, appears first in your finished report. In other words, you work in one order, present in another.

Title

The title, while the first part of a report to strike the eye, is often the last to be written. Books are sometimes written with tentative, working titles, which may be changed several times. The title should reflect the report's content, be brief, and correspond to the report's tone. If the document will be listed in an indexing or abstracting journal, the title should contain the most important key words, as described below in the section on abstracts. The abstract may suggest a good title, as may the conclusions section. Often the publication's style prescribes a document's title length.

Proposal

Although a proposal rarely accompanies a finished report, it is part of the overall effort, and you'll incorporate much of it into the final report. Properly done, a proposal's divisions expand into major sections of the finished report: problem statement, literature review, methods, and bibliography. A thorough proposal will contain most of your project's preliminary work. What will remain is to do the project, report the results, and assess their meaning.

As a request to do or not to do something, the proposal ranges in form from simple to complex, usually in proportion to the time, effort, and money involved. Term paper proposals for class may be quite short, only a paragraph or a few pages; proposals for masters' and Ph.D. theses often run 20 pages or more; proposals for million-dollar research projects may run hundreds of pages.

No matter how simple or complex your project, your proposal should answer these questions:

1. What is the problem I propose to investigate?
2. Why is the problem important?
3. What method will I use to solve the problem?
4. What is my timetable and projected cost?

A Sample Proposal. Let us now turn to a sample proposal and see how good formatting makes it easier for both writer and reviewer.

Chuck, an agronomy student, was interested in finding ways of improving the popcorn grown and marketed by the Agronomy Club. In conversations with his professor, he proposed growing several varieties of popcorn to see if he could find a better one for the club to raise. He soon realized, however, that he would not have time to plant, raise, harvest, and test the corn during the single semester he would be enrolled in his technical communication course. He decided to save that experiment for another time and to concentrate instead on another important factor, the amount of water in popcorn. He knew that water vaporizing under heat causes the kernel to pop. He submitted a written proposal (Panel 13-2), which helped him and his professor to evaluate the project's soundness.

Chuck clearly had a research topic. His proposal's good format made it easy for the professor to criticize. Chuck's professor made several substantive and editorial suggestions on improving the proposal. Part of the revised version appears in Panel 13-3. As you see, Chuck's proposal was now in much more precise form. It stressed what he would do, instead of what he would not, and each section justified and expanded the preceding one. When he finishes the study, Chuck may not be able to conclude that humidity is the only factor in reducing unpopped kernels, but he will certainly know if it is a factor. Moreover, his proposal will not have promised more than it can deliver.

PANEL 13-2. Proposal for a Student Research Paper

PROPOSAL

Title: Reducing "Old Maids"

[Margin note: Vague. Can you be more specific? Also, why be sexist?]

Problem: "Old Maids": Is percent humidity the major difference in reducing the frequency of unpopped kernels in four leading brands of popcorn?

[Margin note: Basic idea comes through. But do you intend to treat the other factors, such as type of corn, length of curing, additives? or other treatments?]

[Margin note: What does this sentence contribute? Why not just drop it?]

Justification: ~~Everyone who has bought and made popcorn commercials~~ [irrelevant as cast] ~~knows of the advertisement war between popcorn companies because "ours has fewer old maids."~~ Is there some magical combination of factors that makes certain popcorns leave fewer unpopped kernels, or is it simply a factor of percent humidity? The answer to this question ~~will~~ [can] ~~have practical significance in that recommendations on~~ [wordy] [help] improv[e]~~ing~~ the quality of the Agronomy Club popcorn. ~~can be offered~~.

[Margin note: Why not phrase it so that neither "magic" nor other factors are proposed for examination? As you say below, humidity is the only factor you will examine.]

In the last five years ~~as listed in~~ the Consumer Index ~~there~~ has [listed] ~~been~~ only one article which "offers tips on buying popcorn. . . ." ~~It is~~ Good Housekeeping, September 1978, p. 274, [asserts] ~~In the article it states~~ that humidity is the real reason why ~~the~~ kernels pop and [advises] ~~that with this in mind it is good to~~ keep[ing] opened packages sealed in tight containers ~~or~~ [so] they will [not] lose their popping ability.

There are no[t] similar articles listed in certain [which ones?] volumes of the Biological and Agricultural Indexes or in the Food Science and Technology Index. I did not check many volumes because the subject ~~material~~ [matter] seems ~~to be~~ of a different nature. [vague]

[Margin note: Why not check with professors who specialize in food testing?]

PANEL 13-2. (continued)

Where did you get these methods? Why only these brands? Why a hot-air popper? Why a GE?	<u>Methods and Materials</u>: Test one hundred seeds of Orville Redenbocker's [*spelling*], Food Club's, the Agronomy Club's and Jiffy Pop popcorn in the same GE hot-air popcorn popper and determine the frequency of unpopped kernels in each sample. Each sample will be randomly selected
It's probably not necessary to blindfold yourself. Why 12 hours? Why 80 degrees? Which correlation test, and why is it appropriate? Your intention of "correlating" humidity with popping suggests a slightly different problem statement. Something like: How does percent humidity correlate with frequency of unpopped kernels?	from a freshly bought product while blindfolded. Next, weigh hundred-seed samples of each product to tenths of a gram and reweigh after 12 hours in a drying oven at 80 degrees Centigrade. Determine average percent humidity in the sample and run a correlation test. Should the correlation be poor, possible alternative conclusions will be drawn. I will have to buy the popcorn products from retail outlets. I <u>will obtain</u> [*why wait?*] permission to use Dr. D. Smith's drying oven and Dr. S. Ladd's weighing balance. <u>Timetable/budget</u>: Tests can be run in one weekend. Popcorn will cost $5 or less. Popper is readily available.
Bon appetit! But this statement is out of place in a science project proposal.	~~AND I intend to eat the results!!~~

Chuck accomplished one other important purpose with his revised proposal. He has written a portion of his final report. When he has completed his research, about all that will remain for him to write will be the results, discussion of conclusions, suggestions for further study, and an abstract. His good work at the start will have paid dividends. Not only does a good proposal improve his science, it virtually halves his reporting task.

> **PANEL 13-3. Revised Student Proposal.**
>
> Tentative Title: "Humidity as a Factor in Unpopped Kernels of Eight Popcorn Brands: A Correlation Study"
>
> Problem: Is there a significant difference in the percent humidity in various commercial brands of popcorn? Does humidity correlate with frequency of unpopped kernels?
>
> Justification: Unpopped kernels frustrate popcorn producers and consumers alike. Unseasoned popcorn may be virtually indistinguishable in taste from brand to brand, so commercial producers compete on the degree to which their kernels pop. Consumers are counseled as to ways of preserving popcorn to ensure its popping qualities. Good Housekeeping (September 1978, p. 274) advises keeping opened packages in tightly sealed containers, asserting that humidity is the key factor in retaining good popping characteristics.
>
> This project seeks to measure differences in humidity of various brands and to discover if the number of unpopped kernels correlates positively with each brand's degree of humidity.

Problem Statement

By the time it appears in a full proposal, the problem statement should be both precise and concise. If you have done your thinking, your problem statement will probably consist of a sentence or two. Possibly it will be a question. By far your most difficult task is to phrase your research problem so that you can actually address it. A vague, general research question produces vague results. Try to phrase a problem statement so that when you hand in your report, you know that you have or have not answered the question and that you have or have not finished the project. The problem statements in Panels 13-2 and 13-3 show a student who proceeds from the vague and general to the specific and manageable.

To refine your problem statement, you face several important decisions. Are you interested in addressing the problem? Do you have the skills required? Has the problem been addressed? By whom? Using what method? You can answer these last three questions by a thorough justification/ literature review.

Justification/Literature Review

As one of the proposal's longer sections, the justification usually reviews previously published studies on your topic. As you examine studies, at first you probably will look at general overviews until you have a specific topic in mind. Soon, though, you will home onto specific studies dealing with your narrowed topic. You will discuss these in your literature review.

You should provide clear evidence that the problem has been addressed. In the rare instance where the problem has not been addressed, the literature search proves that fact. As you examine studies for their successes and shortcomings, you begin to see which methods you might use. Overall, the literature review section proves that you've done your homework and are familiar with previous work on your topic.

The final objective of the literature review is a clear statement of your research question. In some fields, especially the social sciences, the research question may be broken into a series of hypotheses, each stated affirmatively in a declarative sentence. The research then produces evidence that confirms or disconfirms each hypothesis, usually according to a statistical test. Most engineering and science fields, though, phrase the research question either as a declarative statement or as a question and let it go at that.

Be careful not to use general or vague statements in describing your purpose. Avoid vague words such as "investigate," "shed some light on," "examine," "seek answers to," and others. What, exactly, does "investigate" mean? To offer to "shed some light" sets up no standard at all and suggests that you are fuzzy in your own mind about how you will pursue your topic and which types of evidence you will seek.

Methods

Having phrased the problem and examined how others have addressed it, you are ready to describe your own plan. How will you carry out the work? If you plan to test hypotheses, explain the tests you will use. What materials do you need? Which sources will you use—library, laboratory, equipment, animals, humans? Are they available to you? Your methods section contains the meat of your proposal. If you have a clear problem statement and justification, you will handle that part confidently.

Timetable and Budget

Almost never will you have unlimited time or money for a project. A realistic timetable helps you to organize your work and saves you sleep toward project's end. Experienced evaluators look to the calendar and budget section to see if you realize how much work you are letting yourself in for.

Bibliography/Reference List

As soon as you begin to consider a topic, your bibliography or reference list becomes a major part of your work. Do yourself a real favor by adopting the standard system used in your field. (The appendix lists common style guides to standard citation systems.) By recording references in the correct style as you go, you vastly simplify the compilation of the final version.

Many proposals require a bibliography to show that you know the relevant research. This bibliography usually grows as you proceed with the project. Keep a complete record of where you look for information. It will help you judge the thoroughness of your search. As you find more and more information, you adjust your study accordingly.

Two approaches to compiling bibliographies exist. Some professionals advocate listing everything you consult, whether or not you quote it in the body of your report. They argue the importance of establishing the scope of your search for useful information. (Historians tend to apply this system.) A second approach is to list only what you actually refer to. Scientific studies and publications seeking a compromise among the need to document, the need to limit typesetting costs, and the need to limit space usually take this approach. In this case the list of sources is usually called "References," "Works Cited," or a similar title, rather than "Bibliography."

Results

Often labeled "Findings," the results section reports your project's outcome. If you tested hypotheses, report findings about them in the order in which the hypotheses appear in the methods section. If you can work your data into tables, graphs, or other visuals, build your results section around them. Keep your tone neutral and objective.

But scientists and their readers do want more. They want to know what results *mean*. So you need another section, in which you discuss and interpret the meaning of your results.

Discussion/Conclusion

Like most parts of a technical report, this section travels under various names, depending on the field. In some fields, the "Results" and "Discussion" sections are merged into one. You may even find "Discussion and Conclusions." Such variations will give you no trouble so long as you remain clear about the different functions being served. "Results" are what happened or what you observed. "Discussion" is why you think it happened. "Conclusions" are the meaning of what happened.

Tell *what* happened before you begin to tell *why*. Follow that order even if your field prescribes a "Results and Discussion" instead of two separate sections.

Limit your discussion to results you actually have reported in the preceding section. Failure to do so probably constitutes the most common flaw of both student and professional work. Make certain you don't drift beyond the evidence you've presented.

Taken together, of course, your results and discussion form the heart of your report. The Projects at the end of this chapter contain suggestions designed to familiarize you with your field's approach to the various sections of a report.

Recommendations

If you're working in a university or other research setting, you may include a section suggesting fruitful lines of further inquiry. You might call this section "Recommendations for Further Study."

In business and industry, though, a recommendations section states what should be done as a result of your study. What steps should the organization take? Should it buy or not buy, start work or not, build or not build? Like the discussion section, recommendations should flow from both methods and results. In industry, recommendations often form the "bottom line" of technical reports.

Abstract

The abstract is brief, usually not more than 250 words. It is a summary of the project: problem, method, findings. Brevity should not mean vagueness. Include specific results, and choose your words carefully. The increasing use of abstracting services means that many researchers will learn everything about your work from your abstract alone.

The words "abstract" and "summary" have become interchangeable. One form of abstract is called a "summary abstract." Strictly speaking, a summary should come at the end of a report. But given how readers actually read technical reports, put the "summary" first, so that they can read it and decide whether to proceed. "Abstract" tends to be used more in academic circles; businesses use "summary."

The National Science Foundation (n.d.) strikes a middle ground between "abstract" and "summary":

> The summary (about 200 words) must be self-contained and intelligible to a scientifically literate reader. Without restating the project title, it should begin with a topic sentence stating the project's major thesis. The summary should include, if pertinent to the project being described, the following items:
> - The primary objectives and scope of the project
> - The techniques or approaches used only to the degree necessary for comprehension
> - The findings and implications, stated as concisely and informatively as possible

This summary will be published in an annual NSF report. Authors should also be aware that the summary may be used to answer inquiries by nonscientists as to the nature and significance of the research. Scientific jargon and abbreviations should be avoided. (25)

The NSF advice contains virtually all you need to do effective abstracts, but two further points may be helpful. First, publications usually have their own word limits on abstracts, so be prepared to conform to editorial policy. Second, the widespread use of electronic abstracting and indexing services dictates that you pay attention to "key word" concepts, so that computers can refer your published work to researchers looking for studies on your topic. When you submit your work to an indexing or abstracting agency, you will receive instructions on selecting key words.

Citations

Chapter 3 offers advice on citing sources. If, as a beginner, you adopt and learn a standard system for your field, you will encounter less difficulty later on. Most scientific disciplines today dispense with footnotes or endnotes almost entirely, replacing them with in-text citations keyed to a reference list. This textbook uses one of the most common styles, the author-date system. For example, the citation "(Macdonald-Ross 1977b)" refers readers to this entry in the References at the end of the book:

>   Macdonald-Ross, M. 1977b. How numbers are shown. *AV Communication Review* 25: 359–409.

To distinguish between the two publications by Macdonald-Ross in 1977, the articles are labeled 1977a and 1977b. For more on the author-date citation style, see Panel 3-10.

For a different in-text citation style keyed to a *numbered* reference list, see the excerpt from a research article published in the *Agronomy Journal* (Panel 13-4). Note, too, how it follows typical scientific article format. The abstract (in boldface type) is followed by the justification and problem statement and a section or materials and methods. The article then concludes with a section giving the study's results (not reprinted here) and a list of the literature cited.

Keep these points in mind as you decide whether to cite:

- Self-evident or generally accepted points ("common knowledge") need no citation. Knowing the difference between the generally accepted and the less well known is part of what makes you an expert.
- If you cite someone's opinion, label it as such in your text. Beginners often cite opinion as fact and thus mislead their readers.

- You may cite several sources in one citation, as long as they all bear on the same point.
- If you are using footnotes or endnotes, avoid long, discursive notes. If your endnote develops into a short essay, decide whether the information should be included in the text itself. If you decide no, drop the essay entirely.

Progress Reports

Even though you do them before the final report, progress reports often contain many of the final report's elements. A middle stage report on the progress of an ongoing project, the progress report's function is twofold: (1) to report findings to date, and (2) to report changes in the project's scope, timetable, or costs.

Most of us have experienced something like this queasy situation. Having chosen a topic she is interested in, having received her professor's permission to go ahead, Alice has run into all kinds of problems: too few sources, important books missing from the library, conflicting conclusions reached by previous researchers. One day she passes her professor in the hall. "Hi, Alice, how's your paper coming?" They both are in a hurry, and other people are there, so she just says, "Oh, fine."

Alice has just offered about as informal a progress report as it's possible to give. For a while, her professor will go on thinking that she is sailing along, and she will continue to feel desperate. What happens later will almost certainly be unpleasant for one or both of them.

The progress report helps people avoid unpleasant surprises. The National Science Foundation (n.d.) requires a progress report that "should briefly summarize activity during the past year [in the case of multiyear grants], identify any significant scientific developments, and describe any problems encountered."

A progress report may be brief, so long as it covers the following key points:

- What engineering, scientific, or other significant developments have you found so far?
- What problems have you encountered, and what solutions have you found or proposed?
- What unforeseen expenses have developed?
- Are you on schedule? Do you propose changes in the completion date?

If your proposal is on file with your sponsor, your progress report need not repeat material covered earlier. In this case, you can draw instead from what will form the methods, results, findings, and conclusions sections of your final report.

PANEL 13-4. Citation Style of Typical Scientific Journal

Mower Blade Sharpness Effects on Turf[1]

D. H. Steinegger, R. C. Shearman, T. P. Riordan, and E. J. Kinbacher[2]

Abstract

This study was undertaken to determine effects of repeated mowing with a dull or sharp rotary mower blade on Kentucky bluegrass (*Poa pratensis* L.) turfs of 'Park' and 'Baron-Glade-Adelphi'. Mower blade sharpness effects on turfgrass quality, leaf spot (*Bipolaris sorokinianum* Shoem.), thatch accumulation, water use rate, and mower fuel consumption were studied. Field experiments were conducted on a Sharpsburg, silty-clay loam (fine, montmorillonitic, mesic Typic Argiudoll) at the Univ. of Nebraska Field Laboratory located near Mead. Turfgrass quality was reduced by dull mower treatment for both Park and the blend. Leafspot incidence increased on Park turfs mowed with the dull mower, but not on the blended turf which was leaf spot resistant. Thatch accumulation was not significantly influenced by mower blade sharpness. Water use rates under field conditions for Park and Baron-Glade-Adelphi turfs were 1.3 and 1.2 times greater, respectively, for turfs mowed with the sharp mower blade than the dull. The reduced water use rate associated with dull mower treatments was positively correlated to reduced shoot density ($r = 0.88$) and verdure ($r = 0.93$). Gasoline use was 22% greater with dull mower blade treatments than with sharp. This study substantiates the hypothesis that repeated mowing with a dull mower blade reduced turfgrass quality and increased disease susceptibility. However, these results refute the generally accepted premise that dull mower blade injury of turfgrass leaf tissue increases turfgrass water use.

Additional index words: Bipolaris sorokinianum Schoem., *Poa pratensis* L., Water use rate.

Mowing height and frequency are known to influence turfgrass quality and performance (1,2,3,6). Proper mower operation and adjustments also influence turfgrass quality. Beard (1) described impaired turfgrass quality resulting from leaf tissue damage occurring after mowing with a dull mower. He suggested that bruised, mutilated leaf tips that occur after mowing with a dull mower blade increased moisture loss and provided a favorable site for pathogen entry into the plant. Little or no information is available in turfgrass literature on the effects of mowing turfs with dull mower blades. This study was initiated to determine the influence of mower blade sharpness on turfgrass quality, water use, disease susceptibility, and thatch accumulation in Kentucky bluegrass (*Poa pratensis* L.) turfs.

Materials and Methods

This study was conducted from April 1978 to September 1981 at the Univ. of Nebraska Turfgrass Research Facility located at Mead, Nebr. A monoculture of 'Park' Kentucky bluegrass was used in the first of two experiments. Park was selected to represent a common type cultivar that was susceptible to leaf spot (*Bipolaris sorokinianum* Shoem.) disease. The second experiment was conducted on a leaf spot resistant blend of 'Baron-Glade-Adelphi' Kentucky bluegrass. Turfs for both experiments were established in September 1976 on a Sharpsburg silty-clay loam (fine, montmorillonitic, mesic Typic Argiudoll). They were fertilized with 20 g N m^{-2}

---

[1] Published as Paper No. 6710 Journal Series. Nebraska Agric. Exp. St., Lincoln, Ne 68583. Received 29 July 1982.

[2] Professor, associate professors, and professor, respectively, Dep. of Horticulture, Univ. of Nebraska, Lincoln, Ne 68583.

> ## PANEL 13-4. (continued)
>
> season^{-1} (45-0-0) and were irrigated to prevent visual drought stress. Herbicides were applied for weed control. No other pesticides were applied during the course of the study.
>
> ### Literature Cited
>
> 1. Beard, J.B. 1973. Turfgrass: science and culture. Prentice-Hall, Inc. Englewood Cliffs, N.J.
> 2. Madison, J.H., Jr. 1960. The mowing of turfgrass. I. The effect of season interval and height of mowing on the growth of seaside bentgrass turf. Agron. J. 52:449–452.
> 3. ———. 1962. Mowing of turfgrass. II. Response of three species of grass. Agron. J. 54:250–252.
> 4. Shearman, R.C., and J.B. Beard. 1973. Environmental and cultural preconditioning effects on the water use rate of *Agrostis palustris* Huds., cultivar Penncross. Crop Sci. 13:424–427.
>
> *(Source: Reproduced from* Agronomy Journal *75 (May-June 1983): pp. 479–480 by permission of the American Society of Agronomy, Inc.)*

PRINCIPLES FOR FORMAT DEVELOPMENT

Examples of effective document design can be found in such everyday forms as blank checks and driver's licenses. The ease with which such documents both elicit and provide information is a measure of their format and design.

An effectively designed blank, such as the one in Panel 13-5, shows how a lot of information can be fitted into small spaces. Notice, too, how the form's spaces were designed with the typewriter in mind. Document design follows similar logic. First decide the purpose, then devise a format to achieve it.

The logical hierarchy of a document's format can guide you. If you design your own format, follow these principles:

- Let audience needs dictate format.
- Elect a format familiar to the audience. You want readers to spend time on content, not trying to decipher your organization.
- Experiment to find an effective format.
- Strive for simplicity.
- Revise freely.

Format must be adjusted both to audience and to the purposes of the report. Format may be prescribed by a publication or funding source. Different formats are used in different fields. A technical report for a business may have its components weighted differently from those in a technical journal. Panel 13-6 briefly illustrates how an engineering trade journal and an engineering research journal weight report parts differently.

PANEL 13-5. Proposal Form of National Science Foundation

APPENDIX I

PROPOSAL TO THE NATIONAL SCIENCE FOUNDATION
Cover Page

FOR CONSIDERATION BY NSF ORGANIZATIONAL UNIT (Indicate the most specific unit known, i.e. program, division, etc.)	IS THIS PROPOSAL BEING SUBMITTED TO ANOTHER FEDERAL AGENCY? Yes ___ No ___ ; IF YES, LIST ACRONYM(S):

PROGRAM ANNOUNCEMENT/SOLICITATION NO.:	CLOSING DATE (IF ANY):

NAME OF SUBMITTING ORGANIZATION TO WHICH AWARD SHOULD BE MADE (INCLUDE BRANCH/CAMPUS/OTHER COMPONENTS)

ADDRESS OF ORGANIZATION (INCLUDE ZIP CODE)

TITLE OF PROPOSED PROJECT

REQUESTED AMOUNT	PROPOSED DURATION	DESIRED STARTING DATE

PI/PD NAME AND SOCIAL SECURITY NO. (SSN)*	PI/PD PHONE NO.

PI/PD DEPARTMENT	PI/PD ORGANIZATION

ADDITIONAL PI/PD AND SSN*	ADDITIONAL PI/PD AND SSN*

ADDITIONAL PI/PD AND SSN*	ADDITIONAL PI/PD AND SSN*

FOR RENEWAL OR CONTINUING AWARD REQUEST, LIST PREVIOUS AWARD NO.:	SUBMITTING ORGANIZATION IS ___ IS NOT ___ A SMALL BUSINESS CONCERN (see CFR Title 13, Part 121 for definitions).

*Submission of social security numbers is voluntary and will not affect the organization's eligibility for an award. However, they are an integral part of the NSF information system and assist in processing the proposal. SSN solicited under NSF Act of 1950, as amended.

CHECK APPROPRIATE BOX(ES) IF THIS PROPOSAL INCLUDES ANY OF THE ITEMS LISTED BELOW:

☐ Animal Welfare	☐ Human Subjects	☐ National Environmental Policy Act
☐ Endangered Species	☐ Marine Mammal Protection	☐ Research Involving Recombinant DNA Molecules
☐ Historical Sites	☐ Pollution Control	☐ Proprietary and Privileged Information

PRINCIPAL INVESTIGATOR/ PROJECT DIRECTOR	AUTHORIZED ORGANIZATIONAL REP.	OTHER ENDORSEMENT (optional)
NAME	NAME	NAME
SIGNATURE	SIGNATURE	SIGNATURE
TITLE	TITLE	TITLE
DATE	DATE	DATE

(Source: National Science Foundation, Grants for Scientific and Engineering Research, NSF 81–79 (Washington, D.C.: U.S. Government Printing Office))

DOCUMENT FORMAT

PANEL 13-5. (continued)

APPENDIX III — **SUMMARY PROPOSAL BUDGET**

ORGANIZATION			FOR NSF USE ONLY		
			PROPOSAL NO.	DURATION (MONTHS)	
				Proposed	Granted
PRINCIPAL INVESTIGATOR/PROJECT DIRECTOR			AWARD NO.		

A. SENIOR PERSONNEL: PI/PD, Co-PI's, Faculty and Other Senior Associates (List each separately with title; A.6. show number in brackets)	NSF FUNDED PERSON-MOS.			FUNDS REQUESTED BY PROPOSER	FUNDS GRANTED BY NSF (IF DIFFERENT)
	CAL.	ACAD	SUMR		
1.				$	$
2.					
3.					
4.					
5. ( ) OTHERS (LIST INDIVIDUALLY ON BUDGET EXPLANATION PAGE)					
6. ( ) TOTAL SENIOR PERSONNEL (1–5)					
B. OTHER PERSONNEL (SHOW NUMBERS IN BRACKETS)					
1. ( ) POST DOCTORAL ASSOCIATES					
2. ( ) OTHER PROFESSIONALS (TECHNICIAN, PROGRAMMER, ETC.)					
3. ( ) GRADUATE STUDENTS					
4. ( ) UNDERGRADUATE STUDENTS					
5. ( ) SECRETARIAL-CLERICAL					
6. ( ) OTHER					
TOTAL SALARIES AND WAGES (A+B)					
C. FRINGE BENEFITS (IF CHARGED AS DIRECT COSTS)					
TOTAL SALARIES, WAGES AND FRINGE BENEFITS (A+B+C)					
D. PERMANENT EQUIPMENT (LIST ITEM AND DOLLAR AMOUNT FOR EACH ITEM EXCEEDING $1,000; ITEMS OVER $10,000 REQUIRE CERTIFICATION)					
TOTAL PERMANENT EQUIPMENT					
E. TRAVEL 1. DOMESTIC (INCL. CANADA AND U.S. POSSESSIONS)					
2. FOREIGN					
F. PARTICIPANT SUPPORT COSTS					
1. STIPENDS $					
2. TRAVEL					
3. SUBSISTENCE					
4. OTHER					
TOTAL PARTICIPANT COSTS					
G. OTHER DIRECT COSTS					
1. MATERIALS AND SUPPLIES					
2. PUBLICATION COSTS/PAGE CHARGES					
3. CONSULTANT SERVICES					
4. COMPUTER (ADPE) SERVICES					
5. SUBCONTRACTS					
6. OTHER					
TOTAL OTHER DIRECT COSTS					
H. TOTAL DIRECT COSTS (A THROUGH G)					
I. INDIRECT COSTS (SPECIFY)					
TOTAL INDIRECT COSTS					
J. TOTAL DIRECT AND INDIRECT COSTS (H + I)					
K. RESIDUAL FUNDS (IF FOR FURTHER SUPPORT OF CURRENT PROJECTS GPM 252 AND 253)					
L. AMOUNT OF THIS REQUEST (J) OR (J MINUS K)				$	$

PI/PD TYPED NAME & SIGNATURE*	DATE	FOR NSF USE ONLY		
		INDIRECT COST RATE VERIFICATION		
INST. REP. TYPED NAME & SIGNATURE*	DATE	Date Checked	Date of Rate Sheet	Initials - DGC
				Program

NSF Form 1030 (10-80) *Supersedes All Previous Editions*   *SIGNATURES REQUIRED ONLY FOR REVISED BUDGET (GPM 233)

PANEL 13-6. Business versus Academic and Scientific Reports

	Technical Report for a Business	Technical Report for a Journal
Abstract	Written for broad readership. Includes commercial objectives.	Written for peers. Highly specialized language.
Introduction	Rationale based on both technical and commercial considerations. Related literature given in brief or placed entirely in appendix.	Survey of related literature.
Procedure	Given in brief. Full account placed in appendix.	Fully detailed.
Results	Given in brief. Full account placed in appendix if necessary.	A crucial section.
Discussion	Combines technical and business aspects. Uses a separate section for recommendations.	Recommendations rare.
Attachments	Appendices give details of related literature, procedure, results.	Reference list obligatory.

(Source: Anne Eisenberg, Effective Technical Communication (New York: McGraw-Hill, 1982), p. 193. © 1982 McGraw-Hill. Reproduced with permission.)

REVISING A BUSINESS PROPOSAL FORMAT

Jim, who wrote the proposal in Panel 13-7, is a mining engineer who knows his coal mining. He works for an international chemical company with 88,000 employees. Jim's proposal is brief, but its format needs improvement. Jim is clearly aware of his audience, Monty, who is interested in one question: "Is the project financially feasible for the company?" Jim has a lot going for him, including expert knowledge, awareness of why he has been sent to survey the coal field, and appreciation for the company's objectives. In other words, he is an expert in his field. However, he might have used format more effectively.

Bearing in mind how readers read, Jim could have used his title or subject heading to phrase the key question: "Should the company pursue the project?" He then could have presented his recommendation so that Monty would not have been in suspense throughout the memo. His new format:

> Question—go or no go?
> Recommendation—no go.
> Reasons why not—six, as listed.
> Counterargument?—listed, so supervisor will know.

Panel 13-8 contains our revision of his report. The writing is the same as in the first version, but the improved format has clarified the message.

PANEL 13-7. Business Proposal

To: Monty   Date: January 20, 1987

From: Jim   cc: Fred
                                                        Mary

Subject: Coos Bay Project, Oregon

    Coal in the Coos Bay Field has been mined at intervals since 1855. The coal bearing units are a part of the Clarno Formation of Eocene Age.

    The only minable coal bed found within the study area was the Beaver Hill bed. It ranks as Subbituminous B, has an average BTU of 9,516. It ranges in thickness from 4'6" to 7'5" and has average moisture of 16.6%, with an average ash of 10.1%. The thin coal beds overlying the Beaver Hill bed were found to have lower BTU values (8,159—9,000) and are considered economically unminable.

    The structure of the Coos Bay Field consists of an elliptical synclinal basin. This basin is approximately 30 miles long and 11 miles wide, with a trend of approximately north—south. The Beaver Hill bed dips anywhere from 10° to 40° within the synclinal basin. Mining the Beaver Hill bed in the past has been underground, using both drift and slope methods. The Beaver Hill bed is found at depths anywhere from surface outcrops to 1,500 ft. within the basin.

    The only advantage the Coos Bay Field has is its proximity to the Japanese market. The railroad runs adjacent to the property, and the coal could be railed to Portland where it could be shipped directly to the Japanese market.

    There are disadvantages to the Coos Bay Field which make it unappealing for the company to further pursue this prospect. There is only one low grade seam (9,516 BTU), with an average thickness of 5—6 feet, which is feasibly minable. The Beaver Hill bed is found to have a relatively high moisture content and a moderate percentage of ash. I feel that mining within this synclinal basin could cause water problems, given the sloughs and swampy nature of the surrounding topography. Mining within a syncline also has a tendency for roof problems. Caving becomes more of a problem when mining in a syncline compared to an anticline.

    From these given disadvantages, I feel that the company should not further pursue the Coos Bay prospect at this time.

PANEL 13-8. Revised Business Proposal

To:　　　　　Monty　　　　　　　　　　　　Date: January 20, 1987

From:　　　　Jim　　　　　　　　　　　　　cc:　　Fred
　　　　　　　　　　　　　　　　　　　　　　　　　Mary

Subject:　　Should the Company Pursue the Coos Bay, Oregon,
　　　　　　Prospect?

Recommendation: Not at this time.

Reasons: The Beaver Hill bed is the only minable bed in the Coos Bay Field, which is an elliptical synclinal basin, 30 miles long and 11 miles wide, running north—south. The Beaver Hill bed dips from 10° to 40° within the basin and is found anywhere from surface outcrops to 1,500 feet. Its thickness ranges from 4'6" to 7'5". A number of disadvantages exist:
1. The seam is low grade, Subbituminous B, with 9,516 average BTU.
2. The Beaver Hill bed has a high moisture content, averaging 16.6%.
3. The ash content is moderate, averaging 10.1%.
4. Sloughs and the swampy nature of the topography might cause water problems.
5. Mining within a syncline tends to encounter roof problems, and caving becomes more of a problem.
6. The thin coal bed overlying the Beaver Hill bed has lower BTU values, 8,150 to 9,000, and is economically unminable.

Coos Bay coal has been mined at intervals since 1855. Beaver Hill coal has been mined underground by both slope and drift methods. The coal-bearing units are part of the Eocene Clarno Foundation. Coos Bay's only current advantage is proximity to the Japanese market. The railroad runs adjacent to the property, and the coal could be railed to Portland and shipped directly to Japan.

HIGHLIGHTS

1. As you make the transition from school to on the job writing, you will find format playing an increasingly important role.
2. People selectively read technical and scientific material according to their needs and interests. Effective formatting accommodates such reading habits.
3. Components of technical reports include: title, abstract, problem statement, justification, methods, results, conclusions, citations, among others. If you understand the purpose each serves, you can understand how and why components can be combined.
4. Properly organized material is arranged so that readers can extract information according to their needs.

PROJECTS

1. Your professor may collect several proposal forms from your university's office of sponsored research. Examine several, one from the National Science Foundation, one from the National Endowment for the Humanities, and one from an industrial firm soliciting research, or other, similar organizations. Compare and contrast the problem statement section of each. Format your results effectively.
2. At your library, find two or three technical reports or research journal articles from your field, either basic research or applied research. Critique the problem statements in each. How specific are they? Can you suggest ways of improving them?
3. Follow up Project 2 by evaluating the results sections of the technical reports. Do you find authors generalizing beyond either their problem statements or the evidence they present in their results sections?
4. For the same or a similar collection of research reports as in Projects 2 and 3, examine the abstracts. Which seem to be summaries and which abstracts? Were you able to find a "summary abstract" such as the National Science Foundation asks for? Did the writers construct their abstracts from key sentences of the important sections of their reports, or did they use some other method?
5. Your professor may have previously written student proposals, reports, or published reports. Choose one, and critique it in detail (as Panel 13-2 critiques the student proposal). Spot ways in which the document could be improved.
6. As a variation on Project 5, critique a classmate's proposal or report for this class.

FOR MORE HELP

Pinelli, T. E., V. M. Cordle, M. Glassman, and R. F. Vondran, Jr. 1984. Report format preferences of technical managers and nonmanagers. *Technical Communication* 31(2): 4–8.

In one of the more detailed studies in recent years, the authors report on the reading habits of managers and nonmanagers.

——, et al. 1984. Report-reading patterns of technical managers and nonmanagers. *Technical Communication* 31(3): 20–24.

From the same data as in the article above, the authors report research results on reading patterns.

Souther, J. W., and M. L. White. 1977. *Technical report writing*. 2nd ed. New York: Wiley.

The text includes results of Souther's classic Westinghouse study.

14
Word Processing and Printing

OVERVIEW

In Chapter 14, we
- ☐ Discuss word processing components
- ☐ Suggest how to work with word processing operators
- ☐ Suggest points to consider when you do word processing
- ☐ Explain why professionals need to understand the printing process
- ☐ Introduce the basic steps in printing

By the mid-1980s, using computers had become an everyday activity for many technical and scientific professionals. From scientists at the Lawrence Livermore Laboratory in California, to Kansas beef farmers, to technicians at Bell Laboratories in New Jersey, computers had become an integral, important tool for communicating about technical and scientific subjects.

In his newspaper column, "Trend Notes" (1985), John Naisbitt, author of *Megatrends,* reported that an estimated 12 million Americans worked on video display terminals—computers—in 1985, and he projected that 50 million people would be working on them by the year 2000.

Computers are used primarily for performing calculations and word processing. Increasing numbers of professionals use computers to prepare illustrations. If you're not already using a computer to prepare written assignments, you may safely assume that one is in your near future. The following discussion seeks to make your transition to computers easier. If you already use one, it suggests ways of using it more effectively.

WORD PROCESSING

As universities buy the units and software, more and more students across the country are working on computers. At some universities, computers are required of all students; at others, computers are assigned as students arrive on campus. By 1983 more than 1,000 Colorado State University students each semester, who were enrolled in composition classes, prepared their assignments on word processors. They used a modified version of Bell Laboratories' Writer's Workbench. Professors Kathleen Kiefer and Charles Smith worked with Bell Laboratories to develop a version of Writer's Workbench for classroom use.

Once students enter a draft of their papers in the computer, the Bell programs check for spelling errors, active/passive voice, organization, diction, specific terms, punctuation, grammar, style, abstract terms, and other elements of writing. The students then revise their papers and run a clean copy for their professors.

Using word processors entails learning to use the hardware—the computer and other tangible objects—as well as the software—the instructions that drive the hardware. The hardware usually consists of the keyboard, computer, video display terminal or screen, disk drives, soft and/or hard disks, printer, and, possibly, modem. Software consists of the instructions—called programs—that format pages; set margins and tabs; indent paragraphs; add page numbers; justify margins; delete words, sentences, and paragraphs; and insert and move words, sentences, and paragraphs. Other software programs check spelling, grammar, style, and other writing components.

With each passing month, more sophisticated systems and programs are announced and advertised, so it becomes difficult to keep abreast of all the changes. The following discussion considers word processing programs and their fundamentals, not specific key strokes and functions.

Dozens of professionals whom we talked with as we compiled this book—technical, scientific, and writing—strongly endorsed the switch to word processing as an aid to writing. They pointed out that word processing can improve the appearance of manuscripts, lessen apprehension about making changes, and provide clean copies without the necessity of retyping.

Working with a Word Processing Operator

As a professional, you or an operator may enter drafts into a computer. For some operators, initially entering copy into a word processor sometimes proves slower than typing. But once material is entered, making changes and printing clean copies are quicker and easier. After initial entry, correcting a moderately revised, 100-page manuscript may take little more than two hours.

Word processing operators may average six or so pages an hour, typists 12 or more pages an hour. Two factors slow word processing operators. First, they must type commands to manipulate the copy. Second, they may suffer more eyestrain, fatigue, and stress than typists. Therefore, many word processing operators take a 5- to 15-minute break every hour and a 30-minute break after three to four hours of working.

Some professionals argue that not everything should be entered on word processing units. For instance, Nancy Wilson (1985), word processing coordinator at Colorado State University, suggests using a word processor for technical reports, bulletins, handouts, journal articles, proposals, resumes, and documents that require revision. She discourages using word processing for one-of-a-kind correspondence or for filling out forms. Lionel Munn (1982), a Canadian member of the Society for Technical Communication, suggests not giving word processing operators manuscripts needing more than 40 percent revision.

In contrast, dedicated word processing users argue for entering everything on a word processing unit because of the ease of making changes, checking spelling, and correcting mistakes. For a small office, a word processor and letter quality printer can more than pay for themselves in saved secretarial costs.

As a student and as a professional, you'll probably find yourself working with word processing operators. When you do, try not to complicate their work. For initial entries, you can help yourself and the operators by

- Asking for their guidelines on manuscript preparation.
- Following those guidelines.
- Typing, if possible, the original draft. A typed manuscript is much easier to read than a handwritten manuscript.
- Writing legibly in ink, if you provide a handwritten manuscript.
- Making a photocopy of your draft before you submit it.
- Using standard copyediting symbols (see Panel 10-3).
- Not writing in margins or on page backs.
- Cutting and pasting only originals that have *not* been entered on a word processor.

When revising and copyediting a manuscript that has been entered in a word processor unit:

- Use standard copyediting symbols (Panel 10-3).
- Use a bright red or other distinctly colored ink when copyediting, so that an operator can scan the manuscript rather than read it to find changes.
- Don't write changes vertically in margins or on page backs.
- Use insertion notes. Note the insertions in the manuscript and refer to a second page on which you've typed the insertions. Label inserts carefully: Insert A, page 20; Insert B, page 20, and so forth.
- Mark through lines, paragraphs, and pages when you want to delete them, but leave them in the original manuscript. Don't cut them out and make the word processing operator hunt through several pages for your changes.

Doing Your Own Word Processing

At this point in the information revolution, all returns are not in on the value of the personal computer in word processing. It seems as if two distinct groups of people have emerged: the "pros," who have computers at home and wear off their fingerprints playing with them, and the "antis," who pretend that personal computers are just another fad.

Writing will always be a personal task. Use whichever method gets you results. William Zinsser, the confirmed typewriter addict, now endorses using the word processor:

> The word processor will help you achieve three cardinal goals of good writing—clarity, simplicity, and humanity—if you make it your servant and not your master. Remember that it's only a machine, so don't be afraid of it. You'll even learn to like it. Take it from an American boy who always hated machines. (Zinsser 1983, 112)

Not only can a word processing system free you of the drudgery of retyping entire drafts, but it enables you to revise and improve your drafts easily. Word processing systems have: standard and adjustable formats, block moves, spelling and grammar checkers, functions that delete and add, search and replace, and so forth.

Consider the help word processing gives in formatting a manuscript. You need not worry about spacing, alignment, margins, hyphenation, or certain other details as you type each page. You let the software program do it for you. Some programs have fixed page layouts—formats with specified top, bottom, and side margins, page number location, hyphenation, line spacing, and many more features. Sophisticated programs come with standard settings, but you can easily change them.

To format with most word processing programs, you must first install the software that tells the computer and printer which program you'll be using. Instruction manuals for the computer, printer, and software provide the specifics. In most cases, you'll find yourself experimenting a bit until you have the page format the way you want it. But once you have your page format standardized for the system, it takes over and automatically formats each page, inserts the correct page numbers, and prints out your manuscript when instructed to do so.

Among the other word processing features, block moves can help you easily recast and revise your drafts. With most sophisticated word processing programs, you can move words, sentences, paragraphs, and whole pages from place to place. Each program functions a bit differently, but the end result is the same. Consider a general description of block moves using WordPerfect Version 4.1 (SSI Software 1985). To move a paragraph, you mark its beginning by striking the block identification keys, move a cursor through the paragraph (which is highlighted on the screen), and then strike the move keys. The computer asks if you want to "append," "copy," or "cut" the highlighted block. If you select "cut" and press the appropriate keys, the computer automatically removes the paragraph, closes up the space between the two remaining paragraphs, and stores the cut paragraph in its temporary memory. If you select "copy" and want to move the paragraph to a new location, you move the cursor to where you want the paragraph inserted, push the move keys again, and push the retrieve keys. The paragraph appears on the screen. Once you practice a couple of times, moving paragraphs is quite easy. With writing and revising made so easy, you can experiment with different organizations for the one that best fits your point. And you can do so without worrying about constant retyping.

Word processing programs also make the deletion and addition of words, sentences, paragraphs, and pages relatively easy. Sophisticated systems allow you to delete a letter, word, part of a line, whole line, or page. Most programs call the addition of copy the "insert" function. With that function, you can begin writing in the middle of a sentence or page, and the computer automatically pushes the old copy down and makes room for the new copy.

Still another useful function is the "search" function. The computer can automatically search a file for any letter, combination of letters, or words that you specify. In drafting the manuscript for this book, we moved chapters around, deleted others, and added new chapters. Our task of changing chapter numbers, panel numbers, and figure numbers was made easier with the software's search and replace function. For example, we needed to replace the number "13" with the number "14." To use the search and replace function, we moved the cursor to the beginning of a chapter, selected the replace function, entered "13," and then instructed the computer to scan the manuscript. When it came to a "13," it would stop and ask us to confirm the change. We'd respond, and the process continued until we finished revising the chapter—all in five minutes.

Software companies market spelling and grammar checkers. Each program functions a bit differently, but the end result of each is the same—to flag or mark possible changes. The computer flags possibly misspelled words, and then you make the final decision to correct the word. In some programs, you look up words; others provide a list of similarly spelled words. Although such programs can help you correct mistakes, they aren't foolproof. The sizes of the programs' dictionaries vary widely. Some programs have only 20,000 words; others have 80,000 or more. A further limitation is that most spelling programs can't distinguish between correctly spelled, but misused, homonyms—words pronounced similarly but spelled differently, such as *to, too, two*.

Computers can help enormously, but they cannot do the final checking of a manuscript. They cannot think for you.

When you try word processing, what can you expect? We've observed that:

- Being skilled in touch typing makes word processing easier. If you don't touch type, buy a software program that teaches touch typing, and practice until you've developed touch typing skills. You'll soon be on your way to composing at the keyboard.
- Learning to use a word processor effectively takes time. Don't wait until a few days before a deadline to learn a system.
- Encountering difficulties frustrates many users. So remember, it's only a machine.
- Overcoming fear of machines, if you suffer from it, takes time.

When you start using a word processing unit, you can make your life easier by

- Taking classes and seminars, if possible, to speed your learning.
- Expecting some problems in the software, hardware, or manuals.
- Seeking help from someone experienced with your system if you can't solve a problem within 10 minutes.
- Learning how to transfer—saving—what you've written from the computer's temporary memory to disk. Many beginners write for hours and then discover that they have sent their prose straight into the corridors of time.
- Checking the available space on disks before you begin using them.
- Checking the disk as you write to see how much more copy it can hold.
- Learning how to save your work if you receive a "disk full" message.
- *Learning how to make backup files and duplicate disks.* (If you don't, the day will come when you wish you had!)
- Learning how to recall information if you make an error.
- Learning how to search the computer's memory and disks should you lose what you've been writing. You probably won't find it, but you can try anyway. You might be successful.
- Keeping disks in their protective jackets when they are not in the machine.
- Avoiding bending, folding, or flexing disks.
- Inserting and removing disks carefully from the computer's disk drives.
- Keeping your fingers, dirt, cigarette ashes, soft drinks, and other gremlins away from disks and computer.
- Keeping disks away from telephones, equipment, and magnetic fields that might scramble the information on them.
- Copyediting your hard copy—the printed version—for errors. You'll catch errors you missed on the screen.
- Reducing eyestrain and fatigue by taking breaks (Panel 14-1).

PANEL 14-1. Reducing VDT Dangers

Carol enthusiastically logged in three- to four-hour stretches at her computer. Within days, she began having severe headaches. After having an eye examination, after rearranging her word processor, changing the contrast of the video display terminal's (VDT) screen, and varying her work habits, her problem quickly disappeared.

The problems of word processor users may be minor neckaches, backaches, and stress or chest pains and other major problems. Experts don't agree on the dangers. In the mid-1980s, no definitive answers have emerged on VDTs' potential dangers to users. No nationwide safety standards have been mandated for all users.

What can you do to safeguard your health? Be cautious. Consider the suggestions of a columnist for *PC Week,* Jim Seymour:

- Glance away from the screen every few minutes and focus on a wall, person, or something at a distance.
- Take a visual break at least once an hour—leave your terminal, walk around, or do something other than looking at the screen.
- Have your eyes examined every 6 months.
- Buy single-vision glasses ground especially for the working distance of your computer when eye exams show your eyes need help. (Seymour 1985, 26)*

By 1984, two office workers' organizations were seeking legal requirements for companies with VDT workers. The requirements included:

- Heat and radiation shielding
- Adjustable desks, chairs, and screens
- Devices to prevent glare from office lights; glare makes screens difficult to view
- Mandatory annual eye examinations
- Flexible breaks, with a minimum of 15 minutes every two hours
- An option for pregnant workers to do non-VDT jobs with no loss in pay. (Hendon 1984, 9-B)†

We add a guideline from optometrists: keep your distance. Stay at least an extended arm's length from your VDT screen.

*Source: The Corporate Video: VDT Precautions Pay Large Return for Small Investment. PC Week 2 (15):26. Reprinted from PC Week, April 16, 1985. Copyright © 1985 Ziff-Davis Publishing Company.

†Source: Random Access: Possible VDT Harm Still Being Debated, Rocky Mountain News, *December 15, 1984, 9-B.* Reprinted by permission of the Rocky Mountain News.

As you learn a system, remember that someone designed the software and hardware to work in a particular way. Your task is to learn that particular way. Keep your wits about you, go slowly at first, and understand the fundamentals.

Besides their many advantages for writing, personal computers provide another advantage for technical and scientific professionals—illustrations such as line, bar, and circle graphs, and line art.

Using Computer Graphics

Computer graphics have become more and more common in technical and scientific fields. As briefly discussed in Chapter 8, a variety of software graphics programs are being marketed.

A *Graphics* edition of *PC Week* (November 19, 1985) defined four kinds of graphics software:

Business graphics programs: programs that automatically plot a graph from a worksheet or plot a graph from a file saved from a spreadsheet package.
Freehand graphics programs: programs that allow the user to draw color pictures on the screen.
Paint programs: allow the user to assemble freehand pictures using different color shades and crosshatchings.
Draw programs: by which the user can manipulate individual graphics, figures, and lines to produce a chart or sign. (*Graphics* 1985, S6)

With such programs you can prepare bar charts, signs, organizational charts, line graphs, circle charts, 3-dimensional drawings, time series charts, and other illustrations. The illustrations can be produced on paper, on acetate for overhead transparencies, or on 35mm slides.

You need the right equipment and software to produce illustrations with a computer. To run some graphics software programs, computers require more internal memory than just for word processing, a graphics card, color monitor, light pen, "mouse," digitizers, and a special printer. To produce colored copies and overhead transparencies requires a plotter equipped with color pens and output. In contrast, some software programs for producing simple illustrations can print them on dot matrix or laser printers.

If you have no artistic skills, graphics programs can enable you to prepare a variety of illustrations. Programs usually have a library of symbols or standard illustrations. You select one kind of illustration and then enhance it with other components, data, and words. Like word processing software, graphics software requires time to learn. But once learned, proponents argue that you can produce illustrations less expensively and more quickly than an artist working freehand can.

When you use graphics programs, you need to consider the final form in which the illustration will appear. If it will be a black and white report, don't spend time preparing color illustrations. If the final form will be a high quality publication, you may need to have some illustrations drawn by a professional artist. As time goes on, however, advances in computer systems will improve the quality of illustrations they can prepare.

Not only do computers help with word processing and preparing graphs, but they're making printing and publishing easier. Instead of submitting a bulky manuscript to a printer, you can send copies of disks. You can even use a modem (a telephone connection) between your computer and a printer's typesetter.

THE PRINTING PROCESS

Startling developments in printing have occurred during the past dozen years. Chief among them are computer applications for preparing and distributing copy and for producing finished work. Whichever method you use, the guiding principle remains the same: only humans can prevent errors in the final product.

By understanding the basic printing process and paying close attention to the details, you can:

- Reduce the number and frequency of errors
- Improve meetings and communications with editors, artists, and printers
- Reduce misunderstandings
- Minimize delays and save time
- Avoid last minute rushes
- Meet deadlines
- Save money

The Basic Steps in the Printing Process

Producing a typeset, printed publication from conception through distribution involves careful attention to dozens of details. To illustrate the process, let's assume that you're a product engineer who has designed a lawn mower so that the average home owner can make modular repairs. It's your job to write the product repair manual.

Some 18 months before the assembly line starts up, you talk with Nancy, the company's editor for lawn and garden equipment. She gives you an outline of the typical editorial and production process (Figure 14-1) and a list of printing terms (Panel 14-2).

A month later, you finish your draft manuscript and make five copies. One goes to your boss for review, a second to a fellow engineer, a third to the company's head repair technician, a fourth to the company attorney for a product liability check, and the fifth goes into your files. Two weeks later all reviewers have returned their comments. You incorporate their comments, and now you're ready to submit the manuscript.

You make an appointment with Nancy. For the meeting you bring your manuscript along with a list of needed photographs and illustrations. Nancy assigns an illustrator, Pete, to prepare black and white line drawings that will be reduced to fit the company's standard manual format. It calls for two 18 pica columns, printed 54 lines deep on 8.5 by 11 inch pages.

Nancy explains that the company prints all manuals in 10 on 12 point English Times (name of typeface) body copy, 18 picas wide, with ragged right margins. Chapter heads are set as 24 on 30 point English Times Bold, and subheads are set as 18 on 20 point English Times Bold. She explains that most readers find reading serif typefaces easier than reading sans serif typefaces, ragged right margins easier than reading justified margins. She cautions you not to use body copy of all capital, boldface type, or italic type, because readers have difficulty reading such copy.

Nancy begins editing the manuscript the next afternoon. As she works, she tightens and clarifies your writing and checks for grammar and style.

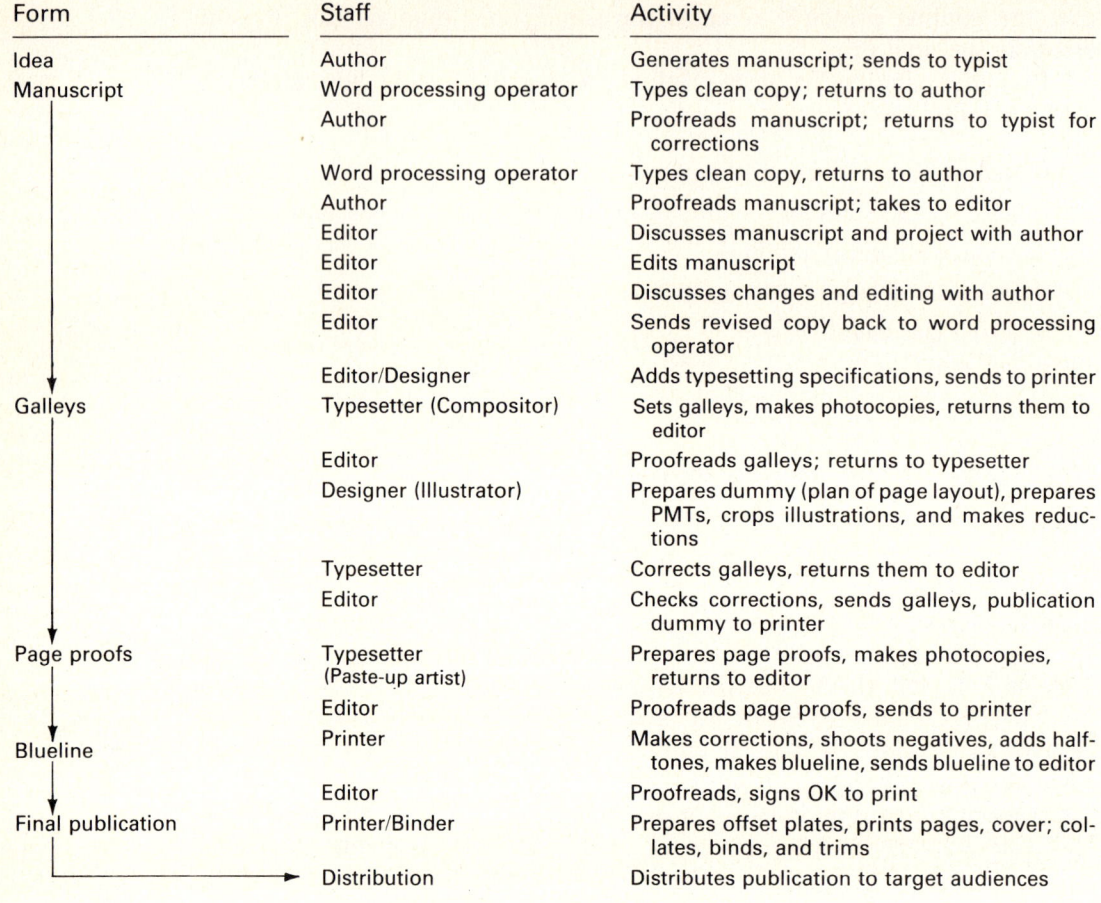

FIGURE 14-1. The editorial and production process.

She's alert to possible content errors. She's knowledgeable about both engine repairs and the requirements of repair manuals. Nancy calls Jackie, an industrial photographer, and arranges for photographs. Jackie provides high quality photographs quickly.

Once she's finished editing, Nancy calls you to clarify several confusing paragraphs and to approve her changes. Then Nancy sends the manuscript to the word processing center for corrections. When a clean manuscript returns, Nancy proofs it and finds only two typographical errors. The word processing operator corrects them and makes copies of the disks.

Before she heads to the company's printing office, Nancy adds the typesetting specifications to the manuscript. When she delivers the manuscript and disks to the printer, she orders 5,000 copies of the manual.

A typesetter transfers the information from the word processing disk to the typesetting computer, which is like a word processing unit. The

PANEL 14-2. Printing Terms

In explaining the printing and editorial processes, we use several printing terms. Here is an alphabetical list:

Blueline: The final proof in a publication's typesetting and layout checked by editor and printer for errors. Its name comes from its blue color.

Body Copy: The type used in the narrative, or body, of a manuscript. Usually body copy is 8, 9, 10, or 12 point type.

Column Width: The width in picas of a column of type.

Cropping: The marking of illustrations to indicate which portion is to be printed.

Display Type: Large type used for headlines, usually 14 points or larger.

Dummy: A mock-up of type, illustrations, and photographs.

Galley: A copy of type set in column widths; used for proofreading and dummying.

Justified: The spacing of lines of type so that right-hand margins are aligned within columns, not ragged.

Leading: The spacing between lines of type. At one time, strips of lead were inserted between lines of type; in photocomposition, leading is added electronically.

Line Art: An illustration containing black, white, and no gray or middle tones; for example, a drawing outlining a part of a mechanism.

Paste-up: The layout showing galley type, PMTs, and windows for photographs pasted in place.

Pica: A printer's unit of measurement. Six picas equal an inch; one pica equals 12 points.

PMT: Photomechanical transfer. A process and the product produced, they enable illustrator and printer to enlarge and reduce line copy or illustrations.

Point: A printer's unit of measurement, it usually refers to type height: 72 points to an inch, 12 points to a pica.

Offset Printing: A popular printing method that uses ink, water, and metal plates. Ink images go from metal plate, to rubber roller, to paper.

Ragged Right: The spacing of lines of type so that the right-hand margins are not aligned.

Sans Serif: Without serifs, or strokes on the ends of letters; a classification of typefaces.

Serifs: The small strokes on the ends of letters in certain typefaces.

Thumbnails: Small artists' sketches of a publication's proposed layout.

Typeface: The style of type used in a publication. Several thousand styles are available. Fortunately, most printers have limited selections.

Typesetter: The individual who operates the photocomposition machine and provides the galleys.

x-height: The height of the lower case x in any given typeface; it usually equals one half the point size of that typeface.

typesetter adds the typesetting specifications, produces typeset galleys, and makes photocopies of them for Nancy.

Nancy asks her editorial assistants to proofread the galleys (Panel 14-3). (Some editorial offices ask authors to proofread galleys, too.) It's the last practical time for changes. When Nancy asks authors to proofread galleys, she reminds them to avoid major changes or rewriting. Too many changes require resetting galleys and boost printing costs. Nancy asks authors to note inaccuracies, typographical errors, and related problems of appearance. Nancy then makes the final proofreading corrections and returns galleys to the typesetter for corrections.

By this time, Pete has prepared the illustrations and the photographer has completed shooting photographs and has provided proof sheets. Pete and Nancy have reviewed the proof sheets and selected negatives from which the photographer prints final pictures. Pete prefers 8- by 10-inch glossy black and white prints, because they produce the best printed results.

Nancy has given a second set of galleys to Pete, who prepares the dummy layout. The dummy shows the paste-up artist how to prepare camera-ready pages. Dummy layouts indicate where copy, illustrations, page numbers, running heads, and photographs go on each page. Pete crops illustrations and indicates the reductions needed to make them fit column widths. He labels the illustrations and clearly identifies them on the layout to prevent errors in placement.

Dummy layouts, photographs, line art, and galleys go to the printer. A paste-up artist waxes the galleys' backs, trims them, and places them on the layout pages. The pages, made of a specially coated paper, allow the artist to lift and realign the galleys without tearing them. Next, the artist prepares photomechanical transfers of the illustrations. The artist reduces the illustrations to the desired column width and size. The artist waxes the PMTs and pastes them up. For each photograph, the artist pastes down a piece of red acetate trimmed to the photograph's size (Panel 14-4).

Afterwards, the artist makes a photocopy of each page for proofreading. Nancy or her assistants proofread the manuscript again, checking page numbers and sequencing, location of illustrations, copy breaks, dropped copy, and other details. They return the page proofs to the paste-up artist, who makes needed corrections. The printer shoots large negatives of the pages and strips the halftones into the clear windows created by the red acetate on the paste-up pages.

From the large negatives, the printer prepares the blueline. It appears exactly as the instructional manual will appear, but it is printed in light blue on yellowish paper. The printer sends the blueline to Nancy for a final check. She and her assistants carefully check the blueline; it's their last chance to correct any errors. But making changes now will prove costly; it requires not only resetting galleys, illustrations, and halftones, and redoing the paste-ups, but also reshooting the page negatives. The

PANEL 14-3. Proofreading

Once your copy has been typeset, you may be asked to proofread photocopies of your galleys. Galleys are typeset columns, with 2- to 4-inch margins, in which you mark your changes. Mark only photocopies, and use a colored pencil. Some proofreading symbols are identical to copyediting symbols (see Chapter 10, Panel 10-3). Proofreading requires two marks for each change, one on or above the line of type and the second in the margin. The marginal mark alerts the typesetter to the error. Proofreading works best if one person reads the original manuscript and a second person marks errors on galleys. Another method is to read the original manuscript into a tape recorder, play the tape back, and mark errors on galleys.

Proofreading Marks

Marginal Mark	Explanation	How indicated in the Copy
1. Marks of Instruction		
ℐ	Delete, take out	He sent the coppy.
stet.	Let it stand	He sent the copy.
sp.	Spell out	He sent ②copies.
tr.	Transpose	He the sent copy.
¶	Paragraph	read. He sent the copy.
no ¶	No paragraph—run in	read.
(sent/?)	Query to the author	He the copy.
2. Marks Regarding Type Style		
ital.	Set in *italic*	He sent the copy.
s.c.	Set in SMALL CAPITALS	He sent the copy.
caps.	Set in CAPITALS	he sent the copy.
c.+s.c.	Set in CAPITALS AND SMALL CAPITALS	he sent the copy.
b.f. or bf.	Set in boldface	He sent the copy.
rom.	Change from *italic* to roman	He sent the *copy*.
l.c.	Set in lower case	He Sent the Copy.
u.+l.c. or c.+l.c.	Set in Upper and Lower Case	he sent the copy.
3. Marks Regarding Defects in the Type		
⊗	Broken letters	He sent the copy.
w.f.	Wrong font	He sent the copy.
ᘐ	Turn inverted letters	He sent the copy.
∥	Straighten marked lines	He sent the copy. He sent the copy.
=	Out of alignment—straighten	He sent the copy.
⚓	Push down space	He sent the copy.

Marginal Mark	Explanation	How indicated in the Copy
4. Marks Regarding Spacing		
⌒	Close up	He s ent the copy.
⌒̸	Delete and close up	He se nt the copy.
#	Insert space	He sent thecopy.
eq.#	Equalize the spacing	He sent the copy.
□	Indent one em	□He sent the copy.
☐☐	Indent two ems	☐☐He sent the copy.
☐☐☐	Indent three ems	☐☐☐He sent the copy.
[	Move left as indicated	[He sent the copy.
]	Move right as indicated	He sent the copy.]
⎴	Raise as indicated	He sent the copy.
⎵	Lower as indicated	He sent the copy.
ld.>	Insert lead between lines	
5. Marks of Punctuation		
⊙	Period	He sent the copy⌄
⌃	Comma	He sent the copy⌄
;/	Semicolon	He sent the copy⌄
:/	Colon	He sent the copy thus
=/	Hyphen	He was copyediting.
⌄	Apostrophe	It is the authors copy.
⌄⌄	Open and close quotes	He said: Send the copy⌄
!/	Exclamation	Send the copy⌄
?/	Question mark	Did he send the copy⌄
⁄⁄ or en	One-en dash	Copy, 17, 34, 227–8
⁄⁄ or em	One-em dash	We went—with copy.
(/)	Open and close parens	It the copy was sent.
[/]	Open and close brackets	He send sic the copy.
∧	Inferior figure	H2O
∨	Superior figure	xy2

(Source: Reproduced by permission from *The Random House College Dictionary, Revised Edition,* Copyright © 1984 by Random House, Inc.)

PANEL 14-4. How Photographs Are Printed

When you submit a black and white photograph for publication, the editor marks—or crops—the photograph, showing the area to be printed. Next, the editor requests the printer reduce the cropped area to fit the column width of the printed page.

To produce that image, the printer creates a halftone, an image with gray (or, less usually, other color) tones halfway between pure white and pure black. To [do this], the printer adjusts the process camera to produce a negative to fit the column width and superimposes a screen over the photograph to break the image into fine dots.

Printers rate screens in lines per inch. Newspaper screens are 65 vertical lines and 65 horizontal lines per inch. Scientific and technical publications often use screens of 120, 175, 200, and 300 lines per inch. As the number of lines increases, dots become smaller, their numbers increase, and quality of image improves.

In offset printing, the printer then strips the halftone negative into the page negative of the typeset copy. Using the page negatives, the printer produces the plates for offset printing. Although other printing processes use slightly different techniques to produce halftones, they all produce images of small dots that give the impression of intermediate tones.

A halftone illustration showing a range of shades. The section outlined in white has been enlarged to show the dot pattern (right). In the shadow areas the dots print as solid black. In the highlight areas the dots drop out completely leaving the white of the paper. In all other areas the dots combine to make shades of gray. For a detailed discussion of halftones, see Allen (1977).

printer provides an approval sheet on which Nancy must acknowledge an error-free blueline.

The printer prepares aluminum offset plates, attaches them to the printing press, prints them, folds, and collates the manual. Next, the printer adds covers, binds, and trims the manuals. The manuals move to the assembly line for packaging with the lawn mowers.

Many publications follow a printing process similar to the one we've just sketched. Details vary, but it's always a many step process with many opportunities to check for errors.

THE CONTINUING COMMUNICATION REVOLUTION

Changes in computers are reshaping writing and printing. More and more typesetting and page negatives are done from word processing disks, eliminating the proofreading of galleys and the pasting up of pages. Software will allow more and more artists or editors to compose pages directly at the computer. Such systems, of course, will require careful editing on the screen.

Electronic publishing and typesetting will become common in many businesses in the next decade. Companies will use electronic, in-house publishing systems to produce reports, sales brochures, price lists, technical documents, and other professional communications. On December 10, 1985, *PC Week* reported that the relatively low cost of computer-based typesetting and publishing will have lured most of corporate America by 1987 (Greitzer 1985). Laser printers will produce both graphics and typed copy in the standard typefaces that once were restricted to printers. Laser printers are fast; they produce pages per minute rather than characters per second.

The ease of operating word processors, the clean copy they produce, and fast printing techniques will encourage many professionals to publish their own manuscripts. But self-publishing carries hidden dangers. Professionals who self-publish will have to review their content carefully, consider their audience carefully, and edit tightly. A word processor, a fast printing job, and a distribution list do not ensure effective communication. Only professionals who are aware of techniques for effective writing can communicate effectively.

HIGHLIGHTS

1. When working with word processing operators, find out how they'd like your manuscript prepared.
2. Learning to use word processing and graphics programs takes time, and you'll probably encounter some problems.

3. Understanding the printing process improves communication with staff, reduces misunderstandings, minimizes delays, saves time and money, and helps meet deadlines.
4. The basic printing process begins as the author generates copy. Then the word processing operator types a clean manuscript for the editor to copyedit and mark for the printer. The typesetter produces galleys and returns them to the editor for proofreading. After proofreading, the editor returns a marked set of galleys to the printer and another set to the illustrator who prepares the dummy layout—a plan for the pasting up of pages. The pasted up pages are photocopied. The photocopies go to the editor for proofreading. After proofreading and after corrections are made, the printer shoots negatives of the page paste-ups, strips in the halftones, and prepares the blueline. It goes to the editor for a final proofreading. The printer then prepares the offset plates and prints the publication.

PROJECTS

1. What word processing facilities does your campus have for students to use? Does your campus have any word processing software for students? If so, which programs does it include? Prepare a 500-word description of the hardware and software.
2. Have you operated a personal computer with word processing software? If so, write a 500-word report describing your experiences in learning the system. What difficulties did you encounter? What factors created these difficulties? How did you overcome them?
3. Does your campus have a publications office? If so, ask a staff member to come to class and explain what that office does. Specifically, ask the staff member to explain: (1) what kinds of publications the office publishes, (2) the processes the office follows in seeing a publication through final printing, and (3) suggestions to professionals who are publishing a manuscript. Write a short report summarizing what the staff member says.
4. Does your campus have a printing plant? If so, what printing processes are used? If possible, arrange a class tour. Write a short report—600 words or less—after your tour, describing the printing process.

FOR MORE HELP

FLUEGELMAN, A., AND J. J. HEWES. 1983. *Writing in the computer age*. New York: Doubleday.
The authors provide a detailed look at word processing systems and how to use them.

Hill, M., and W. Cochran. 1977. *Into print*. Los Altos, Calif.: William Kaufman.
　Hill and Cochran provide an easily read guide to writing, illustrating, publishing, and marketing documents.

Bruno, M. H., ed. 1984. *Pocket pal: A graphic arts production handbook*. New York: International Paper Company.
　The International Paper Company regularly updates this excellent primer on printing and graphic arts.

McWilliams, P. A. 1983. *The word processing book*. New York: Ballantine.
　Although slightly dated, McWilliams's explanations of computer terms and concepts are easy to understand.

Turnbull, A. F., and R. N. Baird. 1980. *The graphics of communication*. 4th ed. New York: Holt, Rinehart and Winston.
　In their classic, Turnbull and Baird provide detailed discussions of the printing process and terms.

Zinsser, W. 1983. *Writing with a word processor*. New York: Harper & Row.
　An excellent book telling of one man's struggles with and triumphs over a word processor.

15
Professional Talks and Slide Presentations

OVERVIEW

In Chapter 15, we
- ☐ Discuss guidelines for preparing and presenting professional talks
- ☐ Explain planning, producing, and giving professional slide presentations

Across the country, professional organizations hold more than 1,000 regional and national meetings annually. Thousands of local, state, and regional professional organizations meet weekly, monthly, or quarterly. Businesses, agencies, and other organizations hold untold numbers of meetings each day, where professionals like yourself give presentations. In their presentations, professionals explain their projects, proposals, progress reports, final reports, and other activities. Audiences vary widely. They may be made up of peers, managers, supervisors, or bosses. Done well, professional presentations enhance your image. Done poorly, they mar it. Panel 15-1 provides a succinct discussion of the fundamentals of giving a speech.

Many professionals add audio-visuals to their presentations, because they know that audio-visuals help audiences to understand their points. Presenters who used visuals won their points 67 percent of the time and were perceived as more professional than those who did not (Oppenheim et al. 1981). The leading audio-visual aids include overhead transparencies and the 35mm slide sets. In 1985, *PC Week* reported, 452.5 million overhead transparencies and 513 million slides were used in presentations (Meilach 1985). Overhead transparencies are popular partly because they can be prepared by computer.

Slide presentations are heavily used, too. Amoco Corporation, previously Standard Oil Company of Indiana, processes 150,000 slides annually (Ruark 1981). Nationwide at least 16 million people own 35mm cameras and some 12 million projectors; they shoot an estimated 1.5 billion color slides annually (Scherer 1978). As many as 70 percent of the slides may be used in professional presentations (Brush and Brush 1978). Slide presentations are popular because 35mm single reflex cameras, projectors, and film are readily available, easily used, and moderately priced.

In this chapter, we emphasize two kinds of oral presentation: the professional talk—the "presentationist" approach—and the 35mm color slide presentation.

THE PROFESSIONAL TALK

Larry Gottlieb (1981, 1984), a public-speaking coach for scientists at the Lawrence Livermore National Laboratory, recommends that professionals use projected pictures, projected words, spoken words, and body language—body movements that enhance their message—rather than merely adding a few visuals to a speech. As in writing, a planned approach helps professionals to develop better presentations.

The professional approach to presentations makes a few key points quickly and provides needed redundancy. In an oral presentation, the audience cannot turn back to check a point. Selective redundancy, built through well organized visuals and skillful speaking, helps an audience understand key points.

The Denver Division of Martin Marietta, an engineering and scientific company, suggests that its engineers, scientists, and technical staff follow an eight-step process for their presentations:

1. Analyze
2. Plan
3. Design
4. Rehearse
5. Produce
6. Practice
7. Deliver
8. Evaluate (Martin Marietta 1976, 4–28)

1. Analyze. Analyze the who, what, when, where, and why of your presentation. Carefully identify your audience, and analyze its characteristics and frame of reference. Let what you learn guide your planning, designing, rehearsing, and presenting.

Evaluate the environment—the "where"—in which you'll give your presentation. Answer the following questions:

- How large is the room?
- How many people will it hold?
- What is the physical arrangement of the room?
- Can you control the lights?

PANEL 15-1. How to Make a Speech

How to make a speech

By George Plimpton

International Paper asked George Plimpton, who writes books about facing the sports pros (like "Paper Lion" and "Shadow Box"), and who's in demand to speak about it, to tell you how to face the fear of making a speech.

One of life's terrors for the uninitiated is to be asked to make a speech.

"Why me?" will probably be your first reaction. "I don't have anything to say." It should be reassuring (though it rarely is) that since you were asked, somebody must think you do. The fact is that each one of us has a store of material which should be of interest to others. There is no reason why it should not be adapted to a speech.

Why know how to speak?

Scary as it is, it's important for anyone to be able to speak in front of others, whether twenty around a conference table or a hall filled with a thousand faces.

Being able to speak can mean better grades in any class. It can mean talking the town council out of increasing your property taxes. It can mean talking top management into buying your plan.

How to pick a topic

You were probably asked to speak in the first place in the hope that you would be able to articulate a topic that you know something about. Still, it helps to find out about your audience first. Who are they? Why are they there? What are they interested in? How much do they already know about your subject? One kind of talk would be appropriate for the Women's Club of Columbus, Ohio, and quite another for the guests at the Vince Lombardi dinner.

How to plan what to say

Here is where you must do your homework.

The more you sweat in advance, the less you'll have to sweat once you appear on stage. Research your topic thoroughly. Check the library for facts, quotes, books and timely magazine and newspaper articles on your subject. Get in touch with experts. Write to them, make phone calls, get interviews to help round out your material.

In short, gather—and learn—far more than you'll ever use. You can't imagine how much confidence that knowledge will inspire.

Now start organizing and writing. Most authorities suggest that a good speech breaks down into three basic parts—an introduction, the body of the speech, and the summation.

Introduction: An audience makes up its mind very quickly. Once the mood of an audience is set, it is difficult to change it, which is why introductions are important. If the speech is to be lighthearted in tone, the speaker can start off by telling a good-natured story about the subject or himself.

But be careful of jokes, especially the shaggy-dog

"What am I doing wrong? Taking refuge behind the lectern, looking scared to death, shuffling pages, and reading my speech. Relax. Come out in the open, gesture, talk to your audience!"

variety. For some reason, the joke that convulses guests in a living room tends to suffer as it emerges through the amplifying system into a public gathering place.

Main body: There are four main intents in the body of the well-made speech. These are 1) to entertain, which is probably the hardest; 2) to instruct, which is the easiest if the speaker has done the research and knows the subject; 3) to persuade, which one does at a sales presentation, a political rally, or a town meeting; and finally, 4) to inspire, which is what the speaker emphasizes at a sales meeting, in a sermon, or at a pep rally. (Hurry-Up Yost, the onetime Michigan football coach, gave such an inspiration-filled half-time talk that he got carried away and at the final exhortation led his team on the run through the wrong locker-room door into the swimming pool.)

Summation: This is where you should "ask for the order." An ending should probably incorporate a sentence or two which sounds like an ending—a short summary of the main points of the speech, perhaps, or the repeat of a phrase that most embodies what the speaker has hoped to convey. It is valuable to think of the last sentence or two as something which might produce applause. Phrases which are perfectly appropriate to signal this are: "In closing..." or "I have one last thing to say..."

Once done—fully written, or the main

(Source: Reprinted by permission of International Paper Company.)

PANEL 15-1. (continued)

points set down on 3" x 5" index cards—the next problem is the actual presentation of the speech. Ideally, a speech should not be read. At least it should never appear or sound as if you are reading it. An audience is dismayed to see a speaker peering down at a thick sheaf of papers on the lectern, wetting his thumb to turn to the next page.

How to sound spontaneous

The best speakers are those who make their words sound spontaneous even if memorized. I've found it's best to learn a speech point by point, not word for word. Careful preparation and a great deal of practicing are required to make it come together smoothly and easily. Mark Twain once said, "It takes three weeks to prepare a good ad-lib speech."

Don't be fooled when you rehearse. It takes longer to deliver a speech than to read it. Most speakers peg along at about 100 words a minute.

Brevity is an asset

A sensible plan, if you have been asked to speak to an exact limit, is to talk your speech into a mirror and stop at your allotted time; then cut the speech accordingly. The more familiar you become with your speech, the more confidently you can deliver it.

As anyone who listens to speeches knows, brevity is an asset. Twenty minutes are ideal. An hour is the limit an audience can listen comfortably.

In mentioning brevity, it is worth mentioning that the shortest inaugural address was George Washington's—just 135 words. The longest was William Henry Harrison's in 1841. He delivered a two-hour 9,000-word speech into the teeth of a freezing northeast wind. He came down with a cold the following day, and a month later he died of pneumonia.

Check your grammar

Consult a dictionary for proper meanings and pronunciations. Your audience won't know if you're a bad speller, but they will know if you use or pronounce a word improperly. In my first remarks on the dais, I used to thank people for their "fulsome introduction," until I discovered to my dismay that "fulsome" means *offensive* and *insincere.*

Why should you make a speech? There are four big reasons (left to right): to inspire, to persuade, to entertain, to instruct. I'll tell you how to organize what you say.

On the podium

It helps one's nerves to pick out three or four people in the audience—preferably in different sectors so that the speaker is apparently giving his attention to the entire room—on whom to focus. Pick out people who seem to be having a good time.

How questions help

A question period at the end of a speech is a good notion. One would not ask questions following a tribute to the company treasurer on his retirement, say, but a technical talk or an informative speech can be enlivened with a question period.

The crowd

The larger the crowd, the easier it is to speak, because the response is multiplied and increased. Most people do not believe this. They peek out from behind the curtain and if the auditorium is filled to the rafters they begin to moan softly in the back of their throats.

What about stage fright?

Very few speakers escape the so-called "butterflies." There does not seem to be any cure for them, except to realize that they are beneficial rather than harmful, and never fatal.

The tension usually means that the speaker, being keyed up, will do a better job. Edward R. Murrow called stage fright "the sweat of perfection." Mark Twain once comforted a fright-frozen friend about to speak: "Just remember they don't expect much." My own feeling is that with thought, preparation and faith in your ideas, *you* can go out there and expect a pleasant surprise.

And what a sensation it is—to hear applause. Invariably after it dies away, the speaker searches out the program chairman—just to make it known that he's available for next month's meeting.

George Plimpton

Today, the printed word is more vital than ever. Now there is more need than ever for all of us to *read* better, *write* better, and *communicate* better.

International Paper offers this series in the hope that, even in a small way, we can help.

If you'd like additional reprints of this article or an 11" x 17" copy suitable for bulletin board posting or framing, please write: "Power of the Printed Word," International Paper Company, Dept. 9, P.O. Box 954, Madison Square Station, New York, NY 10160   © 1983, INTERNATIONAL PAPER COMPANY

INTERNATIONAL PAPER COMPANY
We believe in the power of the printed word.

Printed in U.S. on International Paper Company's Springhill® Offset, basis 60 lb.

- Can you control the heating and cooling system?
- Will a podium, screen, and other equipment be available?
- Will the equipment be movable?
- What time of day will you give your presentation?

Let your answers guide how you prepare your presentation and select your audio-visual equipment. Remember that warm, stuffy rooms, heavy lunches, and other conditions can put your audience to sleep.

Analyze the "what" of your presentation. Considering the vast amount of information that you have on your topic, what do you want your audience to know or do after your presentation?

2. Plan Your Presentation. Most professional presentations benefit from being succinct. Gottlieb (1982) says that audiences are often wide awake when speakers begin. But their attention drops steadily, and many members are fighting sleep 10 minutes into the presentation. By the 20-minute mark, some, he jokes, have a "near death" experience.

When preparing your presentation, plan to talk from your notes. Some professionals write their notes on 3- by 5-inch or 4- by 6-inch note cards. Others use standard typing paper or the edges of the mounts for overhead transparencies. Talk from your notes, and you'll sound natural.

Normal speaking rates run between 120 and 180 words a minute. Thus a 10-minute presentation contains between 1,200 and 1,800 words. So limit your presentation to a few key points. To determine your key points, ask yourself which one, two, or three main ideas you want the audience to recall from your presentation. As you outline, consider what questions your audience might ask. Answer them in your presentation. Structure your presentation around the key points. Use a working outline that incorporates the visuals, the printed words, and the spoken words (as in Panel 15-2). Be sure that your visuals, printed and spoken words, and presentation style stress your main points.

When organizing, remember the old military bromide: "Tell 'em what you're going to tell 'em, tell 'em, and then tell 'em what you've told 'em." Let your introduction provide an overview, the body present main points, and your conclusion restate the main points.

Provide redundancy in the way you coordinate your visuals, printed words, and spoken words. Use cues to tell the audience when you move from one idea to the next. Say, "I've covered point number one, and I'd like to move onto my second point. . . ."

Research shows that audiences recall best the points made early and late in a presentation (Smith 1982, 251–8). In your conclusion, plan to stress the main points from the middle of the presentation.

Should you use humor? Be careful. What you consider humorous, your audience may not. Used inappropriately, humor can ruin your credibility and negate your message. Make sure that humor fits the topic, communicates your point, and doesn't sound forced. Remember, your primary purpose is to communicate, not to entertain.

PANEL 15-2. Outlining a Professional Presentation

Suppose that you're working for a computer manufacturer and you've developed a new computer and software that produce visuals for engineers, scientists, and technical specialists. You've been given 10 minutes to explain your product to a local professional organization. Here's a suggested outline for the opening comments, visuals, printed words, and notes for the spoken words. Note how you present the sequence—from overview to details.

First Point

Consider your general idea, the visuals and printed words, and the spoken words.

General Idea

XYZ Computers will solve your graphics problems; here's an overview of the unit and its potential.

Visuals/Printed Words

Show computer, example of graphics.
Copy: XYZ Model 253 provides high-quality graphics.

Ideas for Spoken Words

Stress: XYZ Model 253; provide overview; explain the computer, how it works, and its diverse graphics capabilities.

Which visuals might you use? Flip charts? Chalkboards? Real objects, such as specimens? Models? Overhead transparencies? Which visuals you select depends upon your topic and on available time, money, and staff to help you prepare visuals. Flip charts and chalkboard will do when you have little time or money or when you need to record ideas from your audience. But flip charts and the chalkboard may not appear so polished as other visuals. Panel 15-3 suggests guidelines for flip charts and chalkboards.

PANEL 15-3. Guidelines for Flip Charts and Overhead Projectors

Flip charts and the chalkboard provide active graphics. You draw and list points as you give your presentation. Flip charts work best in interaction with eight to 10 people. To record information, use standard flip chart pads or newsprint pads and dark-colored, heavy-tip markers so that everyone can see the copy. Use 1- to 2-inch letters. If you must draw a diagram, the *AAACE Communication Handbook* (1976) recommends that you draw the diagram in light pencil before your presentation so that you can see it but your audience can't. Then trace over the light lines with a heavy felt-tip pen.

Yellow chalk on a green board proves most legible. In most classroom settings, the 4- by 8-foot boards provide plenty of space for drawing with thick, 2-inch-high letters. When you write or draw, avoid obstructing the audience's view. Step aside frequently so that everyone can see your points. If you use diagrams, colored chalks may show the different parts of the diagram. As you move from one point to the next, erase old material.

> ### PANEL 15-4. Adding Photographs to Overhead Transparencies
>
> For realistic overhead transparencies, you can add both black and white and color photographs. To prepare photographic overhead transparencies, you'll need to turn to photographic processors, instructional media services, audio-visual firms, or your organization's photographer. For black and white transparencies, a photographer can print from your black and white negative onto 8- by 10-inch continuous tone films such as Kodak's Fine Grain Release film (7302) or Dupont's Cronalar CT77 (Petersen 1979).
>
> For 8- by 10-inch color transparencies, a photographer can print from color negatives and slides, reflected art, and computer graphics. The photographer need only expose the Kodak overhead transparency material, load it into a color processor, soak the film for 20 seconds, laminate, and peel. In the mid-1980s, Kodak advertised the relatively low cost of less than $5 per overhead transparency.

Real objects work well for small audiences. Models work well only if they're large enough for the audience to see. Overhead transparencies work well because you can prepare them beforehand and use projected visuals and words. As we've mentioned, presentation graphic programs for computers are becoming more and more popular. Recent photographic advances enable you to reproduce black and white photographs and color slides as overhead transparencies (Panel 15-4).

3. Design Your Message. Consider the available time, and then use your working outline to develop your message, including both visuals and words. Begin with rough sketches of your visuals, and develop your narrative around them.

With clear acetate and felt-tip pens, such as Vis-à-Vis pens, you can rough out overheads and test them before you prepare the final version. Limit each transparency to one concept. Gottlieb (1981) recommends the HEIDI approach for scientists and engineers at the Lawrence Livermore Laboratory. HEIDI stands for HEadlines, IDeograph, and Idunit.

*HE*adlines are simple, clear, and contain one main idea. Simple declarative sentences carry messages clearly. Put them at the top of a visual. Headline sentences should be less than 11 words, should make an assertion, and should have a "what" (subject) followed by a "so what" (predicate) (Figure 15-1). *ID*eographs are concise, simple pictures: sketches, block diagrams, charts, cartoons, photographs, graphs. Such visuals help reinforce the message of the headline. *I*dunit then ties the presentation together. It consists of four visual units: (1) a title visual containing the presentation's main message, (2) background visuals to familiarize your audience with the subject, (3) body visuals giving the heart of your message, and (4) a conclusion visual repeating the title visual's message (Figures 15-1 and 15-2).

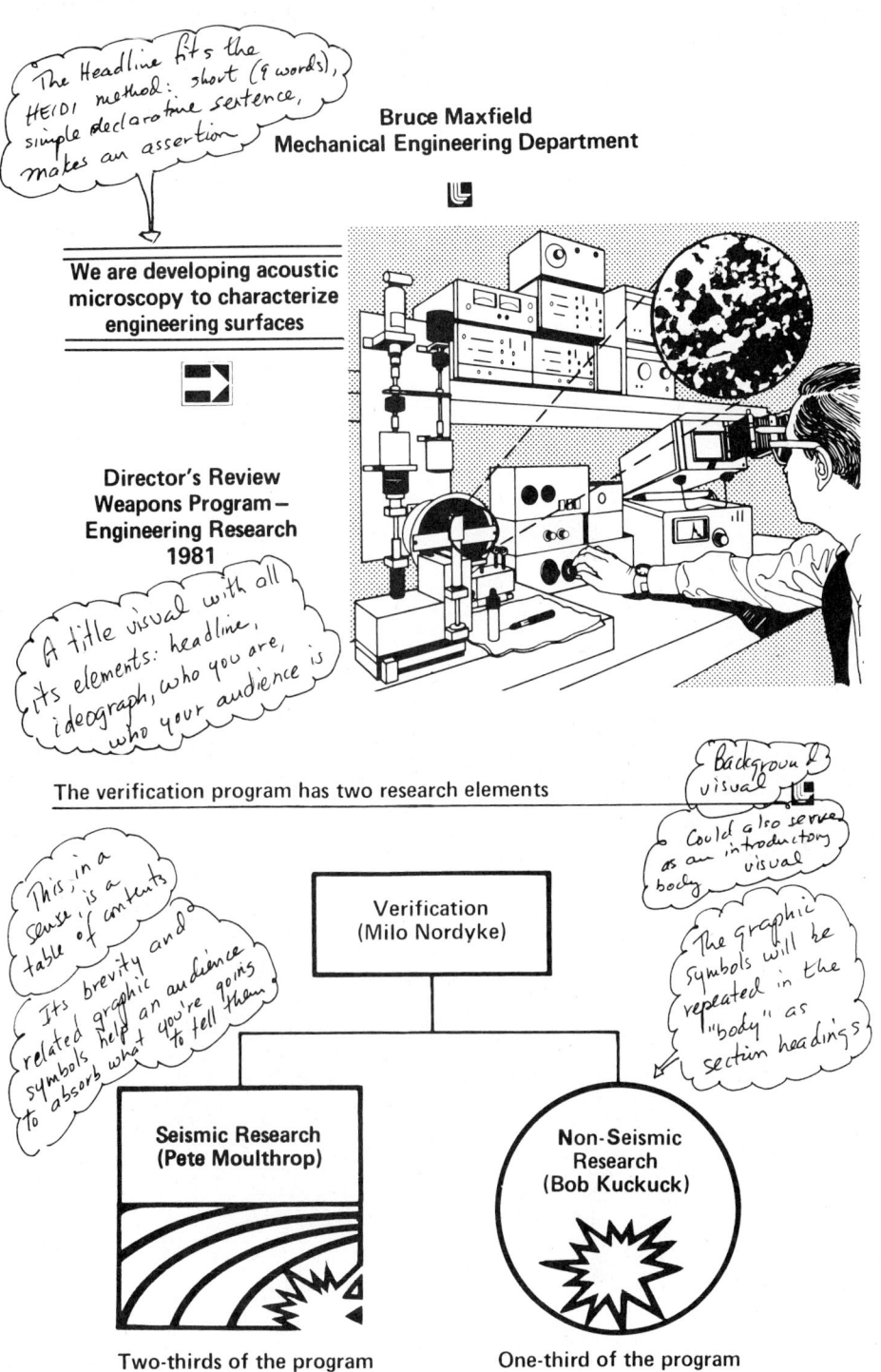

FIGURE 15-1. Introductory transparencies representing the HEIDI technique. *(Source: Reprinted by permission of Larry Gottlieb and Lawrence Livermore National Laboratory, Livermore, Calif.)*

Legible letter height on screen depends on distance from last row

We are developing acoustic microscopy to characterize engineering surfaces

FIGURE 15-2. Examples of body and conclusion transparencies. *(Source: Reprinted by permission of Larry Gottlieb and Lawrence Livermore National Laboratory, Livermore, Calif.)*

> **PANEL 15-5. Make Your Copy Large Enough**
>
> When you prepare a professional presentation with copy blocks, make the copy large enough for everyone to read easily. The letters on the transparency should be at least ¼ inch high, or 1 inch high on the screen for every 25 feet of viewing distance.
>
> For quick, inexpensive copy, the Orator Language Arts ball for IBM electric typewriters produces both capitals and lower-case letters:
>
> ## Orator Type Face
>
> Equipment such as the 3M transparency composer and the Kroy composer make highly legible copy for overhead transparencies:
>
> ## HELVETICA REG. 18
>
> If you have access to typesetting equipment, set your copy set in 24-point serif type, such as English Times:
>
> ## English Times 24
>
> Or use 24-point capital and lower-case presstype to prepare copy:
>
> ## Test Visuals

If you must use lists, tables, and copy blocks, try to add illustrations, even simple ones, to help carry your message. Whatever your budget, make your copy big, bold, and simple (Panels 15-5, 15-6, and 15-7). Avoid computer programs that produce thin letters your audiences will have trouble reading. If appropriate and your budget permits, use color to add interest and appeal to your overheads.

4. *Rehearse Your Presentation*. Using your draft notes and rough visuals, rehearse your presentation before an audience that will give you constructive feedback. Pay especially close attention to your timing, quality of oral presentation, visuals, and message effectiveness. Your rehearsal will help you check your presentation and revise it before you prepare final materials.

5. *Produce Your Visuals and Notes*. You can prepare professional looking overheads on a typewriter, with presstype or other art supplies, typeset copy, or a computer. You'll find that many organizations have staff graphic artists, hire freelancers, or use computers to prepare transparencies. Some schools have laboratories where students can use such equipment.

6. *Practice Your Presentation*. Practice your presentation with your final visuals. First, review the presentation several times in your mind, and then practice aloud. If possible, practice in the room where you will give your presentation. If not, practice in a similar room. Familiarize yourself with the equipment, podium controls, and room lights that you will use

PANEL 15-6. Producing Overhead Transparencies

Prepare the master original on white paper with the copy and drawings illustrating your point.

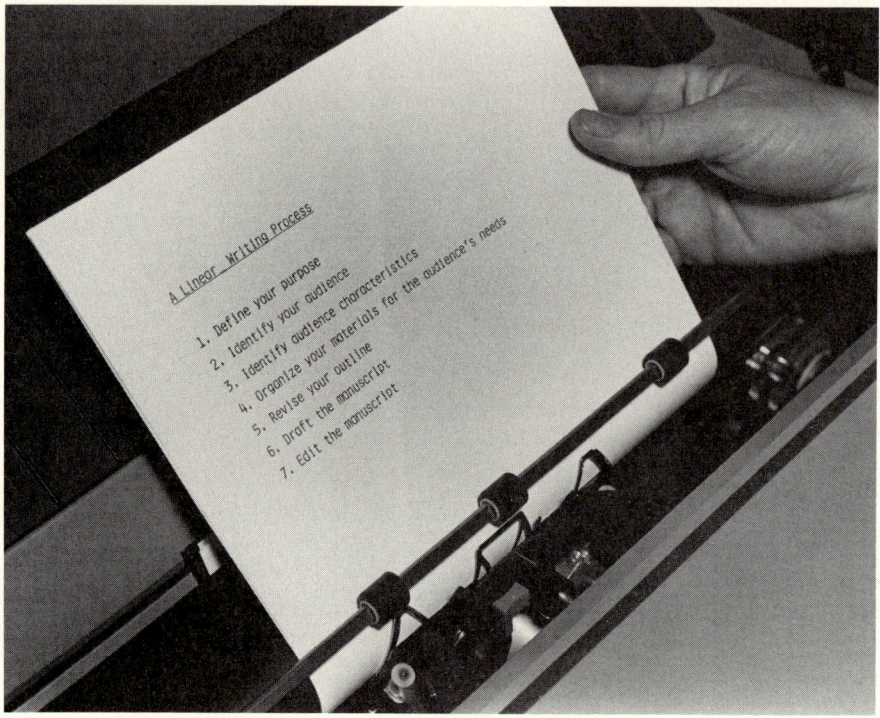

during your presentation. If you will need a microphone, practice with one. If not, practice speaking loudly so that you can be heard throughout the room. Use all of your visuals, and talk through your notes or outline. Then practice your presentation in front of a few people who can provide constructive criticism. When you speak, enunciate clearly so that your audience can understand you. Make eye contact with people in the audience. Move naturally, and don't force your gestures. If possible, videotape your practice presentation. If you can't use a videotape, use a tape recorder. Then review the tape, and you'll spot areas that need more polishing.

Practicing, as recommended here, *helps you overcome stage fright. It boosts confidence and leads to smooth, polished, professional presentations.*

PANEL 15-6. (continued)

Many organizations have transparency machines that use heat process. Make a photocopy of your original; run the photocopy and the transparency material through the machine. Never run originals through such machines. The machines may damage the original. An alternative method of making transparencies is to use certain photocopy machines. Simply load the photocopier with transparency materials, insert your original, and make your copy.

But do plan for disasters. Projector bulbs burn out, equipment breaks down, and people drop their notes. Carry spare projector bulbs and know how to change them, have backup equipment, and number your notes so that you can quickly rearrange them.

7. *Deliver Your Presentation.* Before leaving your office or home, *make sure that you have your notes and visuals.* If you're traveling by bus, train, or plane, *carry your visuals and notes with you.* Don't ship them separately or put them in your luggage and check it. You may arrive for your presentation and find that your luggage, notes, and visuals haven't.

When you begin your talk, speak clearly and forcefully. Maintain eye contact with the members of your audience. As you speak, note your

PANEL 15-7. Hewlett-Packard's Guidelines for Designing Effective Overhead Transparencies

How to Design Effective Overhead Transparencies

If you're using overhead transparencies, here are some simple steps that will add to your professionalism and increase the success of your next presentation.

Keep it simple
- Keep your message simple. Your transparencies should complement your presentation, not repeat everything you say.
- Use a maximum of seven sentences per overhead.
- Limit each of those sentences to no more than seven words.

```
          Agenda
• Review plan
• Present timetable
• Discuss changes
• Approve revised plan and timetable
• Assign new duties
```

Letter style, spacing, and size
- Use capitals (uppercase) and small (lowercase) letters. All-capital lettering is more difficult to read than text-style lettering, which is a mix of uppercase and lowercase.
- Italics, like capital letters, are difficult to read.

```
Upper and Lower Case

ALL CAPITALS

Upper and Lower Italics
```

- Choose spacing that makes the words easier to read. The letter "n" is a good way to judge correct spacing. Leave the space of one "n" between words. Leave two "n's" between sentences. Leave three "n's" between lines (as measured from baseline to baseline).

Leave the space of an 'n' between words. Leave two between sentences and three between lines.

- Use a maximum of three letter sizes on one overhead.
- The image area of an overhead transparency should be approximately 7-1/2 by 9-1/2 inches.

- Keep the format of your overheads consistent. Although your transparencies can be horizontal or vertical, the horizontal format is usually considered best for presentations.

Viewing considerations
- Recommended maximum viewing distance for an overhead is six times the width of the projected image. For example, if your image on the screen is four feet wide, the maximum viewing distance is 24 feet.
- Place your overhead on the projector only when you're ready for your audience to see it. Then, take it away as soon as you're finished with it.

Add color, carefully
- Audiences expect to see color in a presentation, but color shouldn't overpower your message.
- Use color to draw attention to a special element of your overhead.

(Source: Material reprinted courtesy of Hewlett-Packard Company.)

PANEL 15-7. (continued)

Special effects for overheads
There are several techniques for making your overhead transparencies more effective.

Framing
Framing your transparencies makes them easier to handle during a presentation and allows you space to jot notes about your overhead.

Mounting frames are available at most office supply stores.

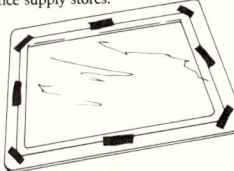

Cardboard arrows
Cardboard arrows can be used to point out important items or to lead your audience through a transparency.

Cut your arrow from lightweight cardboard. You may want to fashion a "handle" for your arrow to make it easier to pick up and move.

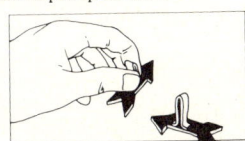

Writing on your transparencies
You can write directly on your transparency without damaging it if you first cover it with a clear sheet of film. Just attach the clear film to all four sides of the frame and you're ready to write.

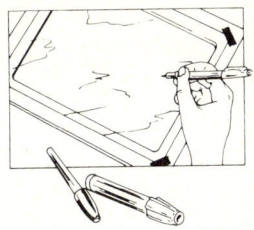

Overlays
There are two ways to attach overlays to the frame of your overhead. In either case, limit the number of overlays per transparency to four. All overlays are hinged into place on the front of the frame so that each layer lines up with the original transparency.

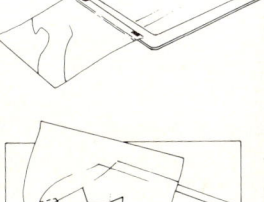

If you always use the overlays in the same order, attach them to the same side of the frame.

If you might not use the overlays in the same sequence each time, hinge each overlay on a different side of the frame. Trim each overlay so that it will fall into place easily.

Masking
With this masking technique, you can progressively disclose information. Cut a sheet of opaque paper into sections that fit the areas of the transparency that you want to reveal. Hinge these sections to the front of the frame, aligning them so that they cover appropriate sections of the transparency. As you present, you can flip back the section you're ready to discuss.

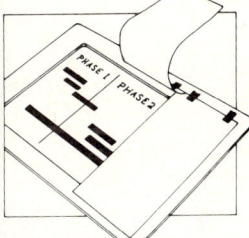

Billboarding
Billboarding allows you to highlight an area of an overhead transparency by adding contrast. Attach a sheet of colored film to all four sides of the back of the transparency frame. Cut out the colored film over the area you want highlighted. Be careful not to cut through the original transparency.

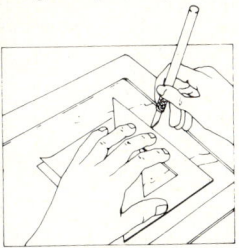

audience's reactions to your points, and adjust your speech as needed. If you deviate from your notes, return to your main points as quickly as possible.

8. Evaluate Your Presentation. During and after the presentation, note the feedback from your audience—attention, questions, and overall reactions. After your presentation, talk with members of your audience. How did they react? What questions did they ask? What kinds of comments did they make? Each time you give a presentation, learn from it, and use what you learn to improve your next presentation.

PRODUCING SLIDE PRESENTATIONS

Well done slide presentations enhance your professional image and keep your audience awake. Poorly done slide presentations can mar your professional image and put your audience to sleep. Poor slide presentations come from a lack of ideas and hard work, not a lack of money, argues Donna Matrazzo (1983), a freelance scriptwriter.

Producing effective slide presentations is relatively easy if you:

1. Plan the slide set
2. Prepare the script and slide set
3. Practice the presentation
4. Give the slide presentation
5. Evaluate the slide presentation

As you learn to produce slide presentations, follow the suggested sequence. Later, if you must work out of sequence, you will know how to avoid many common pitfalls.

Planning the Presentation

Begin by identifying and analyzing your audiences, establishing your overall objectives, selecting the slide presentation form, determining its length, using a viable visual style, and preparing the proposal and treatment.

Identifying Your Audiences. Use the strategy outlined in Chapter 5 to analyze your audiences. Pay especially close attention to the audience's visual familiarity with your topic.

Determining Overall Objectives. Once you know your audience and its characteristics, ask:

- What do I want my audience to know?
- What do I want my audience to do?

Let your answers guide you in developing your objectives.

To improve the audience's retention of your objectives, let the narrative support the visuals and the visuals support the narrative.

Selecting the Form. Most technical and scientific professionals use a single good quality projector, such as a Kodak Carousel or Ektagraphic, and either an 80-slide or a 140-slide tray. Many professionals talk over their slides; others use recorded voice, music, and synchronization tracks. The synchronization track advances the slides automatically.

Determining the Pace and Length. A good rule is to keep a slide on the screen just long enough to communicate your point. Too little time and your audience feels you've skipped out before they've understood the slide. Too much time and you'll put them to sleep. Deciding how much time to allow for each slide is really a function of audience analysis. An audience to whom the technical material is new needs more time to absorb the information communicated by a slide than an audience of peers. Professor Cliff Scherer (1982), of Cornell University, suggests 2 to 4 seconds per slide for a fast paced presentation, 5 to 6 seconds per slide for a moderately paced presentation, and 7 to 12 seconds per slide for a slow one.

Don't talk too much. For an 80-slide set, with a slowly paced presentation, plan to speak for about 13 minutes, and limit yourself to 1,200 or fewer words. For a 140-slide set, with a slowly paced presentation, limit yourself to about 23 minutes and 2,000 or fewer words. For moderately and fast paced presentations, cut your narrative accordingly. Write your script for the ear, and use your script as speaking notes, or have a professional narrator prepare a tape of your script.

Selecting Your Visual Style. The visuals you use set the tone and influence your message. Of the visual styles Kodak identifies in *Images, Images, Images* (Kenny and Schmitt 1981), those commonly used for technical and scientific presentations include:

1. straightforward style, with posed or controlled scenes
2. photojournalistic style, with visuals taken as an event happens
3. illustrated style, with artwork and illustrations

Combining visual styles works well for many technical and scientific presentations, because they often present information in graphs, charts, and tables. If you combine styles, let one style predominate.

Preparing the Proposal and Treatment. If your superiors must approve your presentation, prepare a proposal and treatment. Kodak recommends that a proposal and treatment include:

- A statement of objectives
- Theme and premise
- An outline of the content
- A budget
- A timetable or schedule (Kenny and Schmitt 1981, 109–110)

Your objectives should summarize background on your topic, your communication objectives, and your rationale. Your theme and premise come from your communication objective and establish the slide set's primary function—to inform, instruct, persuade, or document. Your outline identifies the major points you'll make, and your budget provides the slide set's estimated costs. Finally, the timetable establishes how long it will take to produce the slide set.

Preparing the Slide Set

Producing a slide set requires outlining its content; perhaps developing a storyboard; scripting the visuals and narrative; obtaining the needed slides; screening the slides against the script; revising the script and slides; perhaps recording the narrative; preparing instructions for using the slide set; and duplicating the slides, script, and narrative.

Outlining the Script. Start by outlining your major points. Think visually and orally at the same time. Use the triple-purpose outline (Panel 15-8) that Donna Matrazzo, author of the *Corporate Scriptwriting Book* (1980), recommends.

PANEL 15-8. An Outlining Approach

INTRODUCTION TO WILDLIFE AND FISH HABITAT RELATIONSHIPS SYSTEM

Visuals and Sounds	Ideas for Narrative	Audience Effect
Start with scenes suggesting proper management	Historically, Forest Service lacked ways to quantify fish and wildlife resources	Build common background and identify problems audience has had
Show well known biologists attesting to problems; show selected problems	Frustrated managers, biologists wanted and needed system to quantify resources	Audience (resource managers & biologists) recognizes common problems
Show positive resources uses, identification	Now a solution to problem—Wildlife & Fish Habitat Relationships System	Now a solution has been developed to solve their problems

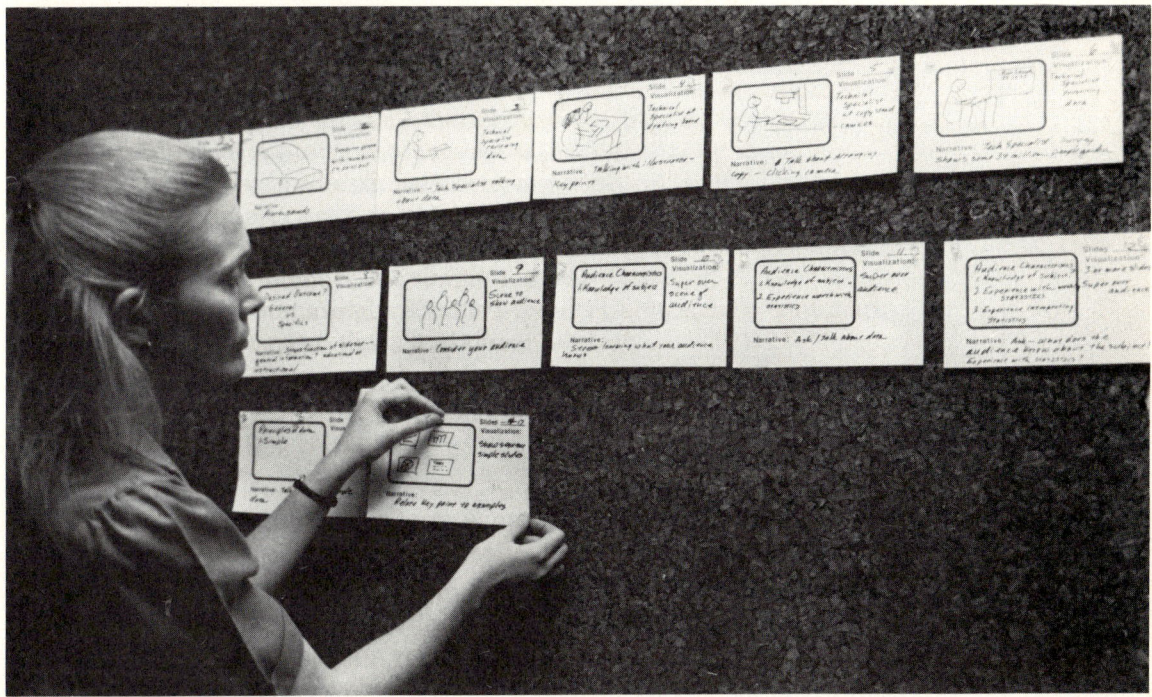

FIGURE 15-3. Using the planning board technique.

Once you have an outline, you can either use the storyboard technique (Figure 15-3) or write a visual and narrative script (Panel 15-9). Some professionals develop the storyboard and then the script; others prefer to script the visuals and narrative directly from their outlines.

Developing a Storyboard. Using a storyboard approach entails drawing sketches and noting the points for the narrative on 3- by 5-inch or 4- by 6-inch note cards (Figure 15-4), one for each slide. Then arrange the slides in sequence. To sequence the cards, pin the cards to a bulletin board, or tape them to a wall (Figure 15-3).

Once you have the storyboard completed, use it as a guide to talk your way through your points. By doing so, you summarize key points, develop a conversational tone, and improve narrative flow. Once you've talked through the script a couple of times, record your narrative with a small tape recorder.

Scripting Visuals and Narrative. If you're using the storyboard, transcribe your taped narrative into a split page script (Panel 15-9), or work from the outline to draft your split page script. Double or triple space your copy so that you have plenty of room for editing. Use the guidelines in Panel 15-10 for writing and editing.

PANEL 15-9. A Split Page Script

The script below comes from a slide set explaining management of slash and logging residue in our national forests. The slide set opens with music, scenes of forest resource management, and a short discussion of resource management. The excerpt below shows the split page script at work: introducing narrators; flagging points—first, second, and third; selecting strong verbs; tying sentences, clauses, phrases, and words to slides.

Visuals	Narrative
23. Slash, 8 × 10 YUM [yarding unmerchantable timber]	NARRATOR: The approach we'll describe can help you evaluate fuel treatments.
24. Siskiyou National Forest sign	The setting of this case study is the Siskiyou National Forest in the Pacific Northwest Region.
25. Four foresters talking/coordinating activities	Here foresters from the National Fuel Inventory and Appraisal Project of the Rocky Mountain Forest and Range Experiment Station worked together with Regional, Forest and District specialists . . .
26. Clear cut and burned area. Little slash on ground	to project, analyze, interpret, and compare the possible outcomes of alternative fuel management decisions.
27. Two foresters looking at printout from terminal; woman operator	The case study's results illustrate several points. First, the technique provides numerical information for making management decisions.
28. Fire scene, heavy log in foreground, fire in background.	Second, the technique improves the ability to predict the size and intensities of fires under different management decisions.
29. Two researchers examining computer printouts.	Third, the process incorporates historical fire occurrence information and weather and fuel information for decision making.
30. Piled slash across lower part of slide	Fourth, the process estimates the size and intensities of fires under different levels of slash treatment.
31. Scenic Cottonwood Pass, Colorado. Mountains, trees	Fifth, the process can be applied in managing timber throughout the West and Northwest,
32. Loading logs on truck	the North-central and Northeast,
33. Slash, brown pine needles in foreground	the Southwest,
34. South pine plantation	and the South.
35. Dave Radloff (grey, yellow & red shirt) reading printout	Here's Dave Radloff, research forester, who coordinated the Siskiyou study.
36. Researcher at computer terminal	RADLOFF: Under the National Fuel Inventory and Appraisal Project's mission, we are developing ways to quantify the evaluation of fuel treatment alternatives.

(Source: Making Fuel Management Decisions, *National Fuel Inventory and Appraisal Project* (Fort Collins, Colo.: Forest Service, U.S. Department of Agriculture, 1980).)

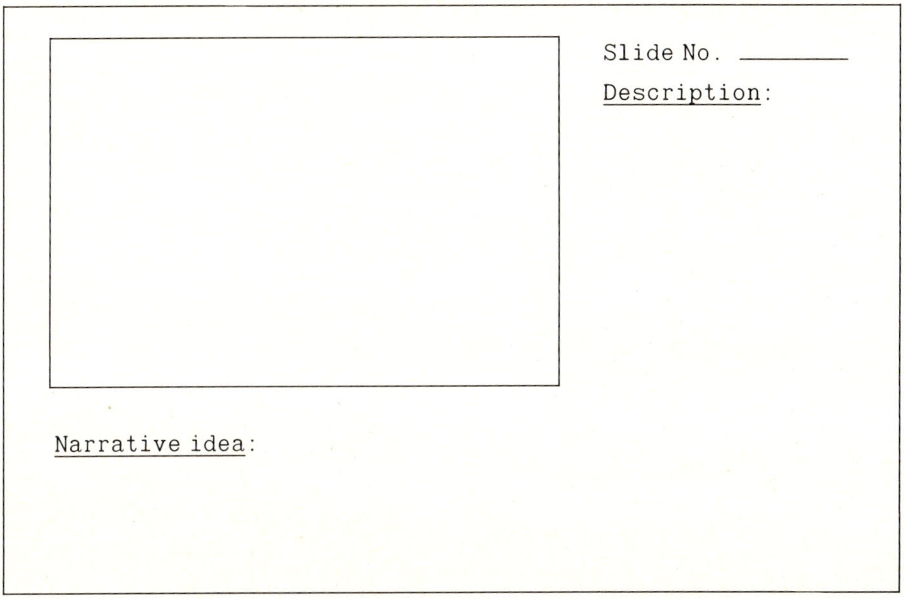

```
                              Slide No. _____
                              Description:




Narrative idea:
```

FIGURE 15-4. A typical planning card for a storyboard.

As you write and edit, remember that your audience hears the words and sees the slides. Use a word, phrase, or clause to communicate your key points. Make the narrative and visuals support each other.

To visualize your script, describe your ideas for slides in words, or sketch your ideas. Your sketches need only give an idea for the slides you will need. When visualizing a slide set, consider the viewers. Are they

PANEL 15-10. Tips on Writing Scripts

A script narrative is writing that will be heard and seen. When you write and edit your script:
- Use simple words.
- Use simple, conversational sentences.
- Use short sentences—12 to 15 words.
- Use strong verbs, in the present tense, if possible.
- Use short, easily understood words.
- Define your terms and jargon with words and slides.
- Avoid adjectives; let the slides modify.

- Make your points easy to follow by saying, "first," "second," "third."
- Spell out difficult words phonetically.
- Test your script by reading it aloud and recording it.
- Listen to your script and ask: Does it flow smoothly? What can I cut? How can I make it sound better?
- Test your narrative and visuals together. Do they support one another? Have you placed proper emphasis on the points you want to make? Do narrative and slides flow smoothly?

Visuals	Audio
	... When you look for diseased, injured, or infected branches and leaves, first scan the tree,
	then move in closer to look for specific problems,
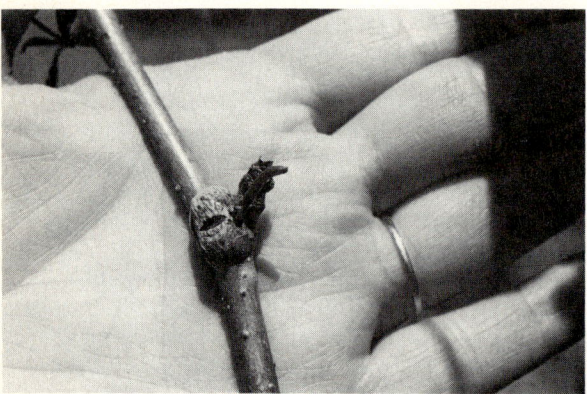	and then look closely at the branches and leaves.

FIGURE 15-5. Progressive disclosure.

visually familiar with your subject? If not, help them develop a visual frame of reference by using progressive disclosures (Figure 15-5).

When scripting the visuals and narrative, avoid a series of slides containing nothing but words if you can get illustrations. Use visuals to add interest to the words. Better yet, add a *few* words to a visual that carries most of the information. Compare the visual appeal of the slides in Figure 15-6.

Obtaining Slides. You can borrow slides, shoot them yourself, have a professional photographer shoot them, or buy them. Use your working script to guide you in collecting needed slides.

Borrowing Slides. If you borrow slides, keep in mind that you'll need slides with the correct exposure, color balance, and composition to carry your message effectively. Some slide collections are excellent, but far more are of poor technical quality. If possible, select original slides shot on professional quality film. Avoid duplicate slides; they create copying problems. Avoid mixing film types in the same slide set. When you borrow slides, handle them carefully, label them with the owners' names, and be careful not to damage or lose them.

Shooting Slides Yourself. When taking slides, shoot more than you'll use in the final slide set. Professional audio-visual specialists estimate their film needs based on a shooting ratio. A shooting ratio represents the number of slides taken for each slide used. Professor Scherer (1982) recommends a shooting ratio of 4:1 for artwork and copy and a 14:1 shooting ratio for candid and field shots. In other words, take four to 14 times as many slides as your final slide set will contain. By doing so, you give yourself dozens of slides from which to select. When you shoot slides, try different angles, and vary your exposures.

Many professionals prefer Kodak Ektachrome over Kodachrome film or other film because they can have Ektachrome processed locally and quickly—usually with 24 hours.

When preparing illustrations, use a horizontal format and the 4- by 6-inch area in the center of 8½- by 11-inch, white paper without watermarks (they show through on slides). Use lettering at least ¼ inch high, or use 18-point type. Limit copy to six or fewer lines. Bold, sans serif typefaces work better than thin, light, serif typefaces. You can hand-letter copy, but presstype or typeset copy produces more professional looking slides (Figure 15-6).

For quick and inexpensive copy blocks with an electric or electronic typewriter, use a Gothic, 12-pitch ball or daisy wheel and a carbon ribbon. To produce strong, clear impressions, place two to four sheets of paper behind the paper on which you're typing. Limit the typing area to a 2- by 3-inch area and six double-spaced lines. Proofread the copy carefully, and look for incomplete strikes as you type each line.

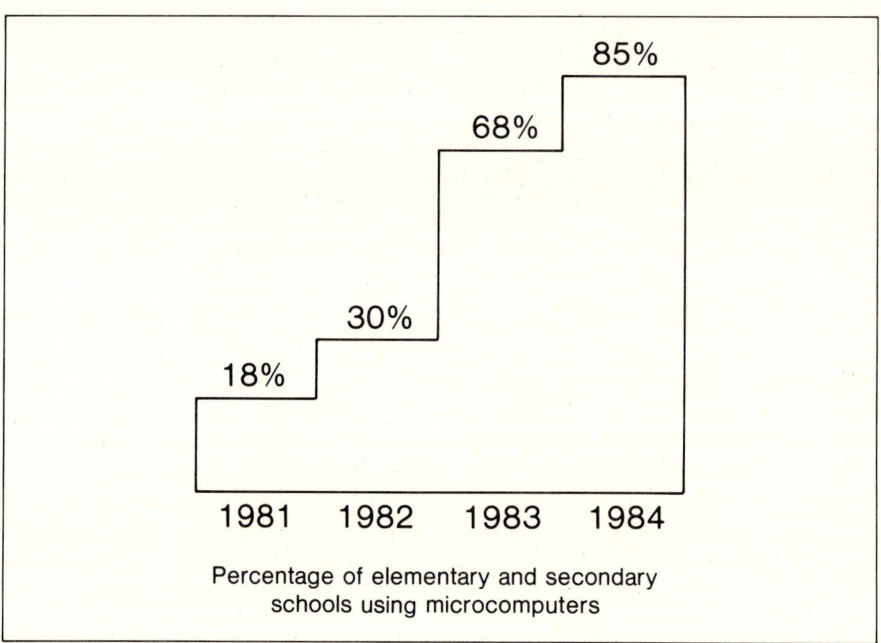

FIGURE 15-6. Compare the interest and quality of the data and title slides on the left and right. A small investment in producing quality visuals improves your professional image.

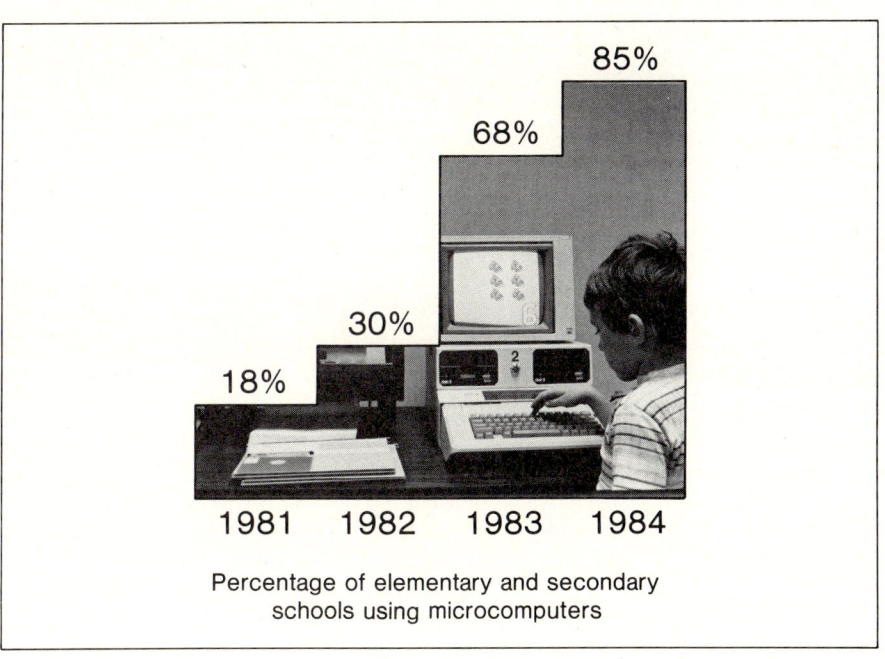

Percentage of elementary and secondary schools using microcomputers

Don't photograph artwork and drawings from existing publications. They seldom reproduce well, either because they contain too much information, they're of the wrong proportion, or the copy is too small to read. They're usually copyrighted, too. *If you want to base a slide on an existing illustration, obtain permission to use it, simplify it, and have it redrawn.* For more information, see Kodak's *Planning and Producing Slide Programs* (Bishop 1984) and *Legibility—Artwork to Screen* (Eastman Kodak 1980).

As described earlier in the section on overhead transparencies, you can prepare originals with art supplies, a computer, or you can hire a graphic illustrator. Whichever method you use, remember that illustrations appear sharper when drawn large and then photographically reduced.

Hiring a Photographer. If you have a funded project, you may be able to hire a professional photographer. When working with a photographer, follow the guidelines in Panel 15-11. Professional photographers have the equipment needed to produce high quality slides and to handle difficult shooting assignments.

Buying Slides. Many professional photographers submit their slides to commercial photographic services that sell limited rights to use the photographs. *Writer's Market* and *Photographer's Market,* each published yearly, list such services. Some state and federal agencies sell slides. To obtain slides, contact the service, identify your needs, and ask about their collections and the rights they sell. Most suppliers base their fees on uses to which slides will be put.

Screening Slides. Once you have collected your slides, begin by removing those that are under- or over-exposed, out of focus, and off color. Next, remove slides with poor composition and content. Set your rejects aside, but don't discard them. You may find yourself returning to them by default. After screening, try to have two or more slides for each one you need.

Arrange the slides in the order of the slide presentation so that you can view each slide with the other slides. To arrange the slides, work on a large slide viewer. Two or three 20- to 40-slide viewers also work well. You can improvise a light table with a storm window or large piece of glass. Tape translucent paper under the glass, and shine a floodlight up through the glass.

As you select your final slides, consider the visual flow and the ideas that the slides present and carry.

Revising the Slide Set. Read your script aloud while looking at each slide. Do the combined slides and narrative produce an effective message? Does the narrative support the slides, and do the slides support the narrative? Consider the tone, pace, and flow. If you find rough spots, add, delete, or

> **PANEL 15-11. Suggestions for Working with a Photographer**
>
> If you plan to hire a photographer, Robert Harvey (1984), chief photographer of Colorado State University's Instructional Services, suggests that you
>
> - Call a photographer weeks or months ahead of when you'll need your slides. Schedule a meeting to discuss your needs.
> - Tell photographers what you know about your audience(s).
> - Give a photographer a list of your specific ideas, not of individual pictures or slides. Explain what you want to communicate.
> - Let photographers use their creative skills to help you communicate your message.
> - Explain when you'll need your slides.
> - Discuss costs, payment methods, other financial details, and rights to the slides.

replace slides and rewrite the script. If needed, return to the slides you've set aside and search for suitable replacements, or shoot replacement slides. If your time is short or your budget tight, rewrite the script to avoid the problem.

Recording and Synchronizing the Tape. If you plan to use the slide script as a speaking outline, skip this stage. If you plan to have a recorded script, you can turn to local radio stations; school audio-visual and instructional offices; or commercial recording studios. Such facilities have sophisticated equipment for producing broadcast-quality tapes. They can provide professional narrators, music and sound effects, sound mixing, editing, synchronization, and other recording services. You can learn to synchronize your own slide presentations, and with practice, you can produce excellent shows for very little money.

Duplicating the Slide Set. If your slide show is more than a one-time presentation, you may want to duplicate it. Once you have a master set of slides, script, and recording, the number of copies you will have made depends on the cost of duplicates and on your needs. Remember that making a couple of extra sets initially will prove less expensive than making duplicates later.

If other professionals will use the slide set, prepare detailed instructions on how to use it properly. Include the brands and models of all equipment needed to use the slide set and information on additional questions about the slide set's content. Explain how to reorder slides should they be removed, dropped, or scrambled. Tell your users to test the entire program before showing it.

Distributing the Slide Set. For each slide set you provide, include the slides numbered and inserted in a tray, the script, the tape cassette, and instructions. Before distributing any slide set, test each to make sure no slides are out of sequence and that the taped narration advances the slides properly. When checking the slides, look for slides that are missing, out of sequence, or inserted backwards. For shipping, use a sturdy box or shipping carton. Many camera stores sell special shipping containers that hold a tray of slides, a cassette, script, and instructions.

Preparing for and Practicing the Slide Presentation

Adequate preparation and practice provide the best insurance for a smooth, effective slide presentation. When you plan a presentation, prepare a checklist, such as that in Panel 15-12. Use your list to guide your slide set assembly and check of equipment. Check your equipment before you use it, and never assume any equipment works until *you test it yourself*.

Once you've tested the equipment and checked your slides, make sure that no one tampers with them before your presentation. If you're traveling, always carry your slides with you. Never ship them ahead or place them in your luggage or baggage. If equipment is provided at your presentation site, arrive early to test it and locate any replacements that might be necessary.

If you'll be talking over the slides, practice talking aloud from the script. But don't read the script. Use it as an outline or as notes. Practice advancing the slides while talking. If you're using a recorded narrative and synchronized tape, follow the script and learn how to correct out-of-

---

PANEL 15-12. Checklist for Slide Presentations

Develop a checklist, and include items such as:

____ Projector	____ Projector stand, if needed
____ Extra bulbs	____ Screen, if needed
____ Remote control unit for projector	____ Small pen flashlight
____ Extension cord for remote control unit	____ Slides in tray, sequenced
____ Electrical power extension cord	____ Extra slide trays
____ Multiple plug or gang plug box	____ Carrying case
____ 3-prong electrical plug adaptor	____ Equipment checked & in working order
____ Tongue depressor to remove stuck slides	____ Tray(s) tested on projector(s) being used
____ Duct tape	____ Tape recorder, if needed
____ Screwdriver, pliers, electrical tape	____ Synchronization unit
	____ Script & presentation notes
	____ Pamphlets, brochures, or reports for audience

sequence slides. Know how to change bulbs, restore the slide tray to its proper place, and clear stuck slides.

Presenting the Slide Set

To build a positive, professional image, arrive early, set up your equipment, and test it. Learn how to control the room lights, and make sure that nothing blocks anyone's view. If necessary, rearrange seating. Make sure that you have a light on the podium so that you can follow the script. If you'll be pointing to slides, have a lighted pointer handy.

When you begin your presentation, face the audience, and stand to one side so that you can check the slides with a glance. Use your script as notes to narrate the slide set. Leave your slides on the screen only long enough for the audience to grasp the point you're making. As you end the presentation, turn the lights up slowly so that you don't blind the audience. Then turn the equipment off according to directions.

Evaluating the Slide Presentation

As you give your presentation, take mental notes of audience reactions to your points. How did the audience react to your message? What questions did the audience ask? Did the questions suggest that you stimulated thinking or created an emotional response? Besides the questions raised, read the audience's reactions by its attention and mood.

Some months after you've finished the slide set, review it to see where you might improve it. Then use your evaluation to improve your next slide presentation.

HIGHLIGHTS

1. Good professional presentations communicate a few key points clearly and quickly to the audience.
2. Both the professional talk and the 35mm slide presentation integrate oral and visual communication techniques.
3. The professional talk integrates projected pictures, projected words, spoken words, and body language.
4. The suggested process for professional presentations includes analyzing, planning, designing, rehearsing, producing, practicing, delivering, and evaluating.
5. Slide sets provide an effective audio-visual communication tool for enhancing professional image.
6. The keys to professional slide presentations are planning, producing, practicing, presenting, and evaluating.

PROJECTS

1. Select a topic. Prepare and present a seven- to 10-minute professional talk on it.
2. For the major written paper for your technical communication course, prepare and give a five-minute professional presentation with adequate visuals.
3. Evaluate the presentations of your classmates. In what ways was each presentation especially strong? What would have improved each presentation? How well organized was each presentation? How well did your classmates use visual aids? How would you rate the delivery quality of the presentations?
4. Select a topic for a slide set, and prepare a 600-word description of your objectives, the audience, and its characteristics.
5. Prepare a proposal and treatment for the slide set described in 4.
6. With a three-column outline, identify the sound, visuals, and audience reactions to your script.
7. Using your working outline as a guide, draft a split page script. Include descriptions of each slide in your narrative.
8. Obtain the slides, and produce your slide set.
9. Give your slide presentation to your class as if it were a professional meeting.
10. Attend the presentation of a technical specialist or scientist—not a class lecture—and then prepare a short report explaining the approach and techniques the presenter used. What were the strengths of the presentation? The weaknesses? Had the presenter been a colleague, what suggestions would you give to him or her?

FOR MORE HELP

Index to Kodak information. 1987. Rochester, N.Y.: Kodak.
   Kodak lists dozens of useful how-to photography publications in this index, which is published annually.

JOENK, R. L. Public speaking for engineers and scientists. *IEEE transactions on professional communication,* March 1980, 1–60. PC-23.
   The issue provides an excellent review of professional presentations.

MATKOWSKI, B. S. 1983. *Steps to effective business graphics.* San Diego: Hewlett-Packard.
   Matkowski discusses preparing business presentations and provides excellent guidelines on developing visuals.

REYNOLDS, L., AND D. SIMMONDS. 1982. *Presentation of data in science*. Boston: Matinus Nijhoff.
The authors explain how to use artists' and drafting supplies in preparing professional publications and presentations.

SINDERMAN, C. J. 1982. *Winning the games scientists play*. New York: Plenum Press.
Having observed hundreds of scientific presentations, Sinderman suggests effective presentation techniques and identifies common pitfalls.

16
Correspondence: Putting Your Best Self Forward

OVERVIEW

In Chapter 16, we

- ☐ Indicate the types of correspondence technical and scientific professionals regularly do
- ☐ Discuss principles of composing effective letters and memoranda
- ☐ Describe popular formats and explain their advantages and disadvantages
- ☐ Illustrate various approaches to effective correspondence: directness, informality, and friendliness
- ☐ Discuss computerized and guide letters
- ☐ Describe the emerging system of electronic mail

LETTERS CAN SHOW YOU AT YOUR BEST

In professional life, it pays to approach written correspondence with the same care you would use preparing to meet company stockholders, influential customers, or a committee interviewing you for a job.

A letter is what you send when you can't go yourself. Use a misspelled word, an eccentric format, an infelicitous or awkward phrase, and you've done the equivalent of wearing mismatched shoes to a formal dinner. If you're lucky, you'll be forgiven, but your gaffe will never be forgotten.

Conversely, a carefully and correctly written letter opens doors, makes an excellent first impression, and represents the absent you better than any deputy.

MAJOR CATEGORIES OF CORRESPONDENCE

We'll discuss the major types of business letters:

- Cover letters, or "letters of transmittal." They simply, briefly tell the addressee a package contains a report, equipment as ordered, or the like.
- Complaints, from customer to firm or firm to customer (usually about nonpayment); from supervisor to employee and perhaps vice versa.
- Inquiries or requests for information.
- Memoranda about day-to-day business within an organization.

Another common type of letter, the application, we discuss in Chapter 17, Job Searching.

Letter of Transmittal

With its major purpose to signal the arrival of information or material, the cover letter, or letter of transmittal, may be brief. Ten months after it had been established to investigate the assassination of President John F. Kennedy, the Warren Commission sent to President Lyndon Johnson a document running more than 700 pages. The letter of transmittal for the study of that terrible and historic event read in full:

> Dear Mr. President:
>
> Your Commission to investigate the assassination of President Kennedy on November 22, 1963, having completed its assignment in accordance with Executive Order No. 11130 of November 29, 1963, herewith submits its final report.

If you wish to acknowledge special help, if you are conveying information earlier or later than expected, if parts are still to come, or if there are other aspects you wish to call to the recipient's attention, you may use the letter of transmittal to do so. A typical transmittal letter for a technical report appears in Panel 16-1. Notice its brevity, cordiality, and directness.

Claim or Complaint Letters

Much more difficult than a cover letter, the claim, or complaint, letter requires careful thought and planning. When you complain about something, you usually want more than to let someone know you're unhappy. You want action. You want a wrong righted, an injustice corrected, a faulty machine repaired or replaced. Your challenge? To phrase your complaint so that you get what you want.

PANEL 16-1. Letter of Transmittal

December 3, 1986

Professor Willard Schmehl
Department of Agronomy
Colorado State University
Fort Collins, CO 80523

Dear Professor Schmehl:

    I am pleased to submit the attached technical report, "Wind-Distributed Crops of Nutritional Value to Thirteenth Century Native Americans of the Great Plains."

    As a student in your AG 430 class this semester, I have enjoyed the opportunity of carrying out this research project, and I look forward to receiving your comments on my paper.

                                              Sincerely,

                                              Roger P. Murdoch

Enclosure

356 Aylesworth Hall
Colorado State University
Fort Collins, CO 80523

    Robert L. Shurter, longtime director of Humanities and Social Studies at Case Institute of Technology and author of the enormously successful *Effective Letters in Business* (1954), notes four components of a good claim letter:

1. An explanation of what is wrong, including specifics such as model number, dates, sizes, colors, or any other details the reader needs for a recheck.
2. A statement of the inconvenience or loss that has resulted.
3. An attempt to motivate action by appealing to the reader's pride, professionalism, sense of fairness, or honesty.
4. A statement of what you consider a fair settlement, or, if you don't know that, an effort to obtain prompt investigation of your claim.

Undergirding all those points, Professor Shurter argues, should be the attitude that we all are human and make mistakes from time to time. It's the old "catch more flies with honey than vinegar" argument. It usually succeeds (Panel 16-2). Although courtesy and gentleness usually prevail, when they do not, you need to increase the pressure. Keep in mind the

---

PANEL 16-2. Letter of Complaint

January 2, 1986

Fruit Company
P.O. Box 87
Harlingen, TX 78550

Dear Fruit Company:

    I am sorry to have to report that the box of ruby red Texas grapefruit I asked you to send to Mr. and Mrs. Marion Gray arrived in bad condition.

    The Grays tell me that more than a dozen of the fruit were rotten when the box arrived. That was the order I made on November 12 for Christmas delivery of your TRR No. 14. The Grays received it on December 21 at their home, 222 North 17th Street, Manhattan, KS 66502.

    It was not easy for the Grays to inform me that my gift was spoiled. Naturally I am quite embarrassed that it reached them in less than excellent condition. Over the past several years, the exceptional size and sweetness of Fruit Company grapefruit in other Christmas gifts I've sent have brought many compliments.

    Do you think you might send a replacement order? Your gesture would certainly reconfirm my trust in you, and, I might add, the Grays' in me.

    Thank you.

                                    Sincerely,

                                    David G. Clark

P.O. Box 878
Deepwoods, CO 80512

need for restraint even as you keep in mind, as Benjamin Franklin put it, that creditors have better memories than debtors. If you threaten too soon, you will certainly alienate your reader.

In responding to a complaint or claim letter, you must be firm without being offensive. Your organization probably endorses one of two positions. Either the customer is always right, and any claim receives full reimbursement or replacement, or fair adjustments are made as the evidence indicates. The second position puts a letter writer in a more ticklish position, for company and complainant may not agree on what "fair" means.

Professor Shurter (1954) lists four principles to observe when you respond to a complaint:

1. Every complaint has merit to the person making it. No complaint is trivial.
2. Every complaint deserves prompt handling or acknowledgment.
3. Your answer should be factual, courteous, and fair.
4. Argument or criticism rarely put people in a generous mood, so avoid them.

Observing the last principle requires courtesy and fairness, without a weakening of your own position. Suppose that your boss asked you to respond to the following:

> The lawn mower I bought from you 3 months ago is a total disaster. How can you sleep at night knowing your company produces such trash? When I put gas in the tank, it all ran out onto my grass, killing it. I discovered the tank was not connected to the engine. I want my money back plus $100 for my dead grass, you nerds.

After investigating, you find that the assembly instructions accompanying the mower clearly state that the tank must be connected before gas is put in. You suspect that the customer tried to operate the mower without first reading the instructions. Your company's policy is to do what's fair on claims. Given the intemperance of the letter, your first instinct might be to respond to kind, with something like:

> Come off it. You can read, can't you? Try following the instructions for once. If you do, you might be surprised. And forget the hundred bucks. We're not about to replace your weed farm.

Get that out of your system, toss it into the wastebasket, and get to work. If you decide against an adjustment, you might try something like:

> Sir:
>
> We regret very much that you had an unhappy experience with our mower. Like all power tools, it is a complex machine that requires customers to follow the instructions provided. Instruction No. 4 points out that the gasoline tank must be connected before it is filled.

> Company policy does not permit us to refund the purchase price after equipment has been owned more than 30 days. Nor does it allow us to reimburse for damage caused when directions for assembly or use are not followed.
>
> We feel certain that our mower will give you many years of excellent service. So that you may use the machine efficiently and safely, we urge you to follow the use and maintenance instructions carefully. We enclose an additional set of instructions for your use.

That response avoids disputing with the customer, offers sympathy, and firmly maintains the company's position. The letter meets the tests of taking the complaint seriously, avoiding criticism, and detailing the policy against refunds.

In planning correspondence, first think through what you want the letter to accomplish. Then think how to accomplish that purpose. You want to show the reader that doing what you want is in the reader's best interest. To accomplish that you must see things from the reader's point of view.

Letters to Colleagues

Letters to supervisors or to people working under your supervision also require careful consideration of readers' viewpoints. You and your company may survive a poorly done letter to a customer, but clumsy correspondence between colleagues fosters ill feelings that may hamper an organization's effectiveness.

Suppose, for example, that someone under your supervision has been consistently late for work for several months. You've spoken to the person, and he promised to do better but in fact has not. Your supervisor thinks it's time you put something in writing, in case the offender eventually must be fired. You'd hate to lose the person, who is an excellent worker, but other workers are losing their concern for promptness. Your objective is clear: to get the latecomer to show up on time. You don't want him to get angry and quit. You use a sympathetic approach.

> Dear Mike:
>
> As you know, company policy requires all of us to be at our work stations at the start of our shifts. One person's lateness in arriving causes unnecessary delays in getting the day's work under way. Moreover, the example both encourages other workers to lose regard for promptness and reflects badly on your professionalism. We should regret it very much if something as minor as your continuing lateness forced us to take strong action.
>
> Consequently, I wonder what we can do to help remove the cause of your consistently late arrival. Will you give the matter thought, and discuss it with me as soon as possible?

In the real-life situation from which that example came, the employee responded positively. The letter said, in effect, "This is serious. You're hurting us and setting a poor example, but we regard you highly and want to help, if you will let us." It turned out the employee had to take his small child to school most mornings. That done, he arrived at the parking lot too late to find a place near his work. When told of the problem, the supervisor arranged a reserved parking space. Everyone felt good, and the tardiness stopped.

Inquiries or Requests for Information

When you ask that someone do something or provide information, that person presumably has the power to refuse. Certainly that person has the power to act promptly or not. Phrase your request so that you receive a positive, prompt answer. Observing a few principles will help:

- Learn the exact name and title of the person whose help you seek. If you can't do that, address your request to the highest-level person you can find. A note from the company president to a subordinate asking, "Can you help this person out?" becomes a request from the boss.
- Tell why you need the information so that people don't define your request as frivolous and give it low priority.
- State your deadline, if you have one. If you need the information for a term paper, and you must write that section by a certain time, say so. But avoid a demanding tone.
- If your request will require more than a little effort, acknowledge that you realize what you are asking for. Many people will go to great lengths if you acknowledge the imposition.

Memoranda

A letter sent within an organization is generally a memorandum. Except in format, internal memos and external letters differ little. Do one well, and you can do the other equally well. When you prepare on company time a memo that is to be read on company time, you're expected to focus sharply and be brief—both attributes of good business letters. Memo format, described below, is brief in eliminating salutation, complimentary close, and other niceties. When you design a memo, ask:

- Why does the person need to know this information?
- How much does the person know about the situation I am describing?
- How much will the person understand?
- What do I want the person to do with the information I will present?

Memos serve two functions: to record and convey information laterally or vertically within an organization; and to serve as bases for decisions. Asking the first three questions above will help as you *record* information

CORRESPONDENCE: PUTTING YOUR BEST SELF FORWARD

PANEL 16-3. Sample Memorandum

Today's trend is away from extensive use of titles and other lengthy identification within organizations. The assumption is that everyone knows the chain of command.

```
                       MEMORANDUM

     TO:     Susan Jones               January 8, 1987

     FROM:   Norm Smith

     SUBJECT: Breckenridge Tract Survey

          I completed the survey Monday and results arrived
     today. They are enclosed for your review. To summarize, we
     may expect to hit oil at around 8,000 feet, but drilling
     costs will exceed those of our other wells in the area by
     at least 30 percent, because of the formations we will
     have to drill through.

          Our choices include:

          1. Going ahead now, even with currently depressed
             prices in the market. Rig rentals are cheap, and
             we might save enough on rent to make up for the
             extra time involved.

          2. Holding off for six months or so, when an expected
             upturn in oil prices might make drilling more
             attractive.

          3. Selling our lease to someone else. We might not
             get what it's worth, but we could pick up some
             quick cash.

          I favor going ahead now if we can afford it. I feel
     as good about this survey as I did the Jackson County one.
     I'm sure there's oil there.

     Enclosures
```

---

in a memo. Asking the fourth will help when you seek a *decision*. Present what you see as the decision maker's options, with a brief evaluation of each. Far from being thought presumptuous, you will probably be appreciated. Panel 16-3 illustrates a memo seeking a decision.

PARTS OF A LETTER

Letters commonly contain several parts, some optional and some not. Most business letters include:

1. Letterhead identifying the organization.
2. Date, typically stated as month, day of the month, and year. Military and military-like organizations state the date as day, month, and year.
3. Inside address, that of the person to whom the letter is sent.
4. Salutation, a stylized or conventional greeting. A few decades ago, the style was "My dear Jones," but today "Dear Mr. Jones:" is preferred. For women, "Dear Ms. Jones" is acceptable, especially if you don't know whether or not they are married. Many women prefer the neutral "Ms.," married or not.
5. Body of the letter.
6. Complimentary close, again stylized and conventional: "Sincerely," "Yours truly," or perhaps "Cordially," if you know the person well. Avoid the embroidered, insincere style of earlier decades.
7. Signature.
8. Typed name and, usually, a title.

Optional and specialized parts of letters are:

9. Direction of letter to the attention of a particular person.
10. Reference to a numerical file or previous letter.
11. Reference to subject treated by the letter.
12. Enclosure.
13. Distribution list.
14. Postscript, usually added by hand by the writer after the letter has been typed.

HELPFUL HINT

Always keep a copy of your correspondence. No one knows why, but fate always seems to rule that it's the letter whose copy you threw out that is the one you need desperately later.

LETTER AND MEMORANDA FORMATS

The format of letters and memoranda used to be the concern primarily of secretaries and stenographers. Nowadays everyone needs to know proper correspondence formats.

Letters outside an Organization

The major letter formats are block and indented. Each combines elements of the other, and you may see virtually any style from time to time. As

> ## PANEL 16-4. Common Letter Format: Block
>
> January 13, 1990
>
> Mr. Robert Perry
> Field Geologist
> Forty-Niner Mining Company
> P.O. Box 1849
> Golden, CO 80225
>
> Dear Mr. Perry:
>
> I understand your company is providing no-cost training in up-to-date placer mining techniques in return for permission to explore acreage in Jackson County, Colorado.
>
> My family owns three sections of land southwest of Walden on the edge of the Mt. Zirkel Wilderness Area. To our knowledge, the land has never been prospected for either gold or silver.
>
> If your company is interested in discussing opportunities, may we hear from you at your convenience?
>
> Thank you.
>
> Sincerely,
>
>
> Matthew Sutter
>
> Rural Route 7
> Walden, Colorado 80621

their names suggest, the two styles differ in the use of paragraph indentations and the placement of the date and complimentary close. Both block (Panel 16-4) and semi-block with indentations (Panel 16-5) are common. Other variations exist, such as a full block, which stacks every element (date, salutation, signature, addresses) at the left margin, and hanging indentation, which indents each line of the elements a space or two. The block and semi-block forms prevail in these days of high-cost correspondence for two reasons.

With the average business letter costing around $8 to produce in the early 1980s, simple formats in which typewriter tabs could be preset were desirable, because this saved secretarial time and money. The second

PANEL 16-5. Common Letter Format: Semi-Block

```
                    Forty-Niner Mining Company
                           Golden, CO

                                            January 27, 1990

Mr. Matthew Sutter
Rural Route 7
Walden, CO 80621

Dear Mr. Sutter:

    We are quite interested in talking with you about ex-
ploring your land near Walden. Unfortunately, our recent
offer to teach placer techniques resulted in so many invi-
tations to explore that it will be some time before we are
in your area.

    As soon as I can arrange a trip your way, I'll be back
in touch so we can get together and discuss matters. It may
be three or four months before I am in the area.

    Thank you for your interest.

                                            Sincerely,


                                            Robert Perry
                                            Field Geologist
```

reason is aesthetic: a simple, spare format is more pleasing to correspondents.

A third major format has come into use, most often by companies soliciting from people whom they don't know. The simplified form, as it is known, features no salutation or complimentary close and contains a subject line—usually in all capitals—a few lines below the address of the person receiving the letter. The memo format, described below and in Panel 16-3, is becoming popular for outside-the-organization correspondence, because it saves space and avoids ambiguities in matters of address.

Besides economics and aesthetics, a reason for using a common business letter format is to show that the writer is part of the professional community. Most organizations may reward originality in research and development, but none we know of encourages originality in formatting correspondence.

The same unwritten law decreeing no misspellings or grammatical errors decrees the use of an acceptable format.

Letters inside an Organization (Memos)

Many organizations provide printed forms for memos:

```
                    MEMORANDUM
                                        Date:

To:

From:

Subject:
```

Such forms often come in two versions, a half-page and a full page. Various memo formats may be used, but all compress the elements of a letter into a compact and standardized shape. Memo format enables you to work as swiftly as possible.

PAGE ARRANGEMENT AND SPACING

Take pains to place your letter or memo attractively on the page. Doing so may require adjusting margins, if you have a brief text. Instead of a 60- or 65-unit line, you may choose a 40-unit one, to improve the letter's overall appearance. If you use company letterhead, avoid a crowded look.

Usual practice calls for paragraphs to be single spaced, with double spacing between paragraphs and other major divisions, such as between inside address and salutation, between salutation and text, and between text and complimentary close. Panel 16-6 elaborates on spacing. A good secretarial text will cover these and other points, or a friendly secretary can give you your organization's preferred model. Other ways to obtain good examples are to save letters you receive and to buy a guide for placing letters on the page at an office supply store.

PANEL 16-6. Spacing a Letter

Locating your letter attractively on the page may require using wider than usual margins, if the letter is short. Never use extremely narrow margins. Likewise, you may vary spacing among various components of a letter, so long as you never use fewer than two lines between major components. The guide below may help.

                                    Date (2 to 6 lines below letterhead)

Inside Address (2 to 12 lines below date)

Salutation (2 to 4 lines below inside address)

Body (2 lines below salutation)
    (Body is single spaced, with 2 lines between paragraphs)

                                    Complimentary Close
                                    (2 lines below last line of body)

                                    Typed Name
                                    (4 lines below complimentary close)

Other Options: enclosures, copies to, return address (2 to 5 lines below typed name, with 2 lines separating each option)

If your correspondence exceeds one page, identify successive pages by a slugging system of some sort, such as:

Dr. Joann Taylor, 2.   May 26, 1987

Repeat the name of the person addressed, the letter's date, and supply a page number in case the pages get lost or out of order.

PANEL 16-6. (continued)

This example shows an attractive, well-spaced letter that follows generally accepted standards for business letter format. The content is also a guide for setting up and spacing your own professional correspondence.

```
                                        July 1, 1987

Dr. Edward Carpenter
1920-23rd Street
Plainview, TX 79411

Dear Dr. Carpenter:

    Spacing a letter properly requires both judging well
what pleases aesthetically and following generally
accepted form.

    Place the letter on the page attractively, neither
too high nor too low. Extremely short letters may require
using wider margins to achieve symmetry, but a 60- or 65-
unit line usually is best.

    Put the date line 2 to 6 lines below the letterhead.
Single space the inside address 2 to 12 lines below the
date. Put the salutation 2 to 4 lines below the inside
letter's body. Single space the body, but leave 2 lines
between paragraphs. For readability's sake, keep your
paragraphs short.

    The complimentary close goes 2 lines beneath the last
line of the body, and the typed name and title go 4 lines
below the close.

    Options such as enclosures, copies to, and return
addresses should go 2 to 5 lines below the title or name
line, with 2 lines separating each option.

                                        Sincerely,


                                        David G. Clark
                                        Professor

3105 2nd Street
Deepwoods, CO 80512
```

LETTER PUNCTUATION

Various styles of punctuating letters exist in the United States. They are often referred to as open, closed, or mixed, and they all have variations. Outside the country still other forms exist. A safe, middle-of-the-road approach, as applied in Panels 16-1 and 16-4, includes:

- Use no punctuation at the end of date lines, receiver's name, writer's name, or any of the other address lines, with the exception of a period at the end of an abbreviation.
- Use a colon following the salutation and a comma following the complimentary close.
- Punctuate within lines as good usage dictates.
- Punctuate the text of the letter as good usage dictates.

ADDRESSING THE ENVELOPE

These days you have a choice between an aesthetically pleasing address and one that meets U.S. Postal Service (USPS) recommendations. We vote for the latter, because business correspondence needs to arrive as quickly as possible. The USPS suggests:

- Use all capital letters.
- Do not punctuate.
- Use ZIP Code Directory abbreviations.
- Use two spaces between word groups, such as street address and apartment or street number.

Single space the sender's and the receiver's address, and include for both the individual's name and/or company, street or post office box address with suite or apartment number on the same line, and city, state, and ZIP code. Center the receiver's address on the lower half of the envelope. Place your return address in the upper left corner.

AVOID SEXIST LANGUAGE

Chapter 10 contains guidelines for avoiding sexist language. Here we consider the question of how properly to address people whose identity, sex, or both are unknown to you. In these years of institutional concern

CORRESPONDENCE: PUTTING YOUR BEST SELF FORWARD

for equal opportunity and fair treatment, old methods no longer serve. The anonymous but masculine "Dear Sir" or "Gentlemen" risk offense.

Jan Venolia, author of *Better Letters: A Handbook of Business and Personal Correspondence* (1982), recommends several possible salutations. To address people of unknown gender:

    Dear Jan Venolia: Dear J. Venolia:

Instead of the traditional Mr., Mrs., or the sometimes controversial Ms., try a simple M.

    Dear M. Venolia:

When You Don't Know the Addressee

"To Whom It May Concern" sounds like something Ebenezer Scrooge would use to certify the worthlessness of Christmas. Avoid that. Some categories permit a fairly graceful "generic" title.

    Dear Fellow Scientist: Dear Customer:

If you know the job title of the person whom you seek to address, you might use:

    Dear Mail Teller: Dear Personnel Manager:

or, as in Panel 16-2, you might repeat the name of the company.

    Dear Fruit Company: Dear Internal Revenue Service:

If you feel uncomfortable addressing an organization as "dear," consider a memo format, which is becoming acceptable for external correspondence.

To: Fruit Company
          P.O. Box 12
          Harlingen, TX 78550

From: David G. Clark
          P.O. Box 878
          Deepwoods, CO 80512

Subject: A Box of Spoiled Grapefruit

Society is still in transition away from the presumably all-male business world. Eventually, we will have comfortable new conventions. For now, realize that many men and women feel sensitive about these issues.

MAKING YOUR LETTERS "SING"

Have you ever received a letter which began, "As per your request of such and such date" or, "Reference yours of the 24th, month preceding"? You must have read letters full of phrases like, "according to our records," "please be advised that," "allow me to say," or "this will acknowledge receipt of your letter." Why people with otherwise good sense put such stuff into their letters remains one of life's mysteries.

In a piece on writing effective business letters, Malcolm Forbes, president and editor-in-chief of *Forbes* magazine, made a telling point. "Suppose I came up to you and said, 'I acknowledge receipt of your letter and I beg to thank you.' You'd think, 'Huh? You're putting me on.'" (1983, p.1) Forbes's lesson, that we should write as we speak, naturally and directly, is simple to grasp but often difficult to apply. For starters, eliminate the trite, pompous, wordy, meaningless phrases that clutter business and professional correspondence. The samples below reflect only a few sad possibilities:

Crime	*Solution*
"According to our records"	"We find"
"Attached please find"	"We attach"
"At this point in time"	"Now"
"For your information"	omit
"This letter is for the purpose of requesting"	simply ask
"In accordance with your request"	"As you requested"
"For the purpose of"	"to," "for"
"In order to"	"to"

By stripping away such verbiage, you lay bare the thought you seek to express. Your letter becomes direct and concise.

To make your letters sing, letter writing experts agree that you should:

- Keep your paragraphs short.

- Make the first paragraph *say* something.

- Be sure that your facts are accurate, your usage correct, and your language precise.

- Be friendly, and stress the "you" rather than the "I," to show that you understand the reader's perspective.

- Be prompt in responding to someone else's letter; avoid the negative message that silence sends.

- Stop when you've said everything necessary.

- Revise.

FORM LETTERS

Government agencies and companies sending lots of mail often use form letters. The federal Office of Records Management (ORM) points to three advantages of form letters over individually written ones:

1. Promptness—mail can be answered faster.
2. Readability—more time can be spent editing and revising a form letter than a hurriedly written one.
3. Economy—form letters cost less. (Office of Records Management 1973)

When you are saying essentially the same thing in letter after letter, the ORM recommends printing letters, so that you have to add only the name and address. A government study showed that printing is economical if 30 or more copies of the letter are sent in a month and the letter contains five lines of copy. If a letter has 20 or more lines, printing is economical if 10 copies a month go out.

Form letters are inappropriate in some circumstances. The ORM's test for appropriateness is good taste: "Discriminating readers quickly detect the stilted quality in a form letter when a personal letter would have been more appropriate. Letters expressing sympathy, appreciation, commendation, or apology usually fall in this category" (Office of Records Management 1973, 10). Although you may, as an officer in a technical or scientific professional organization, draft and send form letters, you are unlikely to use them as an individual. Never use them in job searching, no matter how uncaring the world may seem.

GUIDE LETTERS

When you have to respond to many different people on one topic, but you need to tailor your responses, guide letters can help. Developing a useful guide letter entails recognizing the points that vary and isolating them in separate paragraphs. Then you select, often by a numbering system, the paragraphs you need to assemble the final product.

The need for a guide letter may become apparent over time. For example, a university department office may receive inquiries from

- high school juniors seeking career information
- high school seniors deciding whether to apply for admission
- college students elsewhere thinking about transferring
- former students wishing to return to school
- people thinking of retiring or leaving their jobs to go to college

The response to each would contain much of the same information on career opportunities, the university environment, tuition, and other expenses. But each would need special treatment, too. High school students might not know much about the semester or quarter system universities use. Transfer students would want to know how to get their transfer credits evaluated. Former students would want to know if the credits they took years ago would still apply. A guide letter having paragraphs dealing with such special points could save much time. Members of Congress and other people with heavy correspondence find guide letters essential.

At least one computer firm, Hewlett-Packard, in 1984 marketed an all-purpose software program using the guide letter principle. Known as "Memo Maker," it provided various statements for salespeople to use in maintaining contact with customers. Using a menu, the "writer" selects the appropriate friendly phrases, even down to inquiries about the receiver's health and family members. Then the computer shuffles everything together and produces a warm and solicitous letter.

ELECTRONIC MAIL

When Arthur C. Clarke finished last-minute corrections of *2010: Odyssey Two*, his sequel to *2001: A Space Odyssey*, he sent them from Colombo, Sri Lanka, to New York via the Padukka Earth Station and the Indian Ocean Intelsat V satellite. These days electronic mail can be sent and received by everyone. MAILGRAM, introduced by the U.S. Postal Service and Western Union nearly two decades ago, Telex, and TWX, even older, are such systems.

A modem, or telephone line, attached to your personal computer puts the world at your fingertips. For example, you can use MCI to send a letter via computer, and MCI will make certain that your letter reaches its distant destination the same day or overnight, depending on which service you buy. Telenet, Tymnet, and other services operate electronic mailboxes that let people deposit their messages electronically for you to retrieve at your leisure.

Electronic delivery systems intensify rather than reduce the need for effective writing. Because basic charges are for time spent on line, conciseness remains essential. Clarity and readability give your letters a competitive edge over those of others, no matter how they are delivered.

HIGHLIGHTS

1. Common types of correspondence include cover letters, responses to complaints, inquiries, and internal letters or memos. Definite principles of effective communication apply for letters of each type.

2. Business and professional letters typically contain eight essential and several optional parts.
3. Although various formats exist for professional correspondence, policy and economics, rather than aesthetics, usually dictate the ones used.
4. Various alternatives to sexist language exist and should be used.
5. The most effective professional correspondence is brief, direct, courteous, accurate, precise, and complete.
6. Form letters are sometimes acceptable; guide letters can be helpful in dealing with large numbers of similar letters.
7. Electronic mail, increasingly popular, requires conciseness and clarity as much as, if not more than, traditional mail.

PROJECTS

1. Write a cover letter for a term paper that you plan to submit to one of your professors.
2. Write a letter requesting help from an information source whose assistance you really need, and mail it. Perhaps you need information for a research paper, or you are investigating graduate schools, or you have questions about career opportunities in a certain firm. Show your professor the letter response.
3. Write a complaint letter about a real situation you recently have encountered. Ask for your money back, an adjustment, or a replacement. Make the request as courteous, specific, and effective as you can. Mail it.
4. Your supervisor cannot attend a certain meeting and asks you for a report on what happens. Write a summarizing memorandum about any meeting or class you choose for your supervisor.
5. Improve the following.

MEMO

To: Don Clark, Division Manager

From: Dave Zimmerman, Section 2

Subject: Relocation of the Mail Room

    I don't know whether you are aware of this or not, but the mail room has been moved from what was a convenient location to an absolutely intolerable spot so far from our office that no one wants to go for the mail anymore.
    So far as I can tell, this was an arbitrary decision by the mail service. They did not consult anyone I've talked to. When I asked the mail carrier why, he said it was to speed up delivery. I think

it was really to let the mail carriers get through work earlier so they can take longer coffee breaks.

Our secretaries now consider a trip to the mail room as punishment. As a result, nobody wants to go, especially on days the mail is heavy. That means our outgoing mail is going to be later, since the incoming mail arrives later.

If you really care for the sections under your management, you'll find some way to get the mail room put back where it was, a few steps down the hall from our office.

FOR MORE HELP

Doris, L., and B. M. Miller. 1983. *Complete secretary's handbook.* 5th rev. ed. Englewood Cliffs, N.J.: Prentice-Hall.
Complete and authoritative, this book is a standard in the field.

Murphy, H. A., and H. W. Hildebrandt. 1984. *Effective business communication.* 4th ed. New York: McGraw-Hill.
This book contains excellent advice on writing a wide range of business correspondence.

Venolia, J. 1982. *Better letters: A handbook of business and personal correspondence.* Berkeley, Calif.: Ten Speed Press.
This well written brief book covers the major points of good letter writing.

17
Job Searching from a Communicator's Perspective

OVERVIEW

In Chapter 17, we
- Explain job searching strategies and resumes
- Discuss how employers find candidates
- Suggest job searching techniques

Job searching gives you a good opportunity to apply your communication skills to solving a real problem. After all, finding just the right job entails, at least in part, solving communication problems. A problem-solving strategy may help you land your first job (Figure 17-1).

A strategy is important for four reasons. First, if you're like most college graduates, you'll search for five, six, seven, or more jobs in your life, and you'll change careers twice or more. Once you've learned it, a good strategy will help you throughout your life. Second, you need to present yourself in the best possible way when job openings are limited. You may face intense competition for any openings you find. Third, during economic recessions that make jobs scarce, an organized strategy will help you to market yourself effectively. Fourth, many first-time job seekers fail to consider *employers'* approaches to hiring. A good strategy can help avoid that problem.

Good communication skills will increase your chances of getting a job offer. Know yourself, what you want to do, where you want to go, and communicate these things to prospective employers.

JOB SEARCHING STRATEGIES

Some job searches are conducted off campus, some on campus. For on campus interviews, rather than mailing a cover letter and resume to prospective employers, you speak with their representatives who visit

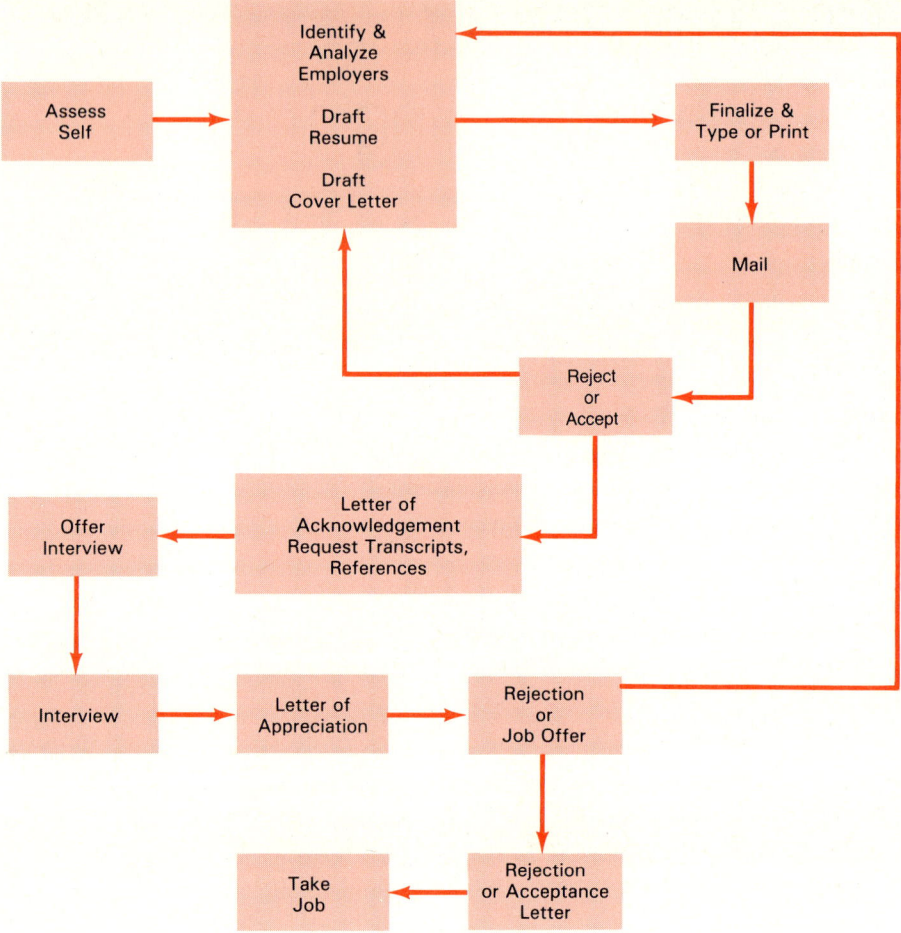

FIGURE 17-1. Job searching off campus.

campus once or twice a year. On many campuses, the career services or placement office coordinates such interviews. On others, departments coordinate the interviews.

You usually sign up a week or more before a campus interview. You provide the career office with a copy of your resume and show up for your interview at the assigned time. If you make a favorable impression, some weeks later you'll be invited to an on-site interview. From that point, the job searching process follows a sequence similar to that off campus, as shown in Figure 17-1.

Some college students conduct their own search. They call potential employers, or they visit an organization and inquire about job openings.

They fill out applications and may be asked to return for further interviews.

Whichever strategy you use, you'll find that it goes more smoothly if you know the resources available and use them to your advantage.

RESOURCES FOR JOB SEARCHING

Resources abound to help you with your job searching. You'll find most resources in your career services office and the department of your major.

Career Services

Most colleges and universities have a career services office to help students and alumni with job searching. Check first with that office. Most have manuals with job-search guidelines and other material. Be sure to ask for a copy of *The CPC Annual,* published annually by the College Placement Council. The *Annual* includes detailed guidelines for conducting a job search as well as employer and employee listings. Be sure to ask, too, about the resources your career services office provides:

- campus interviews with employers from government, business, and industry
- credentials services, personal files that include resumes, references, and related documents
- resumes and correspondence workshops and critiques, where staff members will help you to prepare a resume and application letters
- resume notebooks, a file of resumes collected from other graduates
- alumni job bank
- summer jobs and internship listings
- library, directories, annual reports, books, and other literature
- government job information
- practice interviews, including videotapes and critiques
- job vacancy bulletins in government, business, industry, and education

Department Support

Don't overlook your adviser and other professors in your department when you are job searching. Many know potential employers, and most have been through several job searches. They can provide insights unique to your profession. Many department graduates may call your department when they are looking to fill openings. Your department may have staff members who coordinate job lists and searches.

HOW EMPLOYERS FIND AND EVALUATE CANDIDATES

To find applicants, employers use many techniques. They send job announcements to university departments, other companies, and unemployment offices. They advertise their openings in professional publications, newsletters, trade journals, and, sometimes, newspapers. They notify professional organizations. They have their employees talk with employees in other organizations. They send job announcements to individuals who have inquired about openings. If they have several openings, they may send recruiters to universities. They spare little effort to obtain qualified candidates.

Whether you are applying for a job, a scholarship, or an award, remember that the reader's task usually is screening people out as much as in. Financial columnist Milton Rockmore (1984) reduces the process to barest fundamentals: "It's a game of match-up played against time." He quotes an employment manager at Xerox corporate headquarters:

> I don't have time to sit and read long letters. Forget peripheral information like references, commendations, and stuff like that. Leave it for later in the hiring process. We sweated writing the ad to describe the person we're looking for. Just show us you've got the matching "fingerprints."

A vice president of executive recruitment at Chase Manhattan Bank seconds the Xerox executive. "Show me you fit my specs, and I'll spend time looking at your resume" (Rockmore 1984, 10H).

Just how do you show employers that you "fit their specs"? By addressing only those aspects of the situation that your professional qualifications bear on. Use a straightforward, concise, single-page letter of not more than half a dozen paragraphs, and you have accomplished all that a letter can do. Experts agree on the following points about such a letter:

- Type; avoid fancy, cursive type styles.
- Use good quality, 8½- by 11-inch bond paper. Use white, gray, or buff.
- Steer clear of "cute" beginnings. You never know what sense of humor, if any, the reader possesses.
- Avoid talking about personal matters.
- Tell why you're applying for the job; avoid telling why you want to leave your present job.
- Avoid talk about salary.

You need only to break one or two of the rules to take yourself out of contention. Panel 17-1 shows a loser at work.

PANEL 17-1. An Application that Failed

The applicant may have had right on his side, but his application letter was no place to argue his case. He received no offer for an interview.

    Mr. O. B. Wankanobe
    Chief Forester
    Colorado State Forest Service

    Dear Mr. Wankanobe:

        I would like to apply for a job with the CSFS. For the past three and a half years I have been a forester with the state of _____. I have enjoyed my work here very much, but feel it is time to move on.

        My duties have included those of district forester in the _____ District. I have checked on timbering operations, served on fire patrol, and performed all the tasks expected of me.

        I enjoy being a forester very much and probably would not be writing you at this time were it not that I feel badly mistreated by my supervisor. I got married early this year and at that time asked for a permanent assignment in town, so I could go home to my wife at night. My supervisor first said okay, and for a while everything was all right. Then during fire season this summer he put me back into the field and required me to stay away, often three or four nights at a time.

        My wife began to get nervous at having to be home alone. Every time I got home she would beg me not to leave her alone again. This wear and tear on our relationship left us both upset. My supervisor began to criticize my work late in the summer, and when I tried to explain my side of things, he suggested that maybe I should find a different type of work.

        I still love forestry, and I still want to be a forester. That's why I'm writing you. Please see if you can find some way to help me. Thank you.

                                  Very truly yours,

As employers review cover letters and resumes, interview candidates, and check references, they match candidates against their company's needs. Many questions influence their review:

- Does the candidate have adequate knowledge, background, and skills?
- Has the candidate demonstrated those skills in classes, jobs, or volunteer work?
- When given an assignment, has the candidate followed through with little or no guidance?
- Is the candidate a self-starter?
- Does the candidate know the limits of his or her ability?
- Will the candidate know when to seek help?
- Will the candidate need close supervision?
- Can the candidate follow directions?
- Is the candidate resourceful?
- Does the candidate understand the position's demands?
- Is the candidate trustworthy and honest?
- Will the candidate continue to grow and develop as a professional?
- How well would the candidate fit in the organization and work with other employees?

What turns employers off? The answer varies among employers, but high on the list of negatives are weak letters with typos, spelling, and grammatical errors; ignorance of the organization; self-centeredness; truth stretching or outright lying; inflexibility; inflated expectations about starting position or salary.

JOB SEARCHING TECHNIQUES

Successful job searching begins with an understanding of oneself—education, background, skills, and desires. With that information clearly defined, one can communicate credentials to potential employers.

Assessing Yourself

Take stock of yourself and what you want to do. Question yourself:

- How do my qualifications fit the job I'd like?
- What are my strengths for the position?
- What are my weaknesses?

- What am I doing to overcome my weaknesses?
- How will the job fit into my career plans?

Remember, few people stay with the same organization or in the same job throughout their careers. Thus, ask yourself:

- How will the job fit into my long-term plans? Short-term plans?
- Will the job help me develop new skills?
- Will it be challenging?
- How will I benefit from the job?

Analyzing Employers

A successful job search requires carefully analyzing your potential employers. Apply the audience analysis techniques covered in Chapter 5, and you'll be well on your way to ascertaining critical information about potential employers. Employers are favorably impressed by candidates who know the organization and how it operates and who ask probing questions.

You can learn about employers by reviewing material in career services offices and libraries. Check annual reports, brochures, and any other material you can find. Offices of public and community relations, personnel offices, and organizations themselves provide information on request. Newsmagazines and newspapers carry articles about employers and their products or services.

Check with your adviser to see if any of your school's graduates work for the employer. If so, ask if some would be willing to talk with you about their job. Ask professors about the employer and feedback they've received from graduates.

As you work on your degree, develop professional contacts through job fairs, tours, and other activities. Join the student chapter of professional organizations that you'll enter, attend meetings, and talk with guest speakers. If you can, talk with guest speakers in your classes. Contacts from summer jobs and other activities can help you to analyze employers.

As you gather information, take notes and make copies. For each potential employer, keep your copies in a clearly labeled folder. The file will be a handy reference for writing letters, preparing for interviews, and talking on the telephone.

Preparing Your Resume

Your resume should summarize your background, education, and experience. Most recent college graduates keep their resumes to one page, unless they've had extensive experience. Remember, your resume should prompt an employer to offer you an interview. Employers scan resumes quickly, and if the quick review satisfies them, they carefully reread the resume.

Format your resume to help a quick reader pick out key facts. Set off key sections with heads or captions (Panels 17-2 and 17-3). Use phrases, clauses, and listings. Pack as much information in as you can. But do not crowd your resume. Use strong verbs.

Organize the resume either chronologically or functionally. Chronological resumes summarize background in reverse order, with the most recent activities first. Functional resumes summarize skills. One caution about the functional resume: some employers think job seekers who use it may be hiding something (Sutton 1985). No matter which form you use, account for all your time. If you don't, employers will wonder what you did and why you didn't report it.

The contents of your resume remain up to you. Avoid including points that could cause negative reactions, such as those in Panel 17-4. Most resumes contain:

- Name, temporary (college) and permanent addresses, and telephone numbers. (A permanent address gives employers a way to contact you after you've left college.)
- Professional objectives. (Check your major professor or career service office to see if the career objective goes in your resume or in your letter.)
- Education. Report college degrees, major, universities. List degrees and schools attended in reverse chronological order. If you have an unusual combination of courses, list them by subject matter or title, not by university number. Do not list routine courses. If your grades are strong, you may say so. Provide a frame of reference, such as "3.5 GPA on 4.0 scale."
- Experience. List most recent experience first. Include employer, address, dates worked, titles, and duties. Be brief. Describe activities with strong action verbs. If your jobs do not relate directly to your career plans, report only enough information to demonstrate your ability to work with people and your willingness to work.
- Activities/Interests. Include hobbies, group memberships, leadership positions, sports, interests, and other outside activities. Be careful. Some activities may cause employers to stereotype you and eliminate you from consideration.
- References. Say that they will be "furnished upon request." Tailor your choice of references to each position. Don't use anyone's name as a reference without first asking permission and asking if he or she can give you a good reference. Give people serving as references a copy of your resume. If your career services office provides a reference service, get the necessary forms and have references on file before sending out your resume.

PANEL 17-2. Sample Chronological Resume, Side Headings

```
                         KELLY P. NELSON
                          280 Sail Place
                       Fort Collins, CO 80525
                          (303) 493-8907
```

EDUCATION: Master of Science in Computer Information Systems
              COLORADO STATE UNIVERSITY, Fort Collins, Colorado
              Graduation: August 1985
              GPA: Major-4.0   Overall-3.64   (4.0=A)

              Bachelor of Science in Forest Management
              UNIVERSITY OF MASSACHUSETTS, Amherst, Massachusetts
              Graduation: May 1982
              GPA: Overall-3.64   (4.0=A)
              Special Practicum: Report on America's Wilderness System

EXPERIENCE:
     GRADUATE TEACHING ASSISTANT for Computer Information Systems Department,
          College of Business, Colorado State University.
          Taught 2 classes each semester in COBOL programming. Consulted
          regularly with students on technical problems related to
          programming. Assisted in managing interactive computer lab
          and I/O station. January 1984–May 1985

     FOREST TECHNICIAN for Gunnison National Forest, Lake City, Colorado.
          Seasonal, full-time position.
          Lead crew of 15 technicians responsible for maintaining and
          developing recreation sites throughout the Lake City District.
          May 1982–September 1983

     OTHER POSITIONS:
              Salesman in ski shop, Fort Collins, Colorado, 1982–1983
              Carpenter in southeastern Massachusetts, summers 1979–1982
              Various part-time jobs while attending college.

ACTIVITIES: Member of Colorado State University Data Processing Club
              Member Psi Sigma Pi, National Forestry Honor Society
              Operated Purdue University Forest Management Computer
              Simulation, University of Massachusetts

INTERESTS: Skiing, backpacking, tennis, baseball, jogging, reading

REFERENCES: Furnished upon request.

(Source: Adapted from Placement Manual Spring 1986. Colorado State University Career Services Center, Fort Collins. Reprinted by permission of CSU Career Service and University Communications, Inc., Rahway, N.J.)

PANEL 17-3. Sample Chronological Resume, Internal Headings

ANTHONY PADILLA

CURRENT ADDRESS:
228 Braiden Hall
Fort Collins, Colorado 80521
(303) 491-7843

PERMANENT ADDRESS:
1806 Santa Fe Avenue
Pueblo, Colorado 81006
(303) 542-9900

PROFESSIONAL OBJECTIVE

To work in a social work capacity providing counseling and crisis intervention to youth and families.

EDUCATION AND PROFESSIONAL CERTIFICATION

Bachelor of Arts in Social Work, December 1985.
Colorado State University, Fort Collins, Colorado. Minor in Criminology.

Certification for Sexual Abuse Counselor

SPECIALIZED TRAINING

Active Listening	Treatment of Sexually Abused Children
Values Clarification	Treatment of Physically Abused Children
Group Work	Individual Counseling

PROFESSIONAL EXPERIENCE

Larico Youth Homes
Fort Collins, Colorado
    Intern: Assisted houseparents in two group homes for adolescents. Counseled individuals and families. Fall 1985

Larimer County Social Services
Fort Collins, Colorado
    Volunteer: Counseled abused children referred to Child Protection Team. 1984-1985

OTHER EXPERIENCE

Mountain Cafe
Fort Collins, Colorado
    Waiter: Served meals to a variety of customers, operated cash register, and greeted public. Summers 1982-1984

ADDITIONAL INFORMATION

Student member of National Association of Social Workers
Volunteer for Special Olympics

REFERENCES available on request

(Source: Adapted from Placement Manual Spring 1986. *Colorado State University Career Services Center, Fort Collins. Reprinted by permission of CSU Career Service and University Communications, Inc., Rahway, N.J.)*

Panel 17-4. Some Don'ts for Resumes and Letters

Career services officers recommend that you do not include the following in a resume or letter of application:

- Photographs
- Negative comments about yourself or previous employers, advisers, or colleagues
- High school information
- Reasons for leaving previous jobs
- Salary requirements
- Geographical requirements
- Personal information, such as race, sex, age, political orientation or affiliation, marital status, handicaps, or disabilities
- Personal issues or matters

Affirmative action laws forbid employers to consider or to ask about some of these matters; others are simply irrelevant to your work.

Prepare a draft resume, and circulate it to your adviser, professors, and career services staff members for critiques. Revise accordingly. If you receive conflicting suggestions, and you probably will, weigh the criticisms carefully. It's your resume, and it represents you.

Have your resume typed on an electronic or electric typewriter, letter-quality word processor, or have it typeset. Some advisers advise against typeset resumes as reflecting the printer's polish, not the candidate's. So check into the usual practice in your profession. Proofread your resume several times for typographical and factual errors. Have a friend check it too. Send only your best.

Make multiple copies on a high quality photocopier or by offset printing. Ask that the copy shop do your resume on *bond* paper. Multiple copies run on word processors occasionally insert stray letters or do not print exactly what appears on screen. Proofread each copy carefully. Handle your resumes carefully to avoid smudges or other marks. Keep them away from anything that can damage them. Never send out a poor quality or damaged resume.

Finally, start a file folder for your next job search. Put several copies of your resume in it. Some professionals update their resume every six months or so. Then they're ready to send it off if they learn of a desirable position.

Correspondence

Chapter 16 discusses correspondence principles applicable to job searching. Panel 17-5 suggests contents for the following letters: application, acknowledgment, inquiry about application status, declining offers, seeking information, and accepting positions.

For application letters, write to specific individuals. Tailor the letter to the company and the position for which you're applying. Keep the letter

PANEL 17-5. Correspondence Checklist

Letter of Application

1. Identify the position for which you are applying and indicate how you learned of the firm and position.
2. Indicate why you are applying for this particular position.
3. Describe your main qualifications.
4. Refer the reader to the enclosed resume.
5. Request the next step in the employment process—personal interview, an answer to your letter, etc.
6. Be sure to sign the letter.

Letter of Acknowledgment

1. Acknowledge receipt of offer.
2. Express your appreciation for the offer.
3. Notify the company of the date you expect to make your decision.

Letter of Inquiry of Application Status

1. Request status of application.
2. Recap history of your application.
3. State why you need clarification of status of application.
4. Include thanks for cooperation.

Letter Declining Offers

1. Decline offer.
2. Express your appreciation for the offer and the company's interest in you.

Letter Seeking Additional Information

1. Indicate interest in the company and its offer.
2. Ask for the information you need. Be specific!
3. Express your appreciation for the cooperation you receive.

Letter of Acceptance

1. Accept the offer.
2. Refer to offer letter or document.
3. Tell your travel plans and anticipated arrival date.
4. Express your appreciation and your pleasure at joining the company.

(Source: 1984 CPA Annual. Prepared by Prof. James W. Souther, former Director of University Placement Services, University of Washington. © CPC Annual 1985. Reprinted with permission.)

"you" oriented. On a draft, circle the "I's" with bright red ink. If you see two or more in every paragraph, recast your sentences. Tell how your background fits the company's needs. Don't be self-centered, but do express your needs and desires.

Rather than saying, "I would like to be interviewed between January 15 and 19, when I will be in New York," use a you-oriented approach: "Would it fit your schedule for me to drop by your office for an interview between January 15 and 19, when I'll be in New York?"

Compare, "I read the job description you sent Professor John Jones, and it sounds as though I'm just the person you're looking for," with "The job description that you sent Professor John Jones sounds as though you're looking for an individual with a background like mine."

Write your own letters. Don't find a published example and copy it word for word. When you do, your letters look like all other letters. Be yourself when you write, and sell yourself.

Keep your letters short and succinct. Communicate clearly, cleanly, and quickly. The longer you make your letter, the greater your chances of making a typographical, spelling, or grammatical error. Use only enough information from your resume to entice the reader to review your resume. Finally, proofread your letters carefully.

Keep a copy of your correspondence. Put each letter into your file of background information on the prospective employer to keep your job searching well organized.

Handling Interview Offers and Interviews

Once you've sent out your letters of inquiry and resume, you may receive an offer for an on-site interview.

Double-check the date, time, and place of the interview. Write the specifics down. If you'll fly into another city, check to see if you'll be responsible for making your way to the interview or if someone will pick you up. If time allows, confirm the details in a letter, and express your appreciation. To prepare for the interview, ask what it will consist of and which materials you should bring along. Review your resume, cover letter, and information on the employer. Be prepared to answer specific questions on your qualifications. Dress appropriately. Check with your career services office and major professor for guidance.

Make sure that you understand who'll be paying your expenses for the on-site interview. Most businesses, industries, and private enterprises pay interviewing expenses. Government agencies may or may not. Be polite, but businesslike.

At the interview, you probably will meet with several of the people with whom you'd work. You might be asked to demonstrate skills or give a short presentation. Computer programmers, for example, may be asked to write a short program. Ask your major professor what to expect.

Before the interview, think about which questions you'd like to ask. But let the employer bring up salary, fringe benefits, and other similar points at the appropriate time. Many employers discuss them at the end of the interview.

During the interview, you'll be asked many different questions, and you'll have the opportunity to ask questions of your own. Panel 17-6 contains typical interview questions. Many career service publications list additional questions. Think about what questions you might be asked and how you would respond. Listen to the questions, and think before you begin answering.

After the interview, send a thank-you letter, and express your continued interest in the position.

PANEL 17-6. Typical Interview Questions

- What are your job goals and objectives?
- How would you describe yourself as a person? Give me specific examples that relate to your education, experiences, and training.
- On a rating scale of 1 to 10 (10 being the highest), how would you rate your skills in communication? Persuasiveness? (Cite specific examples to back up your opinions.)
- Tell me about yourself. (This question really means, "Why should we hire you?")
- Why are you interviewing with us? What do you know about our organization? Our needs?
- What appeals to you about this particular job?
- If you could construct your own job in our organization, what responsibilities would you include? Please expand these ideas in relationship to your goals and accomplishments.
- Do you think your grades are a good indication of your abilities? Your potential? Please explain.
- Why did you major in ———? Why did you choose this university?
- What courses in your major did you like the least? The most? Explain.
- What are some of your strengths? Your limitations? Relate your response to your goals and previous accomplishments.
- Tell me a little about your job experiences. How do your experiences and skills relate to the job for which you are being considered?
- What is the most valuable experience gained on your job? Why?
- What do you consider to have been the most significant contribution or accomplishment this year in your job? Why?
- What criteria or parameters are you using to interview employers?
- Where would you like to be in our organization five years from today?
- To achieve your professional goals, what must you do to improve your chances of success? Explain.
- How do you spend your spare time?
- How have you dealt with stress, pressures, and conflict in the work setting? In college?

If you've returned to school after several years of working or military service, you might be asked:

- What prompted you to return to school? Why the ——— degree?

(Source: Adapted with permission from Arthur R. Eckberg, Stalking the Elusive Job, in Spectrums of Opportunities (Chicago: Roosevelt University Career Planning and Placement Office, 1986), p. 3.)

Accepting a Job and Handling Rejections

After the interview, the wait begins. Employers often interview several candidates for a position over a period of several weeks. How long should you wait for an answer? It depends on the position and the points covered in the interview. Employers often tell you when you can expect to hear from them. If you don't hear in two to three weeks, a polite letter often prompts a response.

But don't stop searching for other positions while you're waiting to hear from another employer. The employer may select another candidate. When you do receive a rejection letter, always send a polite letter thanking the employer for interviewing you and expressing your continued interest should another position develop.

Just as the best employers follow up a telephoned job offer with a written confirmation, so should you accept or reject a job in writing, even if you have conveyed your decision by phone. If you accept, your letter will go into your personnel file along with the written offer. If you reject a job or are turned down, either you or your would-be employer may benefit from knowing the reasons.

HIGHLIGHTS

1. Job searching strategies vary, and you'll probably use several in your career.
2. The primary resources for job searching include your school's career services or placement office and professors in your department.
3. Job searching techniques include: (1) assessing yourself, (2) analyzing potential employers, (3) preparing your resume, (4) using professional quality correspondence, (5) handling interview offers and interviews, and (6) accepting and rejecting offers.

PROJECTS

1. Prepare a description of not more than 750 words explaining your career objectives and what you'd like to be doing in two, five, and ten years. Assume that your audience is a search committee of senior employees at an organization for which you would like to work.
2. Assess your background. Prepare a list of your education, work, special activities, awards, and so forth. Don't worry about length; summarize your experiences that are relevant to your job search. A personal assessment is the first step in developing a solid resume.
3. Draft a resume for the first job you want after graduating or for a summer internship. Make at least five copies. Ask for critiques from career services, professors, and your technical communication instructor.
4. Gather as much information as you can on one organization you'd like to work for. Find out how the organization routinely hires. Prepare a report of not more than 750 words describing the employer, the jobs, and how you'd fit into the organization.
5. Draft a cover letter to a specific individual in that organization. Copy your letter, and have it critiqued by someone from career services, professors in your department, and your technical communication instructor.

6. Prepare a resume, and write an application for a job of your choice.
7. Critique each item below, and then edit the sentences as needed:
   a. I am highly qualified for the position of engineering assistant that you announced in your job description.
   b. I have enclosed my resume for your review.
   c. I have found that I work well with my hands but not with older people near retirement.
   d. Your job announcement sounds as though it was written for me.
   e. My brother-in-law mentioned that you have an opening in your office for an entry-level staff member and suggested I write you and apply for the position.
   f. The enclosed resume details my background, education, experience, interests, and provides additional information you will find helpful in considering me for the opening of assistant project leader.
   g. If you would hire me for the position of project leader for your contract in Alaska, I would sure enjoy it. I love hunting and fishing, and I understand hunting and fishing in that area of Alaska is outstanding.
   h. Please consider me for the assistant manager position that you advertised in the career services business and industry newsletter as the position would strongly complement my career objectives.
   i. Your company sounds as though it's the right type of company for me to work for before moving on to a position with ABC Energy Exploration.
   j. I will be vacationing in Southern California during the week of June 5–12 and I would like to drop by your office for an interview. How does 1 P.M. on June 10 sound?
   k. John Jones, a long-time friend, tells me you're in immediate need of an individual with a background like mine. When can you interview me? I can start late this month, if you like.

FOR MORE HELP

BOLLES, R. N. 1986. *What color is your parachute?* rev. ed. Berkeley, Calif: Ten Speed Press.
Bolles presents strategies that have helped thousands find employment. The strategies are appropriate for recent college graduates and experienced professionals.

PART SIX
Ethics, Public Perception, and Evaluating Communication

Behind communication about work lie several fundamentals that affect every stage. Ethics and law play a role in message formulation, in sharing credit among contributors, and in acknowledging sources of help. Communication also contributes to public perception of your company, field, and work. How can you remedy failure and improve on success?

The final three chapters of this book address legal and ethical issues, public understanding of science and technology, and the evaluation of technical communication. These chapters go beyond "how-to" guidelines.

18
Legal and Ethical Aspects of Technical Communication

OVERVIEW

In Chapter 18, we

- Discuss copyright law and professional communication
- Outline common ethical problems faced by technical communicators
- Advise about avoiding and dealing with ethical problems
- Describe product liability law and its implications for technical communication

> law (lô), *n.* 1. the principles and regulations established by a government and applicable to a people, whether in the form of legislation or of custom and policies recognized and enforced by judicial decision.
>
> eth.ics (eth'iks). *n.pl.* 1. (*construed as sing. or pl.*) a system of moral principles. 2. the rules of conduct recognized in respect to a particular class of human actions or a particular group, culture, etc.: *medical ethics; Christian ethics.* 3. moral principles, as of an individual: *His ethics forbade betrayal of a confidence.* 4. (usually construed as sing.) the branch of philosophy dealing with values relating to human conduct, with respect to the rightness or wrongness of certain actions and to the goodness and badness of the motives and ends of such actions.
>
> —*The Random House College Dictionary*

COPYRIGHT LAW AND TECHNICAL COMMUNICATION

Sensational cases of copyright violation occasionally make news. Federal agents may raid a tape and record store or a videotape business, looking for illegally copied tapes and records. Such piracy may divert millions of dollars each year from the rightful owners. Unauthorized duplicating of computer software eventually may reach similar levels. It is unlikely that you will ever be involved in a major copyright infringement case. But because communication products are property, you need to know copyright basics. You need to protect your work from theft and avoid stealing anyone else's work.

Black's Law Dictionary defines copyright as "the right of literary property as recognized and sanctioned by positive law." The United States Copyright Act of 1976 (discussed below) treats "literary property" broadly and takes into account the technological advances of the past century. Individual articles, serials (newspapers, magazines, journals), books, computer software, visual arts such as charts, graphs, microfilm, photographs, films and television shows, works of the performing arts, music, sculpture, plays, choreography, and sound recordings bear copyright, *whether or not they are officially registered with the U.S. Copyright Office, Library of Congress*.

Copyright law has two important concepts. One is that literary products belong to their creators, who have the right to profit from their labors, just like manufacturers, inventors, and anyone else who creates value. The second concept holds that social progress requires that meritorious work be accessible to scholars and researchers who create new knowledge. This is the doctrine of "fair use."

COPYRIGHT LAW OF 1976

In 1976 Congress passed a law placing all copyright matters under federal jurisdiction. Under this law, people hold copyright as soon as they produce anything that is copyrightable. People may still register a work formally with the U.S. Copyright Office, and they certainly should do that when they produce copies for sale.

Whether you formally register or not, you have a federally protected copyright as soon as you meet three tests. Your work must

1. be original
2. have a trace of creative input
3. be "fixed in a tangible medium of expression," that is, be written or recorded in some fashion

In draft versions, a simple statement on the title page that your work is copyrighted will serve notice to anyone that you intend to protect your property rights. If you publish something without intending to hold onto its copyright and later change your mind, you have five years after initial publication in which to register formally. To pursue a copyright infringement suit, though, you must have registered the work. That explains why people who produce books or other important work for sale or lease complete formal registration.

Under the 1976 law, the term of copyright is the author's lifetime plus

50 years. "Works made for hire" and certain anonymous and pseudonymous works may be registered for 75 years from publication or 100 years from creation, whichever is less. A "work made for hire" is defined as that "prepared by an employee within the scope of his or her employment" or a "work specially ordered or commissioned." If you write a user manual to accompany equipment your company manufactures, your company may register it under the "work made for hire" rule. If you and your company agree in writing, you may register it yourself.

Panel 18-1 contains a sample copyright registration form for "nondramatic literary work." The form, two copies of the work, and a $10 fee sent to the Register of Copyrights formally registers your work. Of course, if you never formally register a work, you and your heirs will hold it forever, because the "lifetime plus 50 years" rule never comes into play. However, to preserve your right to legal recourse in case someone takes your property and seeks to profit from your work, register at the time of publication. Book and magazine publishers, scholarly and trade journals, and newspapers routinely copyright their publications.

The copyright notice itself may take one of three forms: © (the letter *C* in a circle); the word "Copyright"; or the abbreviation "Copr." For phonorecordings, the form calls for the letter *P* in a circle, ℗. By using these forms you secure protection in the United States, and, through international agreement, 66 other countries. By adding the words "All Rights Reserved," you obtain protection in another three Central and South American countries.

If you learn of someone who is infringing or about to infringe on your copyright, you have several remedies. First you ask the infringer to stop. If a letter from an attorney gets no results, you can ask a federal court to issue temporary and final injunctions against use of your material. Violation of such injunctions would bring the government into play on your side. You also can ask the court to impound all copies of the material that violated your copyright. If you can show that you have suffered financial loss, you may ask for damages from the infringer. You may ask for statutory damages of $10,000 or more simply because your copyright has been infringed. The 1976 law imposes fines and jail sentences on those convicted of record and videotape piracy and other theft.

LIMITATIONS ON A COPYRIGHT OWNER'S RIGHTS

Copyright protects the way ideas are expressed, not the ideas themselves. No idea, procedure, process, system, method, concept, principle, or discovery, regardless of how it is described, is copyrightable. Protection for processes, methods, discoveries, and ideas embodied in a process or machine may come from patent registration, but that is a distinct method of protecting intellectual property. Copyright protects the form of expression

PANEL 18-1. Copyright Registration Form

FORM TX
UNITED STATES COPYRIGHT OFFICE
REGISTRATION NUMBER

TX   TXU
EFFECTIVE DATE OF REGISTRATION
Month   Day   Year

DO NOT WRITE ABOVE THIS LINE. IF YOU NEED MORE SPACE, USE A SEPARATE CONTINUATION SHEET.

1 TITLE OF THIS WORK ▼

PREVIOUS OR ALTERNATIVE TITLES ▼

PUBLICATION AS A CONTRIBUTION If this work was published as a contribution to a periodical, serial, or collection, give information about the collective work in which the contribution appeared. **Title of Collective Work ▼**

If published in a periodical or serial give: Volume ▼   Number ▼   Issue Date ▼   On Pages ▼

2
a
NAME OF AUTHOR ▼   DATES OF BIRTH AND DEATH
Year Born ▼   Year Died ▼

Was this contribution to the work a "work made for hire"?
☐ Yes
☐ No

AUTHOR'S NATIONALITY OR DOMICILE
Name of Country
OR { Citizen of ▶
     Domiciled in ▶

WAS THIS AUTHOR'S CONTRIBUTION TO THE WORK
Anonymous?   ☐ Yes ☐ No
Pseudonymous? ☐ Yes ☐ No

If the answer to either of these questions is "Yes," see detailed instructions.

NOTE
Under the law, the "author" of a "work made for hire" is generally the employer, not the employee (see instructions). For any part of this work that was "made for hire" check "Yes" in the space provided, give the employer (or other person for whom the work was prepared) as "Author" of that part, and leave the space for dates of birth and death blank.

NATURE OF AUTHORSHIP Briefly describe nature of the material created by this author in which copyright is claimed. ▼

b
NAME OF AUTHOR ▼   DATES OF BIRTH AND DEATH
Year Born ▼   Year Died ▼

Was this contribution to the work a "work made for hire"?
☐ Yes
☐ No

AUTHOR'S NATIONALITY OR DOMICILE
Name of Country
OR { Citizen of ▶
     Domiciled in ▶

WAS THIS AUTHOR'S CONTRIBUTION TO THE WORK
Anonymous?   ☐ Yes ☐ No
Pseudonymous? ☐ Yes ☐ No

If the answer to either of these questions is "Yes," see detailed instructions.

NATURE OF AUTHORSHIP Briefly describe nature of the material created by this author in which copyright is claimed. ▼

c
NAME OF AUTHOR ▼   DATES OF BIRTH AND DEATH
Year Born ▼   Year Died ▼

Was this contribution to the work a "work made for hire"?
☐ Yes
☐ No

AUTHOR'S NATIONALITY OR DOMICILE
Name of Country
OR { Citizen of ▶
     Domiciled in ▶

WAS THIS AUTHOR'S CONTRIBUTION TO THE WORK
Anonymous?   ☐ Yes ☐ No
Pseudonymous? ☐ Yes ☐ No

If the answer to either of these questions is "Yes," see detailed instructions.

NATURE OF AUTHORSHIP Briefly describe nature of the material created by this author in which copyright is claimed. ▼

3 YEAR IN WHICH CREATION OF THIS WORK WAS COMPLETED This information must be given in all cases. ◀ Year

DATE AND NATION OF FIRST PUBLICATION OF THIS PARTICULAR WORK
Complete this information ONLY if this work has been published. Month ▶   Day ▶   Year ▶   ◀ Nation

4 COPYRIGHT CLAIMANT(S) Name and address must be given even if the claimant is the same as the author given in space 2. ▼

See instructions before completing this space.

APPLICATION RECEIVED
ONE DEPOSIT RECEIVED
TWO DEPOSITS RECEIVED
REMITTANCE NUMBER AND DATE

DO NOT WRITE HERE OFFICE USE ONLY

TRANSFER If the claimant(s) named here in space 4 are different from the author(s) named in space 2, give a brief statement of how the claimant(s) obtained ownership of the copyright. ▼

MORE ON BACK ▶
• Complete all applicable spaces (numbers 5-11) on the reverse side of this page.
• See detailed instructions.
• Sign the form at line 10.

DO NOT WRITE HERE
Page 1 of _____ pages

PANEL 18-1. (continued)

	FORM TX
EXAMINED BY	
CHECKED BY	
☐ CORRESPONDENCE Yes	FOR COPYRIGHT OFFICE USE ONLY
☐ DEPOSIT ACCOUNT FUNDS USED	

DO NOT WRITE ABOVE THIS LINE. IF YOU NEED MORE SPACE, USE A SEPARATE CONTINUATION SHEET.

PREVIOUS REGISTRATION Has registration for this work, or for an earlier version of this work, already been made in the Copyright Office?
☐ Yes   ☐ No   If your answer is "Yes," why is another registration being sought? (Check appropriate box) ▼
☐ This is the first published edition of a work previously registered in unpublished form.
☐ This is the first application submitted by this author as copyright claimant.
☐ This is a changed version of the work, as shown by space 6 on this application.
If your answer is "Yes," give: **Previous Registration Number** ▼   **Year of Registration** ▼

5

DERIVATIVE WORK OR COMPILATION Complete both space 6a & 6b for a derivative work; complete only 6b for a compilation.
a. Preexisting Material Identify any preexisting work or works that this work is based on or incorporates. ▼

b. Material Added to This Work Give a brief, general statement of the material that has been added to this work and in which copyright is claimed. ▼

6

See instructions before completing this space.

MANUFACTURERS AND LOCATIONS If this is a published work consisting preponderantly of nondramatic literary material in English, the law may require that the copies be manufactured in the United States or Canada for full protection. If so, the names of the manufacturers who performed certain processes, and the places where these processes were performed **must** be given. See instructions for details.
Names of Manufacturers ▼   **Places of Manufacture** ▼

7

REPRODUCTION FOR USE OF BLIND OR PHYSICALLY HANDICAPPED INDIVIDUALS A signature on this form at space 10, and a check in one of the boxes here in space 8, constitutes a non-exclusive grant of permission to the Library of Congress to reproduce and distribute solely for the blind and physically handicapped and under the conditions and limitations prescribed by the regulations of the Copyright Office: (1) copies of the work identified in space 1 of this application in Braille (or similar tactile symbols); or (2) phonorecords embodying a fixation of a reading of that work; or (3) both.
a ☐ Copies and Phonorecords   b ☐ Copies Only   c ☐ Phonorecords Only

8

See instructions.

DEPOSIT ACCOUNT If the registration fee is to be charged to a Deposit Account established in the Copyright Office, give name and number of Account.
Name ▼   **Account Number** ▼

CORRESPONDENCE Give name and address to which correspondence about this application should be sent. Name/Address/Apt/City/State/Zip ▼

9

Area Code & Telephone Number ▶

Be sure to give your daytime phone ◀number

CERTIFICATION* I, the undersigned, hereby certify that I am the
Check one ▶
☐ author
☐ other copyright claimant
☐ owner of exclusive right(s)
☐ authorized agent of _____
          Name of author or other copyright claimant, or owner of exclusive right(s) ▲
of the work identified in this application and that the statements made by me in this application are correct to the best of my knowledge.

10

Typed or printed name and date ▼ If this is a published work, this date must be the same as or later than the date of publication given in space 3.
_____ date ▶ _____

Handwritten signature (X) ▼

MAIL CERTIFI- CATE TO	Name ▼	Have you: • Completed all necessary spaces? • Signed your application in space 10? • Enclosed check or money order for $10 payable to *Register of Copyrights*? • Enclosed your deposit material with the application and fee? **MAIL TO:** Register of Copyrights, Library of Congress, Washington, D.C. 20559
Certificate will be mailed in window envelope	Number/Street/Apartment Number ▼	
	City/State/ZIP ▼	

11

* 17 U.S.C. § 506(e): Any person who knowingly makes a false representation of a material fact in the application for copyright registration provided for by section 409, or in any written statement filed in connection with the application, shall be fined not more than $2,500.

☆ U.S. GOVERNMENT PRINTING OFFICE: 1982-361-278/58   Sept. 1982—600,000

and gives the copyright owner the right to prevent others from copying the work without permission.

Strictly speaking, anyone who repeats portions of a work without first obtaining permission infringes on a copyright. But to enforce such a rule absolutely would greatly inhibit scholarship, even though it might protect authors' rights. In practice, copyright law attempts to compromise between authors' property rights and the needs of scholars, researchers, and other writers as they seek to generate new knowledge and forms. Three basic limitations on a copyright owner's rights exist:

1. *Licensing,* which permits infringement subject to payment of a royalty or getting signed permission in advance. For example, if your school has a media services unit that helps professors produce slide shows for classes, chances are the unit has a license for using copyrighted music as background for narration. Radio stations that play recordings log their plays, and eventually a few cents per play go to the copyright holders. Professional musicians pay for the right to play copyrighted songs. Authors of books such as this one may pay copyright owners to reprint cartoons or photographs. Authors of scholarly books, with little chance for large commercial success, generally are granted permission without fees. The point is to ask for permission in advance of use.

2. *Photocopying,* which permits infringement so long as there will be no commercial advantage. For example, a professor might copy a brief article for her class to read. But the gray area is large. What about a stockbroker who copies the latest market analysis from a *Forbes* magazine to give to clients? The broker does not sell it to them, but the magazine might argue that the legally correct procedure would be for the broker to buy copies for clients.

3. *Fair Use,* which is not infringement. Section 107 of the 1976 copyright law says, "the fair use of a copyrighted work . . . for purposes such as criticism, comment, news reporting, teaching (including multiple copies for classroom use), scholarship or research is not an infringement of copyright" (Pember 1981, 276). This is the major protection for users of copyrighted materials. In judging "fair use," courts weigh several factors: purpose and character of use (such as commercial or nonprofit educational), nature of the copyrighted work, amount of the work used in relation to the whole, and the effect of the use on the potential market or value of the copyrighted work.

Proceed with caution when you want to include copyrighted material in a publication, slide set, presentation, or other communication. Fair use is not defined by law and is variously interpreted. Quoting excerpts is permissible under fair use, but if you quote 200 words of a 250-word poem, you have infringed. A professor may make three dozen copies of a chapter from a copyrighted book to distribute to students. That one-time use can be considered "inspirational" under the law, but if the professor uses the same chapter next term, the law may deem the professor to have avoided purchase, which deprives the copyright holder of legitimate profits.

Obtaining permission takes time and effort, but it can save grief. Any publication that uses copyrighted material should obtain prior permission of the copyright owner. The editor of a Colorado company's employee magazine wrote an article of which he was quite proud. His publication was on an exchange list with similar magazines in other states. One day he noticed his article in another magazine, published in a state far across the country. His prior permission had not been sought, and even worse, the name signed to the article was not his, but that of the editor of the distant magazine.

Angry and curious, the Colorado editor wrote a letter in which he expressed admiration for the article and asked permission to reprint it. Back came the answer: okay, but be sure to acknowledge your source. That reply so infuriated the Colorado editor that he wrote a letter to the president of that distant company. He enclosed a copy of his original publication, carrying a date months earlier than that of the stolen version. The president apologized and fired the editor who had violated copyright (Danbom 1984). Panel 18-2 contains a typical permission form, used for materials in this book. Notice that the form specifies the nature of the use and the exact material to be used.

Copyright remains an extremely complex subject. It is more than a legal basis for the moral or ethical standards that apply when you use someone else's work. Copyright confirms the property rights that exist in literary effort and production. Existence of copyright law, with its protection for authors, stimulates literary efforts, because authors know they may expect to collect the fruits of their labors. At the same time, through the doctrine of fair use, copyright law encourages the use of copyrighted material in the development of new knowledge.

Those who produce technical and scientific communication are not exempt from observing copyright law. Copyright covers music for slide shows, charts, graphs, photographs from magazines and books as well as those belonging to individuals, interesting graphics for a manual, introductory paragraphs that seem to describe a situation better than you could hope to do, and more. Someone worked hard to compose or develop those. You can no more use them legally than you can borrow someone's car without first obtaining permission.

PLAGIARISM: LITERARY THEFT

Defined as the appropriation or imitation of the language, ideas, or thoughts of another author, and their representation as one's original work, plagiarism is of two types: inadvertent and deliberate.

PANEL 18-2. Sample Permissions Form

Department of Technical Journalism   Colorado State University
303/491-6310 or 6319   Fort Collins, Colorado
                                        80523

February 24, 1986

Mr. William Stolgitis
Executive Director
Society for Technical Communication
815 Fifteenth Street, N.W.
Washington, D.C. 20005

Dear Mr. Stolgitis:

Dr. David G. Clark and I have written a textbook tentatively titled <u>The Random House Guide to Technical and Scientific Communication</u>. Random House, Inc., will publish the book in December 1986. Our book will be hardbound and run about 450 pages.

May I have your permission to reprint the materials indicated below in all editions of our book and derivations thereof for world distribution rights?

It is understood that this material or portions thereof is to be used as examples or to illustrate a point or technique. Full credit will be given to you, if you desire.

I would appreciate your signing and returning one copy of this form to me at your earliest convenience. In signing, you warrant that you are the sole owner of the rights granted herein and that the work does not infringe upon the copyrights or rights of anyone.

Sincerely,

Don Zimmerman
Associate Professor

Material to be reprinted: From <u>Proceedings of the 31st International Technical Communication Conference</u>, 1984, H. E. Vogt, page VC-26, Figure 5. "Wordless instructions--say it with pictures," the illustration series showing the unpacking of the VDT monitor.

Permission granted by _____ Date _____

Form of copyright acknowledgment: _____

_____

Inadvertent plagiarism is frequently committed by people who just do not know better. Unfamiliarity with the existence, much less the complexities, of copyright law may lead someone to assume, for example, that it's legal to photocopy a table from a textbook and to include it in a term paper. When their mistake is pointed out, these people may plead ignorance. They may be warned and forgiven, under the assumption that they have learned a lesson.

But deliberate plagiarism is intellectual thievery. The editor who stole the magazine article described above committed deliberate plagiarism. Every department in every university has its stories of plagiarism committed by students or professors. Not too long ago, plagiarism was punishable in academic circles by automatic dismissal. University honor codes required students to report plagiarism. Even today, plagiarism is grounds for failure in a course or for more serious punishment, depending on how flagrant the case.

Plagiarists Sometimes Get Away with It

One reason that plagiarism is not more widely publicized when it is found is that it is embarrassing. Who relishes admitting having been fooled by a thief? Another reason that plagiarists sometimes escape public denouncement lies in the vagueness of the definition. Similar ideas can occur to several people. Before a charge of plagiarism can hold up, the offending language has to be quite close to the original. Sometimes it is easy to show the resemblance, because plagiarists tend to be lazy (or else they would not steal). But often the people who should check for plagiarism, such as professors or journal editors, are themselves too busy, trusting, or lazy to investigate.

Some plagiarists do get away with their thefts. But the threat of exposure remains, ticking away like a burglar alarm on time delay. One graduate student pirated parts of an old report and put them into his doctoral dissertation, claiming the work was his original effort. He passed and went on to a prominent career in higher education. Fifteen years later, after he had been president of a university for five years, his plagiarism came to light, destroying his career (Broad and Wade 1982).

Two Well-Known Plagiarists

In their remarkable book on dishonest scientific and technical researchers, *Betrayers of the Truth: Fraud and Deceit in the Halls of Science* (1982), William Broad and Nicholas Wade recount many incidents of plagiarism. Elias A. K. Alsabti of Iraq was exposed only after he had plagiarized some 60 scientific papers in medical research journals all over the world. He used stolen papers to win appointments at several medical research centers

> ## PANEL 18-3. An Example of Plagiarism
>
> The *British Medical Journal* in 1980 published the following excerpts from two articles. The original article was published by Drs. K. W. Pettingale and E. H. Tee in 1977 in the *Journal of Clinical Pathology*. A plagiarized version was published in the *Japanese Journal of Experimental Medicine* in 1979. The plagiarists were E. A. K. Alsabti and K. Muneir. Alsabti, who plagiarized some 60 articles during the late 1970s, apparently was so pleased with Pettingale's work that he plagiarized him a second time. The *British Medical Journal* estimates that there are 8,000 medical journals worldwide. Author credentials are difficult, if not impossible, to check.
>
> Original Version
>
> There appear to be differences in the serum protein levels between women with breast cancer and those with benign breast disease. These differences are seen preoperatively at the earliest time of clinical detection of a breast tumour and do not appear to be related to the age difference between the groups. Raised serum caeruloplasmin has been reported previously in patients with both cancer and chronic inflammatory disease (Sternlieb and Scheinberg, 1961; Snyder and Ashwell, 1971), but an increase in β_2 glycoprotein levels does not appear to have been reported before. In previous studies the levels have been either unchanged (Cleve, 1968) or reduced (Snyder and Ashwell, 1971). Our study, however, is confined to one type of cancer at an early stage and is not strictly comparable with the previous studies quoted.
> —*Journal of Clinical Pathology*, 1977
>
> Plagiarized Version
>
> There appear to be differences in the serum protein levels between women with breast cancer and those with benign breast disease. These differences are seen preoperatively at the earliest time of clinical detection of a breast tumour and do not appear to be related to the age difference between the groups. Raised serum ceruloplasmin has been reported previously in patients with both cancer and chronic inflammatory diseases (8, 9) but an increase in β_2 glycoprotein does not appear to have been reported before. In previous studies the levels have been either unchanged (3) or reduced (8). This study, however, is confined to one type of cancer at an early stage and is not strictly comparable with the previous studies quoted.
> —*Japanese Journal of Experimental Medicine*, 1979
>
> (*Source:* British Medical Journal, *July 5, 1980, p. 41. Reprinted by permission.*)

in the United States before he was finally exposed. Panel 18-3 contains an example of Alsabti's work. Notice how close his version is to the original.

The ancients felt their temptations, too. Robert R. Newton argues conclusively in *The Crime of Claudius Ptolemy* (1977) that Ptolemy, whose theory of the solar system was accepted for 1,500 years, stole the observations of Hipparchus, a contemporary Greek. Hipparchus made his observations from Rhodes, which lies five degrees of latitude north of Alexandria, Egypt, where Ptolemy worked. Since Ptolemy does not report any of the stars that are visible in the southernmost five degrees of the sky at Alexandria, Newton concludes that he stole Hipparchus' observations and claimed them as his own.

How to Avoid Even the Suspicion of Plagiarism

The most obvious advice on avoiding plagiarism is the simplest: do your own work. Use plenty of sources, and credit them generously. In handling others' work, it is far better to err on the side of giving too much credit than too little. When someone has discussed a draft with you, acknowledge that help. You will in no way detract from your own genius, and you will compliment the other person.

People often plagiarize charts, tables, graphs, and other illustrations from published sources, under the mistaken assumption that such material is for anyone to use. To avoid plagiarizing, select the specific information you need from a published source, adapt your own version, and credit the source. That probably is fair use. If you want to publish your adaptation, obtain permission from the copyright holder.

ETHICS AND COMPETITIVE PRESSURE

The late historian of science, Derek J. de Solla Price, of Yale University, wrote (1963) of differences between how basic scientists and people in technological research and development view the ethics of publishing. Basic scientists consider publication an end product that pays off in recognition, a sense of accomplishment and contribution, perhaps advancement, and use in further research. People in research and development, on the other hand, see publication as not altogether beneficial, and as secondary to coming up with new products, securing patents, or getting grant money. These differences lead to differences in the openness in their respective publications.

Many scientists are reluctant to describe their work fully. In their publications, methods are described so sketchily that replication is nearly impossible (Broad and Wade 1982). Why? Much of research today is proprietary because of funding sources. Research funded by the federal government is sometimes classified. Privately funded research may similarly be restricted. Scientists who by training and inclination would like to share their research sometimes must limit access to it.

Thus the high cost of research and the scarcity of funds place extreme pressures on professionals and students alike. Robert H. Ebert, former dean of the Harvard Medical School, describes those pressures as they affect medical students. He might have been speaking for all kinds of scientists:

> The spirit of intense, often fierce competition, which begins during the premedical experience to get into medical school [is] encouraged thereafter. Stories of cheating among premedical students are common. . . . Once training is completed and the long hard climb on the academic ladder has begun, there

is intense pressure to publish, not only to obtain research grant renewals but in order to qualify for promotion. . . . In an environment which can even permit success to become a more coveted commodity than ethical conduct, even the angels may fall. (Ebert 1980, 18)

In other words, in science the profit motive often collides with the desire for truth.

OTHER ETHICAL PROBLEMS

People in science and technical professions may violate ethics by concocting data, "cooking" results, withholding adverse information, and not sharing credit.

Concocting Data

Early in 1983, the National Institutes of Health released the results of a yearlong study of the work of a young researcher at Harvard Medical School (Wallis, Schapiro, Wymelenberg 1983). Because of fraud he had committed at Harvard and at Emory University, the young doctor could not receive federal funds for ten years. He had faked dates to make a few hours seem like weeks of work and falsely reported data. Animals that were reported injected had not been injected at all. The *New England Journal of Medicine,* one of the most prestigious medical research publications, acknowledged that it had been taken in by the phony research. That instance of fraud was just one among several similar incidents at Yale, Cornell, and Boston universities, in which researchers succumbed to temptation to keep grant funds flowing in.

Various proposals have been made for tightening safeguards against fraud. These include allocating grants to laboratories, not individuals, on the theory that tighter surveillance of researchers' results would occur. Another proposal is that laboratories become more open in discussing the work of individuals. "Falsehood is necessarily a private matter; open discussion must help to show it up—or even show that falsehood is unnecessary" (Plagiarism, Piracy and Principles 1980). If today's world pressures scientists to work for money, prestige, and other rewards, the best safeguard remains a strong sense of personal ethics.

"Cooking" Results

The father of genetics, Gregor Mendel, supposedly neatened the results of his famous experiments with peas by removing peas that violated the law he was proposing. Like plagiarism, adding or subtracting from results may be either deliberate or inadvertent. People who deliberately change evidence commit willful dishonesty. Anyone participating in an experiment in which test animals are removed during the course of the project should ask some questions. Not to do so might further the commission of a fraud.

Psychologists have well documented our inadvertent tendency to see what we are looking for. If teachers, for example, are told that school children of average ability are gifted, the teachers tend to assign higher grades to those children. Double blind experiments, in which neither researchers nor subjects know which treatment group subjects are in, long have been used to protect against such self-deception. The best defense against the natural tendency to find what we are looking for entails developing and applying a rigorously critical attitude toward our results. (Posing the questions in Chapter 9 on evaluating content is a good place to start.)

Withholding Adverse Information

A close relative of tampering with research results is the withholding of adverse information about a product. This form of fraud poses harm to public safety and tends to be practiced by producing and marketing organizations. Drug and chemical companies that withhold or ignore evidence of the harm their products can do and manufacturers that do not tell employees and consumers about product dangers commit fraud.

In the late 1950s and early 1960s, for instance, a "wonder sedative" known as K-17 by its German manufacturer was prescribed by physicians worldwide. K-17 supposedly had no toxic dose level, no narcotic effect, no influence on circulation, and no risk of overdose. Within a few years, though, some tragic side effects turned up. Women who had taken the drug early in pregnancy gave birth to children whose hands and feet were deformed. Even after many reports of the deformities reached the manufacturer, it tried to withhold research showing a connection between the drug and birth defects in rabbits. By about that time, some 8,000 deformed children had been born. Thalidomide, as K-17 was known to the public, left a terrible legacy. Its manufacturer's attempts to suppress adverse information have been a model ever since of how *not* to do things.

Questions of which information to release and which to withhold can become complicated. Public health as well as jobs, profits, even corporate existence may be threatened. By the mid-1980s, the evidence showed that manufacturers should disclose all adverse information or risk losing all. For example, some Manville Corporation workers alleged that the company knew but did not tell them of the dangers of the asbestos that the company manufactured and sold. Projected losses from the nearly 20,000 lawsuits were so great that the company declared bankruptcy in 1982. If the company did know, but withheld information, that decision was an expensive one.

Sharing Credit for Authorship

If we accept the premise that money (for continued research and as reward for successful work), promotion, and other forms of recognition can result from publication, an important question arises early in any collaboration: how should credit be shared?

Research articles have grown shorter over the past decade, and the number of authors per article has increased. First, many projects are team efforts. Second, playing the promotion game with university and company administrators who measure accomplishment by quantity leads researchers to split their writing into as many pieces as possible.

Various rules apply for assigning responsibility for authorship. Science has not yet reached Hollywood standards, but a billing system exists nonetheless. The first author listed is assumed to be the major contributor. If two authors contributed equally, the first author listed is assumed to be senior in the company or university hierarchy. An exception is articles by professor and graduate student. Sociologist Warren Hagstrom, in *The Scientific Community* (1965), a survey of scientists, reports that most advisers put their students' names first and may not list themselves at all.

The sound practice is for collaborators to agree ahead of publication, and as early in the project as possible, on the order of names to be listed. In-house documents, not meant for outside distribution, deserve similar consideration. Junior authors should question senior authors about how they will receive credit. Without clear agreement, people risk injured feelings.

PRODUCTS LIABILITY

In recent years, technical communicators have become important players in the field of products liability law, because written instructions are often keys to the safe use of a product.

A commuter airliner experienced landing gear problems at Casper, Wyoming, one day in 1984 (Nelson 1984). A crew of maintenance mechanics flew up from Denver, took the landing gear apart, made temporary repairs, and reassembled it. While the plane was sitting on the ground, everything seemed fine. But when the plane lifted off the runway, the landing gear blew apart and forced an emergency landing. When he investigated the incident, Greg Feith of the National Transportation Safety Board discovered what had caused the explosion. The mechanics had installed a retaining ring incorrectly. When the plane took off, nitrogen gas inside the gear blew out the ring.

The mechanics put the retaining ring incorrectly into place, Feith reported, because the maintenance manual gave misleading instructions. Had death or injury resulted from that incident, the company that provided the manual might have been liable for millions of dollars in damages.

Development of Products Liability

A rapidly expanding area of law, products liability was practically invisible until a few years ago. Today, whether you are in manufacturing, construction, chemical development, or any other field, you should be aware of the relationship between your product, instructions for its safe use, and your legal responsibilities. A bit of legal history is required to make the picture clear.

For a long time, a manufacturer could not be held liable for injury to a consumer unless a contract existed between the two. *Winterbottom* v. *Wright,* an 1842 English lawsuit, illustrates the point. A mail coach driver was injured when the coach broke down; he sued the manufacturer, alleging that the coach was defective. The court ruled that the driver could not collect because there was no contract between him and the company. The postal service had bought and owned the coach. It had a contract with the maker. The driver was a third party with no share in that contract:

> If the plaintiff can sue, every passenger, or even any person passing along the road, who was injured by the upsetting of the coach, might bring a similar action.

Winterbottom v. *Wright* was the principal case for nearly a century of American and English law, and the injustice it preserved was variously described as a "thorn in the side of progress," and a "fishbone in the throat of the law" (Ross 1981).

Makers of such dangerous products as poisons and explosives were eventually exempted from the rule, on the grounds that those manufacturers should be able to recognize the dangers and offer protection against them to people who were not purchasers.

Then in 1916 a Scotsman named MacPherson bought a beautiful new Buick (*MacPherson* v. *Buick Motor Co.* 1916). One of its wheels was defective, and while MacPherson was driving his car the wooden wheel crumbled, causing a wreck in which he was injured. Although he had bought the car from a dealer and not from the Buick Motor Company, MacPherson sued Buick. The court agreed with him, ruling that cars are dangerous instruments for which a manufacturer is liable in case of defects.

Today, products liability law has developed much further. It rests on a complex set of assumptions:

- Manufacturers can anticipate and guard against dangers that the consumer cannot.
- Manufacturers may insure against risks involved in use and can treat that insurance as a cost of doing business, but a single consumer would find the cost of such insurance overwhelming.
- The public interest dictates that selling defective products be discouraged and that the responsibility be on manufacturers, wholesalers, and retailers.

- Consumers cannot fully investigate a product's soundness nor prove negligent conduct.
- Trademarks, advertising, slogans, and other marketing strategies tend to inspire trust and reduce skepticism, lulling consumers into uncritical acceptance.

Communication's Role in Products Liability

Generally, a manufacturer can be held liable when a defective product injures a person who has used the product in a reasonable manner. This doctrine of absolute, or "strict," liability places a heavy burden on manufacturers. Moreover, the Consumer Product Safety Act of 1972 made manufacturers almost inevitably accountable to the federal government as well, usually to the Consumer Product Safety Commission. Virtually any injury to a consumer during product use may eventually be held to be due to some defect in the product or *in instructions for its use*.

Therein lies the application of products liability law to technical communication. User manuals, warning labels, and signs are methods that manufacturers use to guard against injury and lawsuits. The manufacturer has the duty to provide adequate warnings against misuse and to give complete directions for proper use.

Attorney Kenneth Ross has pointed out three ways that a product may be proved defective: in design, manufacturing, or marketing. Marketing defects, he has written in the *Journal of Products Liability* (1981), include:

1. Failure to provide any warning of the hazards involved
2. Failure to provide adequate warning
3. Failure to provide adequate instructions for the safe use of the product

Before a manufacturer can be charged with not providing one of these warnings, a legal duty to warn must be present. If the product is dangerous and that danger is not completely obvious to a user, then the maker has a duty to warn. An open or "obvious" danger need not be warned against, but what is "obvious"? True, an experienced carpenter should not have to be told that an electric saw is dangerous. But, argues Ross, underscoring the importance of audience research, *a manufacturer should not overestimate the intelligence and training of a particular user*. All sorts of people buy and use electric saws. Therefore, when in doubt, warn.

Injury that results from unforeseen uses is not the manufacturer's fault, today's law holds, but makers have a duty to foresee reasonable uses. But what is such use, and what is adequate warning? Answers to those questions come, ultimately, from judges, juries, and courts. To make sure that injuries do not occur and lawsuits do not result, you owe your organization all the assistance you can give when you write an instruction manual or warning label.

FMC Corporation, the world's largest producer of soda ash, makes a wide range of agricultural and other chemicals. The company sells machinery to industrial, agricultural, defense, and other users here and abroad. Its central engineering laboratories have developed an extensive product safety label system (FMC Corporation 1985).

Panel 18-4 illustrates how FMC's sign artists go about reducing hazard warnings to their simplest components. The intensity of hazard is communicated by red for immediate danger, orange for potentially severe, and yellow for minor hazards. Warnings that tell how to avoid hazards are cast in as few words as possible. FMC presents the points vertically: first showing the level of danger; then the nature of the hazard; consequences; and, finally, how to avoid the hazard. Simplicity and clarity are readily apparent in the FMC system.

Five considerations prove important in creating adequate warnings (Ross 1981):

1. choosing the correct signal word
2. deciding who the audience is and its education, training, experience, dress, clothing, anatomy, and other details
3. deciding placement of labels, reading distance, viewing angle, and lighting
4. writing an adequate warning label
5. choosing a symbol or pictograph that is effective and that complies with any government or industry regulations

Once the duty to warn is met, the duty to provide instructions comes into play. A typical instruction book, Attorney Ross notes, contains eight sections: (1) introduction; (2) description; (3) receiving, handling, storing; (4) installing; (5) operating; (6) inspecting, maintaining, adjusting; (7) overhaul and repair; and (8) supplementary information, such as technical specifications. Ross suggests including two other sections: (9) a disclaimer asserting that the instruction book does not modify the sale contract and does not take all situations into account, and that the seller's liability is limited. He suggests including: (10) a page describing safety concerns, referring readers to pages in the book dealing with those points, and defining what is meant by "danger," "warning," and "caution," if those terms are used in the book or on labels.

Beyond that, warnings should be written in the clearest and simplest possible language. Manufacturers also must make sure that ultimate users actually get warnings and instructions. This may mean attaching instructions to each item and sending an additional set to management. Too many companies, writes Ross, put the important use documents into an office file to preserve them, instead of making certain that users have them.

Negligence in preparing instruction manuals and accompanying documents can prove as costly as any mechanical or electrical defect. Communications about use and maintenance should be designed as carefully as

PANEL 18-4. FMC Safety Hazard Signs

Pictorial Panel

The pictorial message usually consists of a black pictorial on a white background for high contrast and visual consistency. Additional colors can be used occasionally to assist in the proper interpretation of the pictorial message. For example, the pictorial warning of hot surfaces takes on added meaning when the surface and emanating heat waves are shown in red.

Care must be exercised when developing a two-color pictorial to ensure that the selection and proximity of colors is visually effective and does not dilute the meaning of the intended message.

Instruction Message Panel

The instruction message panel always consists of white lettering on a black background. White lettering on black is considered to be more legible in poor lighting conditions and where darkened areas exist on a product such as in recessed or partially enclosed areas.

The predominantly black message panel also provides high contrast with the predominantly white pictorial panel. The high contrast functions as an attention-getting device for peripheral vision (i.e., the lateral distance between the label and the point where a person's eye is already directed).

Border

The narrow border surrounding the label is always white and functions as a visual barrier to prevent the label colors from blending with the product color. Good contrast between the label and the product is assured at all times.

Color Control

Safety color swatch sheets are located in Section 14. When ordering safety labels, color swatches should be provided to the printer for precise color matching.

(Source: FMC Corporation, *Product Safety Sign and Label System*, Santa Clara, Calif. Reprinted with the permission of FMC Corporation.)

any machine. One does not need to be a lawyer to design effective technical communication. But awareness of legal aspects is basic to competent technical communication.

HIGHLIGHTS

1. Copyright is the means by which literary property rights are secured and maintained. Under the Copyright Act of 1976, copyright is exclusively under federal jurisdiction.
2. Plagiarism (literary theft) is common and easily avoided. Deliberate or inadvertent, it can lead to ruinous results.
3. Other forms of ethical malpractice, such as concocting results, "cooking" data, and hiding unfavorable information, seem common in a world where competition demands success. The costs of unethical practices are paid by society and the individuals who commit them.
4. Products liability law holds manufacturers accountable for injuries sustained while people use a product. Defective products, instructions, and labels that fail to warn properly are potentially negligent.

PROJECTS

1. Ask your university library for a copy of its policy on photocopying copyright materials. What is "fair use" as the policy defines it?
2. Choose a professor who has published copyrighted materials, and ask who registered the work, whether permission was required for any material, who got permission, and how.
3. Does your school have a policy on plagiarism? Look in a catalog, general bulletin, or at the office of student affairs. Does the policy define plagiarism and explain why it is incompatible with the pursuit of knowledge?
4. The instructor's manual for this book describes ethical questions involving technical communication. In class, discuss one or more of these. What would you do in each situation? Are the actions described illegal or unethical? Why or why not?
5. Find a manufactured product's instruction or user manual that contains a warning against unsafe use of the product. Drugs, electrical equipment, and power tools nearly always are accompanied by such warnings. What other kinds of products have liability disclaimers?

FOR MORE HELP

BROAD, W., AND N. WADE. 1982. *Betrayers of the truth: Fraud and deceit in the halls of science.* New York: Simon and Schuster.
Broad and Wade review cases of scientific fraud.

PEMBER, D. R. 1984. *Mass media and the law.* 3rd ed. Dubuque, Ia.: William C. Brown.
Pember's work contains readable chapters on copyright, defamation, invasion of privacy, and other topics of interest to professionals.

SINDERMANN, C. J. 1982. *Winning the games scientists play.* New York: Plenum Press.
The book covers many aspects of the communication of scientific work.

U.S. Copyright Office, Library of Congress, Washington, D.C.
The primary source for information on and applications for copyright.

WEINSTEIN, A. S., et al. 1978. *Products liability and the reasonably safe product: A guide for management, design, and marketing.* New York: Wiley.
The authors provide a readable, detailed look at how suits result from product failure and misuse.

19
Creating Better Public Understanding of Technical Fields

OVERVIEW

In Chapter 19, we

- Describe achieving equivalence of enlightenment between the public and scientists
- Point out the mutual mistrust among science and technology, the public, and the mass media
- Show that despite mistrust, all parties agree that science and technology must be well understood by the public
- Show how professionals can reach the public
- Suggest how professionals can correct public misunderstanding of their work

SCIENCE, MEDIA, AND THE PUBLIC

> The American press as a whole . . . has done an irresponsible job of discussing important technical issues that are not easy for the public to understand.
> —Presidential Science Adviser Dr. George A. Keyworth

Those words of criticism (Keyworth 1985) reveal several assumptions about the social roles of media, technology, science, and government:

- The media have an obligation to report on science and technology.
- For various reasons, media tend not to fulfill this obligation as experts want.
- In a broad sense, the public dictates policy on science and technology, even though it may not fully understand the issues.

- Even presidents require scientific advisers to interpret issues for them.
- Technological and scientific fields increasingly provide interpretation for media and public.

Science writer Jon Franklin (1981) of the Baltimore *Sun* once related the following story to a conference called to consider better ways of communicating university research results. It seems that astronomers at a Caribbean observatory detected the beeping of a pulsar, before such things were known. Deciding to announce their discovery to the media, they sent a researcher to the library with instructions to get the address of the most widely distributed newspaper in the United States. The researcher came back with an address in Lantana, Florida. The astronomers talked with an editor, who immediately sent a reporter to the observatory. The reporter took notes as the scientists played tapes of the beeping noise from outer space and discussed their theories as to what it meant.

"You don't really know for sure what is doing the beeping?" asked the reporter. "That's right," said the astronomers, properly tentative in their conclusions. "Then," said the reporter, "it might even be intelligent life trying to contact other life?" "Well," replied the scientists, "anything is possible." And so, Franklin observed, the most important astronomical discovery of that decade was announced on the front page of *the National Enquirer,* in a huge headline that read, "Space Beings Contact Earth."

This story illustrates some basic truths about how the different cultures of science and media tend to view one another. Neither group understands much about the other's approach to a given topic. Yet each must deal with the other to serve a third group, the public.

PUBLIC SUPPORT IS ESSENTIAL

When it launched the first earth satellite on October 4, 1957, the USSR also launched serious doubts in the United States about the quality of U.S. science. A prominent politician issued a four-step plan to help the United States "win the race for scientific supremacy" (Swinehart and McLeod 1960). Three of the steps had to do with the public: increasing its acceptance of the value of basic research, giving it a more favorable picture of science and scientists, and convincing it that universities deserved more support in training scientists and engineers.

Although the "race for scientific supremacy" may be no nearer resolution today, public support did come. By 1981 nearly 900 American spacecraft had gone into orbit, United States citizens had walked on the moon, and rockets had probed space. Similar evidence appears in the fields of cancer research, environmental conservation, energy development, and others. The public, if properly informed, can understand, appreciate, and support the objectives of science.

The road to what political scientist Harold Lasswell (1948) once dubbed "equivalence of enlightenment" between experts and citizens is built by the mass media. Local and national newspapers, magazines, television, and radio provide a huge "town meeting" for the discussion and formulation of policy. People in science seek public funding and public approval. Communities vote, for example, on whether a local laboratory may create new life forms. A state legislative committee holds hearings on the possible ban of agricultural chemicals. Often the decision lies with people who are not scientific experts.

Who is to represent a field before nonexperts? Who is capable of presenting information that is both accurate and simply stated, so that intelligent people not trained in a field may acquire equivalent understanding? You are, if you will learn the public relations principles for representing yourself and your field effectively.

In "What the Public Needs to Know About Science and Technology" (1981), Jacques Richardson lists scientific problems facing humankind "today and in the foreseeable future." One category of such problems: the development of policy for mastering science and technology. What is the role of science and technology in bridging the gulf between science and public. The "popularizer is challenged daily to explain science or technology to the indifferent or hostile lay person" (Richardson 1981). Although the public supports technology and science, at times the public is hostile to them.

People Love Science . . .

The public does enjoy learning about science. Television shows such as "Cosmos," "The Body in Question," and various National Geographic specials prove that. So does the popularity of magazines devoted to science, such as *Discover, Omni, Smithsonian,* and *Natural History*. A science editor at *Time* magazine found (Bennett 1979) that for half a dozen years in a row, any issue with a science cover story was among the best sellers of the year.

. . . and Resent It, Too

But technology and science can injure people and foster distrust. *Public Opinion Quarterly* (Pion and Lipsey 1981) published a review of two decades of survey research on public attitudes toward science. The authors found that upper middle class and highly educated young people consistently held antagonistic attitudes toward science. A general decline, from 56 to 41 percent, occurred between 1966 and 1977 in people expressing a "great deal of confidence" in science. "Seeds of disenchantment with science and technology are present," the study concluded:

> The fortunes of science appear to be largely dependent on the voices of relatively small numbers of academic and scientific elites and upon the favor of enlightened decision makers (p. 313).

As seeds of disenchantment take root, relations between science and public become more adversarial. Nuclear power advocates marshal physicists and other scientists who support their side. Opponents do likewise. A natural fallout is public distrust, because scientists of seemingly equal reputation appear to be in vehement disagreement. Yesterday's miracle, DDT, with its power to eliminate malaria, today is known to enter the human food chain and appear in mothers' milk, with potentially terrible consequences. Whom is one to believe? The public becomes conditioned to view skeptically all scientific developments.

Science and Technical Professionals Are Suspicious

Two factors inhibit reporting about mathematics research, Dr. Lynn Arthur Steen (1981), a mathematician at St. Olaf's College, told a conference. Those two factors are reporters and mathematicians. Scientists often think of reporters as

- Inaccurate
- Prone to distort or exaggerate
- Incomplete
- Concentrating on personalities
- More interested in results and applications than method

Why do these perceptions occur? Inaccuracy and distortion almost always result not from reporters' conscious intentions to distort, but from efforts to translate scientific terms into plain language and clear examples. For example, a scientist cited as the major source in an article (Ryan 1974) about tumors in cats and dogs complained about its first sentence: " 'Relax, you can't catch cancer from your cat,' says Dr. Blank." "What I stated," the scientist said, "was that the probabilities were small, based on my data; however, my data were not good enough to rule out the possibility of a rare event relationship." The reporter wanted a simple, readable beginning. The scientist resented the reporter's finality. Both had proper objectives for their primary audiences.

As for scientists' other criticisms of reporters, it is worth keeping in mind that most scientists who communicate effectively reduce complexities to everyday terms. Reporters themselves worry about incomplete stories on government, economics, and other significant topics. Likewise, telling a story in its human terms injects life into an otherwise highly abstract topic. Reporters devote attention to personalities in a story as a way to improve readability, and thus increase the audience.

Emphasizing results and applications over methods probably results from the natural desire society has to improve. In the long run, scientists are as interested in that objective as anyone else.

Media Are Skeptical

From many a reporter's viewpoint, technical and scientific professionals:

- Have superior and secretive attitudes. Some scientists retreat into technical language to avoid reporters (Russell 1981).
- Have all the usual human weaknesses, including large egos and the tendency to exaggerate the importance of their findings. For example, Dunwoody (1981) reports that a science writer told her that several researchers had claimed to have found the cause of Sudden Infant Death Syndrome, a still unsolved mystery.
- Assume that all journalists are guilty until proven innocent of all the sins described above.
- Think that journalists should merely report findings in the scientists' own words and avoid any effort to interpret research. A survey of 140 scientists found they agreed that journalists should avoid interpretation and should permit scientists to review their stories before publication (Ryan 1979). In the same survey, 198 science reporters and editors said they opposed such practices.
- Sometimes feel they should be able to dictate whether an interview will be published at all. The role of the press is to find things out and to publish them, reporters argue. Also, since science reporters know that scientists compete for funding and thus are seeking favorable publicity for their projects, reporters now routinely ask about funding sources; they also usually ask other scientists for their views on the significance of a development.
- Are naive about how media really operate. Scientists make little effort to recognize the urgency of reporters' work. A daily newspaper or television show reporter wants information now, not tomorrow or next week.
- Do not understand how errors may be introduced by a headline writer or an editor at the television station.
- May neglect the public interest aspects of their work. Whether or not the public understands the science in question, much of it affects the public. These implications are obvious in fields from weapons research to rush hour road repair.

PANEL 19-1. Scientists, Media, and Public Understanding

When a scientist has a big idea and good evidence to support it, media attention can be intense and sustained. Dr. Walter Alvarez of the geology and geophysics department of the University of California at Berkeley has had several years of publicity concerning one of his ideas. As a graduate student in 1970, he was doing research in Italy's Apennine Mountains, looking for evidence of previous reversals in the earth's magnetic fields. In a former ocean floor, now part of a canyon near the town of Gubbio, Alvarez found a thin layer of reddish-brown clay just at the boundary between the Cretaceous and the Tertiary periods. The clay contained an abnormally high concentration of a rare gray metal, iridium. Alvarez knew that nearly all iridium found on earth is of extraterrestrial origin. Furthermore, this deposit was some 25 times greater than amounts distributed generally, and the boundary between the Cretaceous and Tertiary periods marks one of earth's great extinctions: the passing of the dinosaurs some 65 million years ago.

Working with his father, Luis Alvarez (a Nobel prize winner in physics), and nuclear chemists Frank Asaro and Helen Michel, Alvarez formulated and tested a theory. If so much iridium was present, it must have arrived at once, perhaps in an asteroid. The impact would have raised a huge cloud of dust that would account for the thin reddish-brown layer of clay. The dust cloud could have darkened the sky for months, long enough to stop photosynthesis in many plants. The resulting catastrophe would have demolished much of the food chain. The largest animals would have suffered most and would have starved in a short time.

The Alvarez group announced this theory in 1979. Within three years iridium deposits had been confirmed in 36 sites worldwide at the Cretaceous–Tertiary boundary. No point of impact was found, although the ocean floor west of Portugal appeared a possibility. The actual location may never be found.

The neatness of the Alvarez theory, the mounting evidence of worldwide iridium deposits, and public interest in the dinosaurs' disappearance brought much publicity. Soon Alvarez had accumulated a 3-inch thick file of articles on his work. Articles appeared in *Time*, *National Geographic*, the *New Yorker*, *AmeriCan Way*, *Physics Today*, *Family Weekly*, the *New York Times*, *Christian Science Monitor*, *Mosaic*, and many more. The theory was the subject of a Public Broadcasting Service television production of "Nova" ("The Asteroid and the Dinosaurs," March 10, 1981), and of an "In Search Of" television program episode.

Interviewed for this book, Walter Alvarez (1982) expressed mixed feelings about media attention to his work. Although he believes that the public has a right to know about research, especially research funded by public money, he nevertheless expressed some disappointment in the media treatment of his work. Publicity had brought notoriety and misunderstanding, even among scientists. "I've been attacked at scientific meetings by scientists on the basis of what they have read in magazines or newspapers," he said.

Alvarez was also critical of reporters who expected him to teach them about his field. "If they haven't done their homework, I have no time for them," he said. He added that some reporters think they write about science, when in fact they seem to seek controversy, and try to pit one scientist against another. Some media, he complained, trivialize serious research.

Below are three accounts of the Alvarez group's work. First is the abstract of an article the group itself wrote for *Science* (June 6, 1980):

Platinum metals are depleted in the earth's crust relative to their cosmic abundance; concentration of these elements in deep-sea sediments may thus indicate influxes of extraterrestrial material. Deep-sea limestones exposed in Italy, Den-

PANEL 19-1. (continued)

mark, and New Zealand show iridium increases of about 30, 160, and 20 times, respectively, above the background level at precisely the time of the Cretaceous–Tertiary extinctions, 65 million years ago. Reasons are given to indicate that this iridium is of extraterrestrial origin, but did not come from a nearby supernova. A hypothesis is suggested which accounts for the extinctions and the iridium observations. Impact of a large earth-crossing asteroid would inject about 60 times the object's mass into the atmosphere as pulverized rock; a fraction of this dust would stay in the stratosphere for several years and be distributed worldwide. The resulting darkness would suppress photosynthesis, and the expected biological consequences match quite closely the extinctions observed in the paleontological record. One prediction of this hypothesis has been verified: the chemical composition of the boundary clay, which is thought to come from the stratospheric dust, is markedly different from that of clay mixed with the Cretaceous and Tertiary limestones, which are chemically similar to each other. Four different independent estimates of the diameter of the asteroid give values that lie in the range of 10 ± 4 kilometers. (Alvarez et al. 1980, 1095)*

The Alvarez group took sophisticated measurements, such as x-ray fluorescence, mass spectrometry and irradiation. The National Science Foundation and the National Aeronautics and Space Administration supported the work. But some media, in translating from scientific to popular versions, made the research seem like trivial speculation. This story from the July 16, 1979 issue of *Time* is one example:

Doomed Dino: Puzzling Over Its Demise

When dinosaurs vanished abruptly 65 million years ago, they left an enduring mystery—and created a scientific parlor game. Hypotheses abound to explain the extinction. Brains too small in bodies too large? Emerging mammals feasting on dinosaur eggs? Now comes evidence for another possibility. Geologist Walter Alvarez, probing an ocean canyon near Gubbio, Italy, discovered an abrupt increase in iridium in a limestone layer dating back to the dinosaurs' demise. Probable cause: some mysterious, still unfathomable event.

Alvarez and his team from the University of California at Berkeley were sampling the strata because they provide a rare, undisturbed record of reversals in the earth's magnetic field. Such fluctuations can influence climate, and possibly allow more cosmic radiation to assail the earth's atmosphere. One layer, only a centimeter thick and tracing back 65 million years, showed a sharp excess of iridium, an element 1,000 times more plentiful in otherwordly matter than in the earth's crust. The "spike" in the readings made a sobering point. "It's the first experimental evidence that something quite extraordinary happened then," says Physics Nobel Laureate Luis Alvarez, who gave his son a helping hand. A supernova that could have wiped out the dinosaurs? "A very small probability," says Alvarez *père*. Also possible but improbable: a cloud of interstellar gas or a large meteorite. On with the parlor game.†

Physics Today, a publication that might be characterized as the trade journal of the discipline, carried a story in its May 1982 issue. The story is noteworthy for two reasons. The writer shaped the story toward the audience of physicists by listing physicist Luis Alvarez first, and the writer defined some geological terms, illustrating our point that outside of their own field, scientists often must be treated as intelligent lay people.

Dinosaur extinction due to asteroid? Recent developments appear to confirm what at first seemed an outrageous idea: A meteorite or asteroid a few kilometers in diameter hit Earth about 65 million years ago and caused a major wave of biological extinctions.

---

*Copyright 1980 by the American Association for the Advancement of Science. Reprinted by permission of AAAS and Prof. L. W. Alvarez.

†Copyright 1979 Time Inc. All rights reserved. Reprinted by permission from TIME.

> **PANEL 19-1. (continued)**
>
> The hypothesis was suggested two years ago by Luis Alvarez, his son Walter, and two colleagues at the Lawrence Berkeley Laboratory, Frank Asaro and Helen Michel, to explain the extinctions that mark the end of the Cretaceous period (the age of the dinosaurs) and the beginning of the Tertiary (the age of the mammals). This boundary is fairly clear in the fossil record; there is, for example, a very abrupt change in the marine microorganisms whose fossil remains make up limestone and chalk. In some areas where limestones of the appropriate age are exposed (one fairly famous area is near Gubbio in Italy), the boundary is marked with a thin layer of clay. There are similar, but apparently not as drastic, changes in the terrestrial flora. And, of course, there is the disappearance of dinosaurs from the fossil record. *
>
> Aside from the article in *Science,* over which the Alvarez group had control, various news media shaped vastly different accounts for their audiences. About all the scientists could do was try to make sure the reporters understood the meaning of the research.
>
> *Reprinted by permission of* Physics Today.

Reporters view themselves as sentinels for society and often seek to make technology's impact clear. Reporters increasingly seek connections between technical activity and its social ramifications. Thus, in their eyes, the scientist's or engineer's narrow concentration on projects may produce results but also harm society.

To overcome the difficulties posed by such attitudes, whether you consider them justified or not, requires effort. Outside of your organization, those efforts usually fall under the heading of public relations.

WHAT IS PUBLIC RELATIONS, AND WHY IS IT IMPORTANT?

Every organization has a public relations policy, although less effective groups define public relations as good publicity when things are going well and no publicity when things are going badly. Notions like that are shortsighted, for good public relations are needed most during tough periods. A forward-looking policy can mitigate many a crisis.

Basic public relations theory is quite simple: in the marketplace of ideas, where truth and error and everything in between collide, we all have the right to present our views as effectively as we can. If we don't join in, we leave the field to arguments that may be wrong. Thus every company, every government agency, every organization should represent itself to the public. Even when laws forbid the existence of public relations personnel, as federal law does for the federal government, the rationale that everyone should have access to the marketplace of ideas is irrefutable. The United

States government may have no "public relations" officers, but it does have "public information" officers.

Components of Effective Public Relations

Public relations experts agree that consistently effective organizations bear in mind three major facts:

1. No matter what the issue, several publics exist. Each should be addressed in terms that have meaning for it. Doing so, without appearing to say different things to different audiences, requires careful coordination of effort.
2. Any organization's public relations unit should be placed for easy access to top management. Effective public relations always involves two-way communication—from management to publics and from the various publics to management. Entering the marketplace of ideas means more than trying to persuade everyone to your point of view. It means being open to others' viewpoints, too.
3. Successful public relations entails four basic steps:
    A. Researching the basic problem facing the organization and the attitudes and composition of the publics involved
    B. Planning the total campaign according to the directions revealed by research
    C. Communicating through the media, in the formats research dictates, by the timetable worked out in the planning stage
    D. Evaluating the results and effectiveness of the campaign and making certain that management understands the results, so that further action may be considered

An Effective Public Relations Campaign in Applied Science

The National Hail Research Experiment (NHRE), carried out during the 1970s, illustrates the basic steps in effective public relations (Borland 1982). Administered by the National Center for Atmospheric Research under funding from the National Science Foundation, the NHRE could easily have become a public relations nightmare. It did not, thanks largely to careful adherence to good public relations policy.

From the time cloud-seeding came into widespread use in the United States in the early 1950s, weather modification has been a touchy subject. Many bitter disputes have arisen over questions of whether cloud-seeding should be done at all, much less when and where. No one likes drought, but no one wants a hailstorm or a flash flood either. When the NHRE began, its managers realized that more than atmospheric science would be involved. Social scientists played important roles even before starting cloud-seeding in the target area of northeastern Colorado, southwestern Nebraska, and southeastern Wyoming.

In 1971, Human Ecology Research Services, a Boulder research firm, gathered data during interviews with

- 200 residents of the target area
- a group living on the margin of the area
- a control group living about 100 miles southeast of the experimental area

The groups were interviewed on six occasions: before and after field operations in 1971, before and after operations in 1972, and in August 1973 and 1974. In addition, farm organizations and other community groups were interviewed annually.

Interviewers sought several kinds of information from each respondent: attitudes toward weather, weather modification, and science; extent of belief in whether or not cloud-seeding could produce desirable effects; media most often read, listened to, or watched; level of awareness of the experiment; attitudes toward it; whether respondents preferred local or nolocal control of weather modification; extent to which respondents attributed weather events to the experiment; and demographics. These included age, sex, education, socioeconomic status, religion, and land use—urban, farm, or ranch.

Research helped define the publics, identify their organizations and information-seeking patterns, and formulate the kind of information each group should receive. Well before any cloud was seeded, social research identified various publics in the experiment area: townspeople and rural residents, farmers and ranchers, people with considerable education and those with little. Moreover, researchers knew all these groups' media use patterns.

According to Henry Lansford, public relations director of the National Center for Atmospheric Research during the project, top management "accepted and supported the proposition that public information would be a critical element in the success or failure of NHRE" (Lansford 1977, 1). More than two years before field work began, an information campaign was underway, not just among people who lived under the clouds that were to be seeded. Recognizing that several outside publics existed, NCAR briefed state governors and other officials.

As field operations approached, a form letter and brochure went out to 9,500 rural boxholders in the area and to 40 national and regional agricultural publications. Personal letters went to county agricultural extension agents. Lansford visited them as well as newspaper, radio, and television news editors, and he gave talks to farmers' and ranchers' organizations, to service clubs, and to other groups. He maintained close contact with state officials, especially midway through the experiment when new weather modification legislation was introduced.

PANEL 19-2. Using Institutional Voices

Often a public relations department can help scientists reach the public quickly and efficiently.

In March 1976, prairie dogs in the Comanche National Grassland in Baca County of southeastern Colorado were found to have fleas carrying plague (Danbom 1983). A team from the Fort Collins branch of the Centers for Disease Control (CDC) dusted the prairie dog burrows by the time the news was made public on March 30.

On May 22, a case of human plague was confirmed near Cañon City, 150 miles northwest of the grassland. By that time epidemiologists felt certain that prairie dogs, rock squirrels, ground squirrels, chipmunks, and wood rats not only in Colorado but in several other western states were experiencing plague.

The Colorado Health Department warned summer vacationers to avoid contact with rodents and to use insect repellent to ward off fleas. By June 10, three more rodent cases were confirmed. Governor Richard D. Lamm, on June 19, at the request of the CDC, asked the Environmental Protection Agency (EPA) for permission to use DDT in six affected counties. By that time, the plague threat had made national news.

The EPA granted the DDT exemption, and spraying began in several Colorado counties. The Sierra Club questioned whether the threat to humans were sufficient to meet federal requirements for DDT use. Those requirements call for a showing that significant health problems will occur if DDT is not used. Only one confirmed human case of plague had occurred in Colorado. The controversy brought public health and environmental scientists into conflict, against a backdrop of increasing public anxiety. Plague outbreaks in California, Nevada, New Mexico, Arizona, and Wyoming complicated matters.

But candor of the Colorado Health Department prevented the controversy from leading to panic. Its director of public relations, Rowene C. Danbom, kept the public and the media informed of developments as they occurred. New confirmations were announced, but the small risk to humans was constantly emphasized. Danbom sought to inform the public without hampering the scientists. She scheduled health department and CDC scientists for press conferences and arranged for media representatives to observe a training session on DDT use. The Sierra Club did not object to how the DDT was used once permission had been granted.

Public understanding was enhanced during a tense situation. Without media relations personnel, scientists concentrating on the health aspects almost certainly would have overlooked some public aspects. For one, the health department public relations staff, in addition to keeping local media informed, contacted the Colorado state tourism agency and Denver convention and visitors bureau. Through those offices, tourists and conventioneers were made aware of the latest developments. There were few rumors and no panic.

As the experiment progressed, continued monitoring provided NCAR with shifts in social attitudes, so that points of concern could be addressed as they developed rather than after matters became crucial. Preliminary research suggested formation of a citizen's council on hail research. Accordingly, a dozen residents of southwestern Nebraska and northeastern Colorado, including farmers, ranchers, and a farm implements dealer, formed a council that met several times a year during the experiment. They heard from NCAR scientists and discussed with project leaders preliminary results and future tests.

During field operations, Lansford prepared and sent a weekly newsletter to county agents, news media, and anyone else who requested it. The letter described the week's activities in cloud-seeding and current weather conditions. In response to a suggestion by the citizens' council, by the third summer of operations Lansford was telephoning a one-minute summary of the past 24 hours' activity to four area radio stations, which taped the reports for replay on news programs.

Social science research during the NHRE led Lansford to an interesting conclusion about public relations in science and technology. Early in the project, surveys showed that the more people knew about the experiment, the more likely they were to accept it. Later, this association was less clear. Still later, research showed that many people exposed to the extensive public information activities retained few specifics, but still regarded the experiment favorably. Lansford reasoned that many people, "because we were working so hard to keep them informed about what we were doing, concluded we must be an honest and responsible group of people who could be trusted to deal fairly with them" (Lansford 1977, 2). When it comes to complex questions of science and technology, few public relations efforts can do better.

ACTIVELY IMPROVING PUBLIC UNDERSTANDING

Successful organizations do not wait for the occasional large project to begin practicing effective public relations. They communicate every day with various publics about activities that may benefit or harm those publics. As a technical or scientific professional, you can reach your publics by:

- Responding to media queries about your own or others' work
- Working with your firm's public relations office to arrange for and showcase announcements of your projects
- Doing your own public relations work, either through the mass media or by addressing people directly

Helping Media

Journalists want the most knowledgeable sources they can find, and they want them right away. Engineers, extension agents, horticulturists, and other experts may be called by reporters. Perhaps a local reporter wants to localize a national or international news story. Being interviewed is flattering, but some guidelines will be helpful. Most companies do not mind their employees speaking to the media, as long as the employees speak with company authorization. Educational institutions encourage such communication; public agencies range from actively encouraging to discouraging. Learn your organization's policy.

Corporations and other large organizations sometimes have their public relations departments produce guidelines for interviews with reporters. A major producer of health care products, Smith Kline & French Laboratories (now SmithKline Beckman) produced such guidelines for its technical and scientific personnel (Smith Kline & French 1964). We abridge those guidelines and add some of our own.

1. Jot down the reporter's name, organization, and the questions you are asked. If any doubts arise later about who said what, you have a means of refreshing your mind. For interviews that you know are controversial, it's fair to use a tape recorder. Your interviewer probably will.

2. Arrange personal interviews so that you have few interruptions. If your organization has a public relations staff, someone from that office may wish to attend. It also may be useful to have an associate handy to search the files if you need something quickly.

3. In asking for the interview, the reporter should tell you what topics will be covered and may even give you some preliminary questions. But don't expect to get all questions ahead of time. Reporters like to observe reactions to questions. Although a reporter is a trained fact-gatherer, he or she probably knows little about your field and may proceed less directly than you expect. The big questions usually are: How did you get into this research or project? What did you do? What did you learn? What are its implications?

4. Expect radio and some newspaper reporters to use short telephone interviews. On the phone, be as concise as you can. A radio reporter tapes your responses but can broadcast only the bare essentials. Print reporters also can take down short sentences more easily than long ones. Give short answers to television reporters, too. Effective politicians and others frequently quoted learn to speak in 15- to 30-second "bites" that a television editor can stack or arrange in any order. Keeping your answers short helps get your facts, not someone else's interpretation, on the air.

5. Remember TV's need for good visual material. The camera needs an interesting, moving background. Prepare visuals that explain your work.

6. Take care with your rhetoric. Barbara Cox and Charles Roland of the Mayo Foundation in Rochester, Minnesota, reviewed (1973) all articles and letters about psychoactive drugs that appeared in two medical journals during one year. They found many emotional words, such as "alarming," "abuse," "addiction," and "epidemic." When scientists use such words, they impeach their credibility, and readers grow apprehensive.

7. Respond promptly to questions. Reporters need answers right away. Answer questions by day's end.

8. Be prepared. James E. Lukazewski, president of the public relations consulting firm of Brum & Anderson ExecuCom, Inc., stresses that point in his firm's "Having Effective Media Interviews" (1984), a brochure for executives. The brochure advises executives to list the questions they would love to respond to, if only someone would ask, and to draft answers of 125 to 150 words. It also advises them to list the questions they never want to be asked and to answer these in 125 to 150 words. Before the CBS program "60 Minutes" was allowed to enter the New Jersey plant of the Manville Corporation, company public relations manager Neal Amarino (1985) prepared a list of some 179 questions he thought Dan Rather might ask. They were tough questions, because Amarino knew that "60 Minutes" was interested in the protection, or lack of it, the company provided for its asbestos workers. Amarino rehearsed company executives. Rather asked 171 of the questions on the list. The resulting story was much more complete and favorable to the company than it otherwise would have been.

"No Comment"

Although the best policy is to respond to media inquiries, because a story is being done anyway and one ought to express one's point of view, "no comment" is sometimes next best. *The Public Relations Journal* (Detwiler 1979) asked public relations professionals when they say "no comment." Their replies:

- When they are not qualified
- When they are angry or when the issue is controversial
- When they do not have proper clearance from their organization
- When they have not had time to consider the full public impact of their comment
- When their company's competitive edge would be blunted
- When the issue is legal or regulatory
- When their company is under investigation by a law enforcement agency or when contracts are being negotiated

Relying on the Public Relations Department

If your organization has a public relations department, it can help you to reach the proper audiences with news of your work. If you're with a large organization, announcement of your work will be scheduled to mesh with other public announcements. Public relations departments also schedule announcements and press conferences to meet media deadlines. To make early evening news shows, news conferences usually are scheduled for morning or early afternoon. For announcements not requiring news conferences, they may arrange coverage in Sunday or Monday morning papers, which usually have more space than other editions. Supplying pictures of your work reduces reporters' work and increases your chances of good coverage.

The public relations department also can train you to feel more comfortable with reporters. The better you are at being interviewed, the more valuable you are to your company. Many companies use engineers or scientists as spokespeople because they project authority. But engineers and scientists often lack ability in dealing with media and in public speaking.

POPULARIZING YOUR WORK YOURSELF

In 1959 E. B. White published in the *New Yorker* a series of what he called "bulletins tracing man's progress in making the planet uninhabitable." Shortly he received a letter from a well known marine biologist urging him to look at environmental pollution. Having read one of the biologist's books, White urged her to undertake the task herself. She did. In 1962, Rachel Carson's *Silent Spring* appeared (Lear 1970). The ensuing controversy over pollution raged for years, but Carson's scientific credentials gave her views weight that a nonscientist's would have lacked.

Agricultural researchers, county extension agents, forest rangers, U.S. Soil Conservation Service employees, engineers, and technical experts with other public agencies continually face the question of how to communicate their message to the public. Small daily newspapers, weeklies, radio stations, and nonmetropolitan television stations are glad to have their reports, but they rarely have the staff to translate from scientific to popular versions.

Animal scientists, agronomists, and wildlife biologists often produce research with immediate application to livestock and crop production or to the conservation and management of wildlife. Many such people write for state farm and ranch publications, conservation journals, and popular magazines. Most newspapers and radio and television stations engage scientists to report regularly through columns or special programs on conservation, water pollution, crop management, and other topics. Scientists who write for media in this fashion save time and effort, because expert and journalist become one. As a scientist you know what to say, and as journalist you know how to say it.

A complete course for professionals interested in writing popularized science articles is beyond our scope here, except for a few a guidelines. One route open to anyone is a college course in newspaper and magazine feature writing. Such a course will give practice in structuring articles and writing for general audiences, and it provides criticism from the instructor and students.

Another alternative is to seek help from a sympathetic journalist who can suggest what the local paper or station might publicize about your

PANEL 19-3. Popularizing Science for Better Public Understanding

Technical and scientific reports often may be translated into plain language accounts that interest wider audiences. Here are two versions, scientific and popular, of a research project. Note how the writers define and then address their audiences.

Here are the title and abstract of a research article that appeared in *Ecology*:

ENERGETICS OF A SUBURBAN LAWN ECOSYSTEM
John Howard Falk
Math/Science Group, University of California, Berkeley

Abstract. A study of the energetics of a suburban lawn was conducted in 1972–73 in Walnut Creek, California USA. Several major components of the annual primary and secondary production were measured, including man's role as manager and experimenter in the system.

The system was extremely productive with net productivity of 1,020 g/m^2 per yr compared to cornfields with productivity of 1,066 g/m^2 per yr and exceeding tall grass prairie values of around 1,000 g/m^2 per yr. Homopterans, with maximal values of 19 mg/m^2 were plentiful; other typical grassland species, like Araneida, were scarce, representing only 1% by weight of the total invertebrate population. Food utilization per unit area by suburban birds considerably exceeding natural grassland bird utilization (46 kcal/m^2 per yr vs. 1.01–2.33 kcal/m^2 per yr); lawns are ideal foraging sites for open area adapted, flock-feeding species.

Man was the dominant consumer in the community, accounting for 10% of the herbivory and nearly 100% of the scavenging. Energy inputs (labor, gasoline, fertilizer, etc.) amounted to 578 kcal/m^2 per yr, equalling or exceeding corn production for a comparable net productivity, but not necessarily utilitarian return.

Key words: California, energetics, lawn; grassland management; lawn ecosystem, urban and suburban; productivity, lawns. (Falk 1976, 141)*

*From *Energetics of a suburban lawn ecosystem,* by J. H. Falk, *Ecology,* 1976, 57, 141–150. Copyright © 1976 by Ecology. Reprinted by permission.

Using Falk's research article, New York Times News Service reporter Bayard Webster produced a popularized account for a general newspaper audience.

"YOUR LAWN'S FULL OF ENERGY"
By Bayard Webster
© N.Y. Times News Service

NEW YORK—To a suburban homeowner, a lawn is a patch of grass requiring raking, mowing, fertilizing, and watering at considerable expenditure of time, energy, and money that usually brings, in return, a dividend of aesthetic pleasure.

But a California biologist who conducted a year-long study of a typical suburban lawn ecosystem determined that green patch—a teeming community of hundreds of species of plants and animals—is as productive in terms of net energy as a field of corn. The study was made by Prof. John H. Falk of the University of California at Berkeley, who also found that a suburban lawn, contrary to the opinion of many perspiring grass mowers or leaf rakers, produced almost three times as much energy as the physical energy man exerted in caring for it.

The energy, measured as living and dead material in terms of calories, is not in a readily usable form; but its measurement helps to shed new light on man's impact on an ecosystem.

Using a variety of equipment, including a rotary lawn mower to clip the grass, a vacuum cleaner to collect insects, and his kitchen oven to dry collected specimens, Falk analyzed his lawn in a way that few, if any, scientists have done. The results of his study, believed to be the first of its kind, are reported in the current issue of the scientific journal *Ecology.**

*Wisconsin State Journal, *May 23, 1976.* Copyright © 1976 by The New York Times Company. Reprinted by permission of The New York Times Company and the *Wisconsin State Journal.*

field. We know one extension service agent who began in just this way. After being interviewed several times by a local journalist, she tried writing articles. Success followed, and the local paper soon offered her a regular column. Now other extension agents supply material for her column, and she has branched far beyond her first subjects.

HOW TO COMPLAIN

Perhaps a columnist or editorial writer has taken you to task. Perhaps someone who does not understand your science has given you a local version of Senator William Proxmire's "Golden Fleece Award." Should you ever complain about how you've been treated by a news medium? Do so only if you bear in mind that the medium has the last word. Here is a checklist for action when you have been hit by a media "cheap shot": a form of misinformation disseminated to gain an advantage for the source (Hutchcroft 1983).

- Don't respond to every attack; pick your issues carefully.
- Writing a letter to the news medium may not be the most effective way to reach the general public. Often that's like arguing with the umpire. You may feel better, but the call stays. The best channel—letter, phone call, visit, advertisement, publication of your own, or some combination—depends both on the audience for the response and the organizational goal to be served. Consider these well.
- Decide whether the attacker has reached a "teachable moment." Response at that time can pay off well.
- If you decide to respond, do so as promptly as possible.
- Keep your response to your area of professional competence and to the point in question. Avoid branching out into general areas of complaint.
- Be cautious in fighting others' battles. You may not know the whole situation.
- Avoid personality conflicts between the responder and the offending source.*

If you are the victim of a botched, distorted, inaccurate media report, you may owe it to yourself to respond, not only to call attention to the unfairness of the treatment but to protect your interests. But practitioners advise that rebuttals may lower public confidence in a communication medium and do not necessarily raise public confidence in the institution

*Source: Adapted from *Responding to Media Cheap Shots: Observations on the CAST Experience*. Paper No. 16. Ames, Iowa: Council for Agricultural Science and Technology. Used by permission.

making the rebuttal (Detwiler 1979). Moreover, you cannot reassemble the original audience and reach them with the facts. For these reasons, professionals try to get it right the first time.

HIGHLIGHTS

1. Professionals frequently must deal with news media to communicate with various publics. Relations between these two "cultures," science and media, sometimes are hurt by misunderstandings and hard feelings.
2. The importance of science and technology to society requires all parties to work hard at informing the public, which often must make decisions on less than full understanding of technical issues.
3. Reporters must be accurate, but you should aid their understanding and thereby increase public understanding of your work.
4. Effective public relations build public trust through consistent communication.
5. The basic steps in public relations are (1) research into the situation's components, (2) planning of the communication campaign, (3) communication of a message, and (4) evaluation of results.
6. Responding to media inquiries is one way to increase public understanding of science and technology. Interviews require careful planning.
7. Popularized versions of scientific work can raise public awareness.
8. If you or your organization feel victimized by media treatment, careful complaining sometimes helps.

PROJECTS

1. Choose a recent important development in your field, and trace its progress from initial report in a research journal, through trade journals, popular magazines, and newspapers. To find published accounts, use the major index of your field, the *Science and Technology Index* for trade journals, newspaper indexes such as the *New York Times Index,* and the *Reader's Guide to Periodical Literature.* How do the various publications design their reports to suit their readers' backgrounds? Do you find any distortions?
2. Your instructor may invite to class a public information writer from your school's public relations office. Ask about the kinds of problems posed by scientists, news media, and research subject matter in getting news across to the public. How does the public relations office try to solve those problems?
3. Speak with a researcher at your school about dealing with the news media. What advice does that person offer?

4. Your instructor may invite a local reporter to discuss problems in reporting on technology and science. What advice does the reporter have for getting information accurately published?

FOR MORE HELP

BOWEN, M. E., AND J. A. MAZZEO, eds. 1979. *Writing about science*. New York: Oxford University Press.
    The editors have collected a series of articles by well-known scientists that illustrate writing for different audiences and using different techniques.

CENTER, A., AND F. WALSH. 1985. *Public relations managerial case studies and problems*. 3rd ed. Englewood Cliffs, N.J.: Prentice-Hall.
    The authors report more than three dozen case studies in public relations, many with technical bases.

CUTLIP, S., A. CENTER, AND G. BROOM. 1985. *Effective public relations*. 6th ed. Englewood Cliffs, N.J.: Prentice-Hall.
    Long the standard in public relations, the book gives valuable advice on how to practice public relations.

FAZIO, J., AND D. GILBERT. 1981. *Public relations and communications for natural resource managers*. Dubuque, Ia.: Kendall-Hunt.
    A book on a specialized field in public relations.

FRIEDMAN, S. M., S. DUNWOODY, AND C. ROGERS. 1985. *Scientists and journalists: Reporting science as news*. New York: Macmillan.
    The authors present pieces by science reporters and scientists on how science is reported today.

NEWSOM, D., AND A. SCOTT. 1984. *This is PR: The realities of public relations*. 3rd ed. Burlingame, Calif.: Wadsworth.
    The authors provide a readable and practical guide to public relations.

RIVERS, W. L. 1984. *News in print: Writing and reporting*. New York: Harper & Row.
    An excellent text on writing magazine and newspaper articles.

20
Evaluating Your Communication

OVERVIEW

In Chapter 20, we
- ☐ Discuss the role of evaluation in professional communication
- ☐ Explain the differences between informal and formal evaluations
- ☐ Suggest an informal evaluation strategy
- ☐ Briefly review formal evaluations

In earlier chapters, we described how professionals in the computer industry evaluate their manuals and integrate their evaluations into the communication process. As more and more professionals recognize the importance of effective communication, they evaluate their communication efforts and seek to improve the quality of feedback to their messages. A few examples illustrate their varied approaches.

Getting health information across effectively is a difficult venture, observes Rose Mary Romano (1982) of the National Institutes of Health, Office of Cancer Communications. Because the office's communications cover complex, technical, and sometimes controversial subjects, the staff pretests them on sample audiences. The staff knows that effective communication can change health-related behaviors and reduce the risk of certain illnesses. Pretesting saves money, reduces the risk of alienating and misdirecting consumers, and lessens the need to produce other communications later on.

Recognizing the importance of knowing how and why people buy magazines, the Publishers Clearing House hired a research firm (Lieberman Research 1977) to conduct a national survey of magazine readers. Information from such surveys and from trade magazine readership studies guides publishers as they reshape existing magazines, drop others, and start new ones.

Researchers Lloyd Bostian, Phillip Tichenor, and J. Paul Yarbrough (1977) have conducted evaluation workshops for extension communication specialists. The workshops help the specialists discover ways of improving the cooperative extension service's publications and other efforts to reach the public.

Our point? Evaluate your communication efforts. By doing so, you improve them.

EVALUATION TECHNIQUES—SEEKING FEEDBACK

Most professionals begin with informal evaluations and may progress to formal ones. Informal evaluations are usually subjective assessments of communications' quality and effectiveness. Formal evaluations are highly structured, organized, and empirical. Students and professionals use informal evaluations more frequently than formal evaluations. The discussion that follows provides both a strategy for informal and an overview of formal evaluations.

AN INFORMAL EVALUATION STRATEGY

To evaluate your communication, begin with general questions, such as:

- How can I improve the process?
- How can I improve the product?
- How can I enhance the communication's effectiveness?

Then develop objectives specific to your communication.

Gathering Information

The more common informal evaluation techniques include personal observation and reviews; peer reviews; and audience interviews.

Personal Observation and Reviews. If you were evaluating a communication process, you might note the problems that you had run into. Maybe reviewing a manuscript took two weeks, obtaining photographs took a month because the photographer had five jobs ahead of yours, and printing took four weeks. Observations like these give you insights for the next time.

Ask yourself how you might have improved the communication. Could you have tailored it more effectively to your audiences? Could you have improved the writing by revising more? What changes would you have made if you had had more time?

PANEL 20-1. Instructions to Reviewers

                Editorial Office
                U.S. Fish & Wildlife Service
                1025 Pennock Place, Suite 212
                Fort Collins, CO 80524

INSTRUCTIONS TO REVIEWERS

Although a review of a technical manuscript could proceed in many ways, we suggest the following outlines as one kind of frame that will help to pull it all together.

I. Rapid overall evaluation

    First—Go over the entire ms. quickly and ask: Is the subject appropriate for the intended outlet? If it is obviously outside of the scope of interest of the reading audience, the review can stop at this point. Do you have any suggestions for more appropriate outlets?

    Second—If the subject is appropriate, is the ms. a serious candidate for review? I.e., is it well organized, documented, and reasonably well written? If not, it can be rejected and returned to the author(s) with a copy of "Instructions to Contributors."

II. Organization, logic, and substance of ms.

    A. Is the ms. a significant contribution to the literature? Does it really add new information or concepts to the body of scientific knowledge? Are the questions raised in objectives worth answering?

    B. Are the methods used the best possible approach to the problem?

    C. Is (are) the author(s) competent to use these methods?

    D. Are the conclusions sound; do they adequately support the data?

    E. Does the ms. have integrity and completeness? Are there any aspects of the subject that have been overlooked?

III. Ms. mechanics

Although the details of style and format are the responsibility of the redactory or copy editor, the

PANEL 20-1. (continued)

reviewer will find it useful to keep these points in mind in judging the quality of a candidate manuscript.

    A. Title—should be 10 words or less. Key words should come first if possible. Avoid the use of vague introductory wording.

    Poor: "The effects of hunting on population structure, reproduction, and mortality in hand-reared ducks in southern Minnesota."

    Good: "Harvest and population characteristics of hand-reared Minnesota mallards."

    B. Abstract—Does it exceed 2–3% of length of article? Is it informative and adequate?

    C. Do cross check of literature citations in text versus listing in Literature Cited.

    D. Pick 4–6 references at random; check the accuracy of documentation against original book or periodical in library. If author quotes are included, check these word for word.

    E. Search for first mention of vernacular name in text; scientific name should appear immediately after.

    F. Check spelling of scientific names.

    G. Scan text for alertness of author to details of style and format.

    H. Study Literature Cited—have CBE rules for abbreviations been observed?

    I. Are references to tables and figures included in text?

    J. Are the table and figure captions adequate? informative?

    K. Does the author's name (Do authors' names) appear on upper margin of each text and table page?

IV. Concluding evaluation

    Lastly, remember that authors need encouragement to continue writing. Be incisive and straightforward in your criticism but never use sarcasm or ridicule. Be absolutely fair in your judgments and helpful in your recommendations.

(Source: Editorial Office, U.S. Fish and Wildlife Service, Fort Collins, Colo. Reprinted by permission.)

Let's say that you've prepared a report recommending your company purchase a new piece of equipment. Does the report contain a phrase that confuses or puzzles readers? Have you raised a question in the reader's mind but not answered it? Will your audience say, "I really don't understand your point"? Spotting words or phrases readers may not understand is one of the most difficult tasks for a writer. But time away from your manuscript helps, as do conscious efforts to read your manuscript from the reader's perspective.

To evaluate the effects of your communication, remember your goals. To illustrate, suppose that you rewrote a users' manual for a lawn mower because your company was receiving complaints that the first manual wasn't clear about operating the machine safely. If your goal had been to reduce customer complaints, you might ask the customer relations representative whether the number of calls had decreased. You might examine the customer representative's record of calls to compare the number before and after the revision.

Peer Review. By seeking reviews of your communications, you'll benefit. No matter where your career leads, you will have your work reviewed; you'll be reviewing others' work, too. If you haven't developed a thick skin about your writing, you should try to grow one. Peer reviews can sting.

Some reviewers work subjectively, others in a more structured fashion. When U.S. Fish and Wildlife Service biologists submit their articles to the agency's editorial office, each reviewer follows a set of instructions (Panel 20-1). Such guidelines provide consistency. Other organizations also establish guidelines for reviewers. Reviewers often suggest changes, deletions, and additions. They may accept, reject, or accept a manuscript subject to revision.

When you're reviewed, you can accept the reviewers' decisions or challenge them. If you challenge them, be prepared to support your reasoning and rationale, and be prepared for the consequences of your decision. Many publications will not accept manuscripts unless you revise them according to reviewers' or editors' guidelines.

In business and industry, you'll revise your communication to conform to your superiors' suggestions or guidelines. By doing so, you usually benefit from their knowledge of the factors influencing your communication.

Audience Interviews. After you've distributed your communication, you can seek audience reaction to it through interviews. To prepare for the interview, follow the guidelines in Chapter 4. Apply the same principles and techniques as when interviewing for information, but change your interview's focus.

If possible, interview people in several different positions who will be using your communication. For example, interviewing secretaries, scientists, and writers who use the manual to operate a computer will give you insight into the problems they encounter. Many companies routinely send manual writers out to visit customers to pick up firsthand evaluations.

Interpreting the Information. As you use informal evaluation techniques, interpret your information cautiously. You will learn about your communication, but you may not obtain definitive answers to your questions. If you need definitive answers, turn to formal evaluation techniques.

FORMAL EVALUATIONS

Students and professionals occasionally may work with people who use formal evaluation techniques or use them themselves (Panel 20-2). Formal evaluations usually require structured, quantitative, empirical information. "Structured" means efforts planned to assess the communication objectives. "Quantitative" means number gathering—such as counts and frequencies. "Empirical" means conducting controlled field experiments, collecting numerical data, and using statistical analyses that permit valid, reliable inferences.

PANEL 20-2. Suggested Evaluation Techniques

Here we suggest formal evaluation techniques. The technique you can use depends upon the available funds, time, and resources.

Communication	Technique
Ideas, concepts, slants for communications	Focus groups, nominal group technique, peer review, in-depth interviews, Delphi technique
Lectures, classes, seminars, and workshops	Course evaluations, tests, controlled field experiments, peer reviews
Newsletters, journals, magazines, technical reports	Readability and Cloze tests of manuscript, readership surveys, controlled field experiments, peer reviews
Manuals, how-to pieces, instructions	Knowledge tests, performance tests, in-depth interviews, nominal group techniques, focus groups, peer reviews
Films, slide shows, videotapes	After a showing: self-administered questionnaires, focus groups, in-depth interviews, nominal group technique, peer review
Public information campaigns	Controlled field experiments, surveys, unobtrusive measures, and other observational techniques

Like informal evaluations, formal evaluations can be used during rather than after the communication process. To obtain valid evaluations, establish criteria or objectives before you begin the communication process. The National Institutes of Health reports criteria for evaluating health communications:

- Attention—Does the message attract and/or hold the audience's attention?
- Comprehension—Is the message clearly understood? Are the main ideas conveyed?
- Personal relevance—Does the target audience perceive the message as personally relevant?
- Believability—Is the message and/or its sources perceived as believable?
- Acceptability—Is there anything in the message that may be offensive or unacceptable to the target audience? (Romano 1982, 1)

These questions revolve around a larger one—how *effective* is the communication? Finding ways to answer them will help you improve as a technical communicator, but discovering the answers can be difficult.

Evaluating the Communication Process. If you were regularly involved in producing a manual, you might become concerned about the time required to produce it. You might evaluate the time necessary for different stages of the manual's production. You might ascertain where problems were occurring. Once delays were identified, you could take steps to overcome them.

You also can test your ideas by drafting manuscripts to try out on target audiences. Advertisers regularly pilot test their ideas with focus groups, surveys, and field experiments before advertising extensively.

Evaluating Communications. You can assess the readability of printed materials. In *The Measurement of Readability* (1963), George Klare defines readability as a writing style's ease of understanding. Researchers have investigated more than 200 different readability formulas and more than 150 different variables that influence reading comprehension. These formulas, usually multiple regression formulas, assess comprehension. (See Panel 20-3.) Most express readability as the number of years in school required for reading and comprehending a passage. The formulas are based on the average number of words per sentence, the average number of syllables per hundred words, and other variables. Some researchers (Felker 1980; Selzer 1983) now question readability formulas and writing prescriptions based solely on them. The researchers argue that measured variables are not the only factors that cause reading difficulty.

At best, readability formulas provide indications of how difficult others might find your writing. Other techniques for assessing comprehension test people. In the Cloze test, subjects are asked to read materials and are then tested on them (Taylor 1956; Bormuth 1966). The test consists of passages from the reading with a blank for every *n*th word. Subjects are asked to fill in the blanks and are scored on the percentage of correct responses.

Evaluating Communication Effectiveness. To assess communication effectiveness, audience members must react to the communication. Barbara Winbush and Glenda McDowell, technical editors at IBM in Rochester, New York, test IBM manuals. In a 1980 issue of *Technical Communication,* Winbush and McDowell reported on using inspection tests, written tests, task-oriented tests, and attitudinal tests.

For the inspection tests, readers examined the manuals and marked sections which gave them difficulties. In written tests, users read the manual and completed a test on their knowledge about the equipment and its parts. The written tests included true/false, multiple choice, matching, and completion questions. For task-oriented tests, the editors gave users the manuals and then asked them to perform tasks in simulated environments. Observers watched and recorded users' actions and their problems. For the attitudinal tests, the editors asked users to rate the manuals on specific characteristics. By testing six manuals, Winbush and McDowell learned about problems such as insufficiently developed objectives and topics, ineffective headings, inadequate explanations, and confusing organization. With the problems clearly identified, the editors revised the manuals accordingly.

Questionnaire evaluations provide empirical information on the audience's use of a communication. Panel 20-4 (pp. 416–417) contains an evaluation used by the U.S. Soil Conservation Service to assess a staff workshop. The questionnaire provides quantified assessments. The basic approach can be adapted to other communications.

Most formal evaluations focus on cause and effect. They require knowing that the communication in question, and nothing else, changed information levels, attitudes, or behavior in a target audience. They require the ruling out of alternative explanations with controlled field experiments, numerical data, and statistical tests.

Controlled field experiments require randomly assigning individuals or groups to treatment and control groups. The treatment group receives your communication and the control group doesn't. Then you measure the factors or variables related to your communication objectives. Measurement techniques include surveys, in-depth interviews, unobtrusive observations, such as watching people when they are unaware of your presence, and other techniques. By comparing the treatment and control groups, you can rule out alternative explanations and determine the communication's effectiveness.

To illustrate, suppose you're an engineering major, and program chairman of your school's student organization for engineers. Over the last

> ## PANEL 20-3. The Fry Readability Score
>
> **Expanded Directions for Working Readability Graph**
>
> 1. Randomly select three (3) sample passages and count out exactly 100 words each, beginning with the beginning of a sentence. Do count proper nouns, initializations, and numerals.
>
> 2. Count the number of sentences in the hundred words, estimating length of the fraction of the last sentence to the nearest one-tenth.
>
> 3. Count the total number of syllables in the 100-word passage. If you don't have a hand counter available, an easy way is to simply put a mark above every syllable over one in each word. Then when you get to the end of the passage, count the number of marks and add 100. Small calculators can also be used as counters by pushing numeral 1, then pushing the + sign for each word or syllable when counting.
>
> 4. Enter graph with *average* sentence length and *average* number of syllables; plot dot where the two lines intersect. Area where dot is plotted will give you the approximate grade level.
>
> 5. If a great deal of variability is found in syllable count or sentence count, putting more samples into the average is desirable.
>
> 6. A word is defined as a group of symbols with a space on either side; thus, *Joe, IRA, 1945*, and *&* are each one word.
>
> 7. A syllable is defined as a phonetic syllable. Generally, there are as many syllables as vowel sounds. For example, *stopped* is one syllable and *wanted* is two syllables. When counting syllables for numerals and initializations, count one syllable for each symbol. For example, *1945* is four syllables, *IRA* is three syllables, and *&* is one syllable.
>
> (Source: E. Fry, Fry's Readability graph: Clarifications, validity, and extension to level 17, Journal of Reading 21 (1977): 249.)

four years, your adviser has observed a steady drop in the number of students attending the organization's one-day seminar, "Communication Techniques for Engineers." You discuss the issues with your adviser and decide to try a personal appeal along with the regular letter and brochure announcing the seminar.

You prepare an information letter and simple brochure that you'll send to your school's 400 engineering majors. To test the personal appeal, you decide to add a handwritten, personal note to 100 majors. You randomly select 100 majors, address each by first name, and write, "Dean Jones reports attending the communications seminar has almost guaranteed jobs to graduates. Why don't you sign up for this year's seminar?" You keep a record of the 100 who received the personal notes—the treatment group. You see that 20 of the treatment group and 15 of the 300 remaining majors

EVALUATING YOUR COMMUNICATION

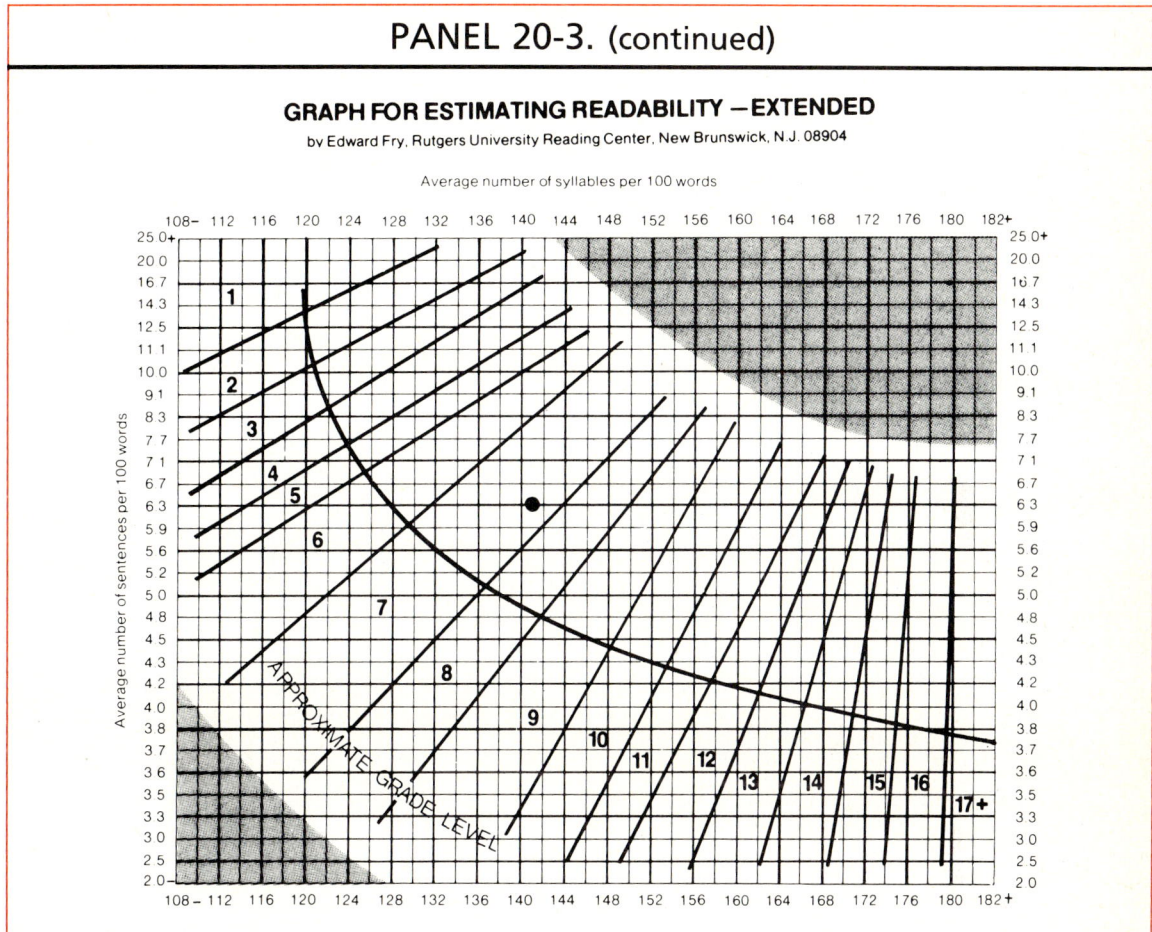

PANEL 20-3. (continued)

GRAPH FOR ESTIMATING READABILITY — EXTENDED
by Edward Fry, Rutgers University Reading Center, New Brunswick, N.J. 08904

registered. Thus 20 percent of those in the treatment group and 5 percent of those in the control group registered.

The conclusion? The personal note *appears* to have increased responses. Although you cannot state definitively that the personal note caused the higher response rate, you can suggest trying the personal note again next year.

Communications seldom lend themselves to ideal experimental designs. The design must fit the communication setting and circumstances. To develop an adequate evaluation design, seek help from university researchers. Professors in journalism, sociology, psychology, education, and other social sciences can help you. You also can consult the more than 40 years of communication research literature. Much of that literature bears on technical and scientific communication. Check the social science literature

PANEL 20-4. A Workshop Evaluation

In October 1983, the Denver, Colorado, Office of the Soil Conservation Service, United States Department of Agriculture, conducted a five-day communication workshop for field staff. Workshop participants completed the following evaluation.

EVALUATION OF TRAINING

SECTION I General Information
NAME (Last, First Middle Initial)   Date of Training
                                                                        , 19

SECTION II Total Course Evaluation   Rating (Circle one)

Organization of Course	1. Well Organized	2. Adequate	3. Poorly Organized
Level of Difficulty of Course	1. Too Advanced	2. Appropriate	3. Too Elementary
Subjects Appropriate to the Job	1. Significant	2. Partially	3. Insignificant
Training Facilities	1. Excellent	2. Fair	3. Poor
Lodging Facilities	1. Excellent	2. Fair	3. Poor
Recommendation to Colleagues	1. Highly Recommend	2. Recommend	3. Do Not Recommend

Comments on Questions in SECTION II

SECTION III Narrative
  Strong Subjects of Course

  Weak Subjects of Course

  Suggestions for Improvement

SECTION IV Instructor Evaluation
  Style and Delivery: Did the instructor exhibit a dynamic personality, effectively use voice to express degrees of thought, emphasis, and sincerity, and use gestures to convey ideas? Was a pleasant atmosphere for learning created? Were visual aids used properly and large enough for all to see?
  Learning Objective Accomplished: Each instructor must provide a learning objective which describes the goal of each unit of instruction. The objective identifies what the student should have learned at the conclusion of a lesson. Was a learning objective presented to the students prior to the lesson presentation, and was this objective accomplished at its conclusion?
  Responsiveness to Participants: Did the instructor properly involve the group? Did all participants feel included and was there a chance for questions, participation and discussion? Was there group activity?
  Knowledge of Subject: The instructor cannot teach what he or she does not know. Did the knowledge imparted appear current? Did the instructor possess in-depth knowledge and understanding of the topic?

PANEL 20-4. (continued)

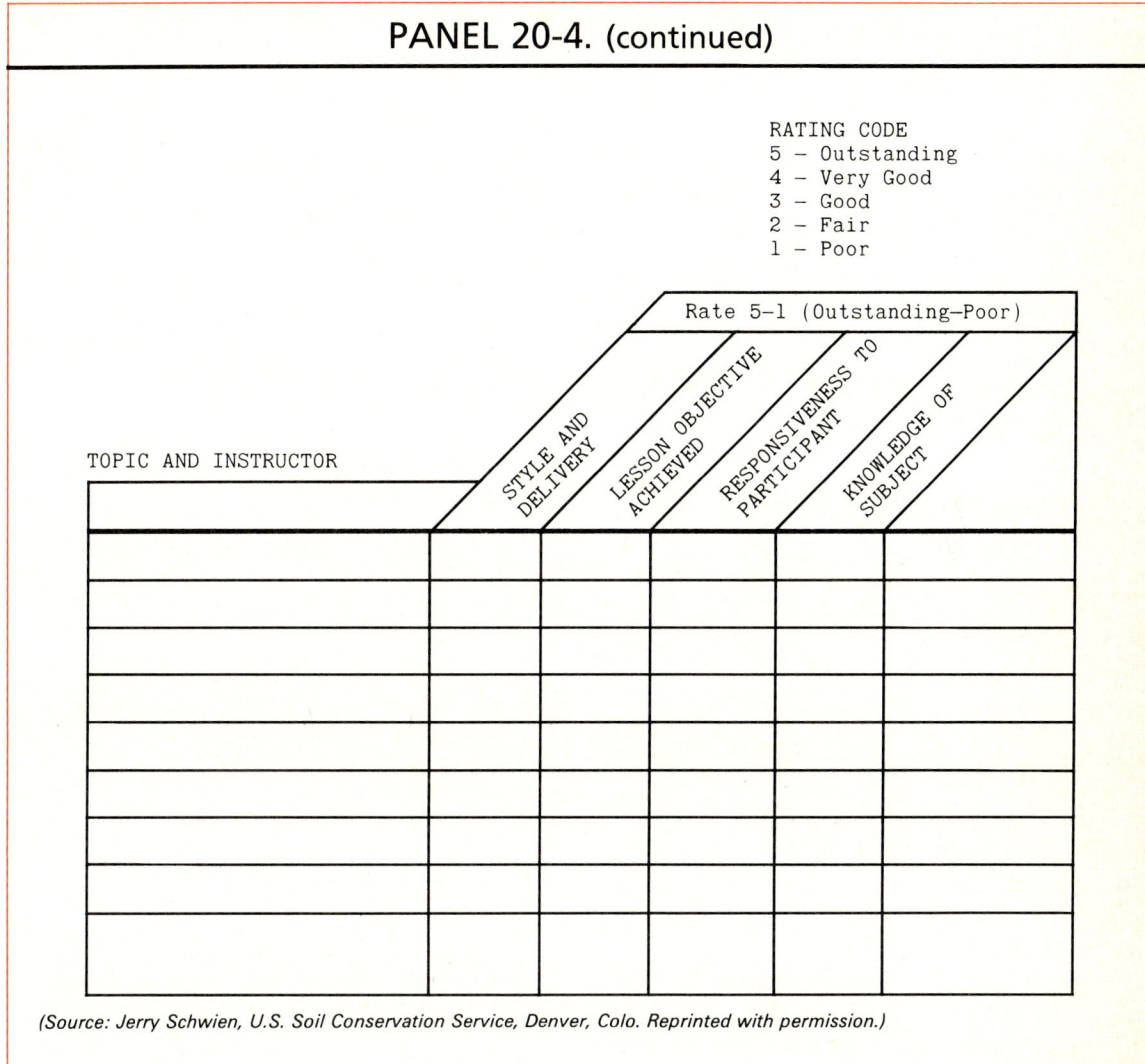

(Source: Jerry Schwien, U.S. Soil Conservation Service, Denver, Colo. Reprinted with permission.)

under these topics: communication effects, diffusion of innovations, and technology transfer.

A Precautionary Note

Some forms of evaluation are more reliable than others. For example, the self-administered questionnaires included in magazines, journals, instructional manuals, and other publications can lead to faulty conclusions. The response rate usually remains well below 50 percent, insufficient for analysis. The respondents may poorly represent all users. Furthermore, you have no way of calculating how closely the responses represent the true population response. To put your evaluations in perspective, ask:

- What limitations must I place on my evaluation?
- How did limited money, time, resources, and skill influence my conclusions?
- How well did those who evaluated the communication understand or represent the target audience?

As you make the transition from in-school communication to professional communication, remember that some evaluation—even though limited—provides more feedback than no evaluation. Evaluations help you learn more about your communication settings, audiences, and messages—and what you need to do to improve your messages. So evaluate carefully, improve your communications, and advance your career.

HIGHLIGHTS

1. Professionals should evaluate their communications.
2. Informal evaluations subjectively assess communication quality and effectiveness.
3. Formal evaluation techniques are highly structured, organized, empirical assessments of a communication's quality and effectiveness.
4. An informal evaluation strategy involves establishing the objectives for your evaluation, gathering information, and interpreting the information.
5. Techniques for gathering information include personal observations and reviews, peer reviews, and audience interviews.
6. Interpret the information from informal evaluations carefully.
7. Formal evaluation techniques include peer reviews, in-depth interviews, focus groups, nominal group techniques, the Delphi process, and surveys.
8. Readability tests contribute to evaluations of communications, but interpret results carefully.

PROJECTS

1. Informally assess the process you used to develop your last written assignment for your communication course. Develop questions specific to your assignment, and answer them. Write a brief report summarizing your evaluation.
2. With the readability formula presented in Panel 20-3, score the assignment in Project 1. Consider the assignment's audience, and

discuss the readability score's relevance to that audience. Write a brief report on the readability score and its relationship to the audience.

3. Consider your last speech or oral presentation. Informally assess your preparation for the speech, your presentation, and the audience's reaction. What might you have done differently? What would you do differently in your next speech? Write a short report on your informal analysis.

4. Swap a major assignment for your technical communication class with a classmate, and evaluate each other's assignment for (1) content, (2) communication effectiveness, and (3) overall appearance.

FOR MORE HELP

RICE, R. E., AND W. J. PAISLEY. 1981. *Public communication campaigns.* Beverly Hills: Sage.
   An excellent review of technical mass media campaigns and the implications of communication campaigns. The points serve as guidelines for evaluating technical communication.

WILLIAMSON, J. B., D. A. KARP, J. R. DALPHIN, AND P. S. GRAY. 1982. *The research craft.* Boston: Little, Brown.
   The researchers detail collecting data through sophisticated statistical analysis.

WIMMER, R. D., AND J. DOMINICK. 1983. *Mass media research.* Belmont, Calif.: Wadsworth.
   Wimmer and Dominick identify social science techniques applicable to evaluating scientific communication.

Appendix: Style Manuals and Writing Aids

To begin your personal communication library, you'll need a good college dictionary, basic style manual, standard grammar reference, and, possibly, a word book. The following lists selected references.

Style Manuals

American Chemical Society. 1978. *Handbook for authors of papers in American Chemical Society publications.* Washington, D.C.: American Chemical Society.

American Psychological Association. 1983. *Publication manual of the American Psychological Association.* 3rd ed. Washington, D.C.: American Psychological Association.

The Associated Press stylebook and libel manual. 1982. New York: Associated Press.

Buxton, D. R., R. C. Dinauer, D. A. Fuccillo, J. J. Mortvedt, and C. O. Qualset, eds. 1984. *Publications handbook and style manual.* Madison, Wis.: American Society of Agronomy.

CBE Style Manual Committee. 1983. *CBE style manual: A guide for authors, editors, and publishers in the biological sciences.* 5th ed. Bethesda, Md.: Council of Biology Editors.

The Chicago manual of style. 1982. 13th ed. Chicago: University of Chicago Press.

Steenrod, N. E., J. L. Doob, L. Carlitz, F. A. Ficken, and G. Piranian. 1980. *A manual for authors of mathematical papers.* 7th ed. Providence, R.I.: American Mathematical Society.

U.S. Government Printing Office style manual. 1973. Washington, D.C.: U.S. Government Printing Office.

Webster's standard American style manual. 1985. Springfield, Mass.: Merriam-Webster.

To find the appropriate style manuals for your technical or scientific field, look at the guidelines in the major publications in your field, or ask your professors. They can tell you the primary style guides and the dates of the most recent editions.

ANSI Guidelines

The American National Standards Institutes, 1430 Broadway, New York, N.Y. 10018, coordinates standards for technical and scientific industries and professional organizations in the United States. ANSI also provides information on the International Standards Organization.

The ANSI standards are voluntary; some employers use them. Selected ANSI standards appropriate to technical and scientific communication include the following:

ANSI Z39.14-1971. *Writing abstracts.*
ANSI Z39.16-1979. *Preparation of scientific papers for written and oral presentation.*
ANSI Z39.18-1974. *Guidelines for format and production of scientific and technical reports.*
ANSI Z39.22-1981. *Proof correction.*
ANSI Z39.23-1974. *Technical report number (STRN).*

Other ANSI standards cover drafting and library practices. For a detailed listing of the standards, write to ANSI and ask for its catalog and prices.

Grammar and Usage Guides

Bernstein, T. M. 1978. *The careful writer*. New York: Atheneum.

Crews, F. 1984. *The Random House handbook*. 4th ed. New York: Random House.

Crews, F., and S. Schor. 1985. *The Borzoi handbook for writers*. New York: Knopf.

Hodges, J. C., and M. E. Whitten. 1986. *Harbrace college handbook*. 10th ed. New York: Harcourt Brace Jovanovich.

Strunk, W., Jr., and E. B. White. 1979. *The elements of style*. 3rd ed. New York: Macmillan.

Dictionaries

The American heritage dictionary of the English language. 1982. Boston: Houghton Mifflin.

Ellis, K. 1976. *The word book*. Boston: Houghton Mifflin.

McGraw-Hill dictionary of earth sciences. 1984. New York: McGraw-Hill.

McGraw-Hill dictionary of life sciences. 1976. New York: McGraw-Hill.

McGraw-Hill dictionary of scientific and technical terms. 1980 3rd. ed. New York: McGraw-Hill.

The Random House college dictionary. 1980. Rev. ed. New York: Random House.

The Random House dictionary of the English language. 1981. Unabridged. New York: Random House.

Webster's new world dictionary of the American language. 2nd ed. 1978. Cleveland, Ohio: Williams Collins & World.

Webster's third new international dictionary of the English Language. 1976. Unabridged. Springfield, Mass.: Merriam.

Books on Writing

Chandler, H. E. 1978. *The how to write what book*. Metals Park, Ohio: American Society for Metals.

George, D., G. A. Olson, and R. Ray. 1984. *Style and readability in technical writing*. New York: Random House.

Flower, L. 1981. Problem solving strategies for writing. New York: Harcourt Brace Jovanovich.

Fluegelman, A., and J. J. Hewes. 1983. *Writing in the computer age*. Garden City, N.Y.: Doubleday.

Plotnik, A. 1982. *The elements of editing*. New York: Macmillan.

Williams, J. M. 1981. *Style: Ten lessons in clarity and grace*. Glenview, Ill.: Scott, Foresman.

Zinsser, W. 1980. *On writing well*. 2nd ed. New York: Harper & Row.

———. 1983. *Writing with a word processor*. New York: Harper and Row.

References

Adams, J. L. 1976. *Conceptual blockbusting*. San Francisco: San Francisco Book Company.

Adams, H. 1918. *The education of Henry Adams*. Rpt. New York: Random House, 1931.

Allard, T., H. Duzet, J. Overholt, F. Salinas, and J. Vreeland. 1982. Group interview, IBM plant, Santa Teresa, Calif.

Allen, A. 1977. *Steps toward better scientific illustrations*. Lawrence, Kan.: Allen Press.

Alvarez, L. W., W. Alvarez, F. Asaro, and H. V. Michel. 1980. Extraterrestrial cause for the Cretaceous-Tertiary extinction. *Science* 208: 1095–1108.

Alvarez, W. 1982. Interview with authors.

Amarino, N. 1985. Personal communication.

American men and women of science. 1983. New York: Bowker.

American Chemical Society. 1978. *Handbook for authors of papers in American Chemical Society publications*. Washington, D.C.: American Chemical Society.

American Psychological Association. 1983. *Publication manual of the American Psychological Association*. 3rd ed. Washington, D.C.: American Psychological Association.

American Association of Agricultural College Editors. 1976. *Communication handbook*. 3rd ed. Danville, Ill.: Interstate Publishers.

American statistics index 1974 annual and retrospective edition. 1974. Washington, D.C.: Congressional Information Service.

Anderson, P. V., R. J. Brockmann, and C. R. Miller, eds. 1983. *New essays in technical and scientific communication: Research, theory, practice*. Farmingdale, N.Y.: Baywood.

Anderson, W., and D. Cox. 1980. *The technical reader: Readings in technical, business, and scientific communication*. New York: Holt, Rinehart & Winston.

Andriot, J. L., ed. 1984. *Guide to U.S. Government publications*. McLean, Va.: Documents Index.

Apple's new chief says concern needs discipline. 1983. *Wall Street Journal*, July 20, 38.

Applied science and technology index. 1986. Bronx, N.Y.: H. W. Wilson.

Associated Press. 1982. How do you spell nitpicking? *Fort Collins Coloradoan*, City Final Section, September 18, 1.

———. 1983. Scientists expect huge crop of mosquitoes in nation. *Rocky Mountain News*, May 13, 56.

Babbie, E. 1973. *Survey research methods*. Belmont, Calif.: Wadsworth.

———. 1982. *Social research for consumers*. Belmont, Calif.: Wadsworth.

Bailey, T. N., A. W. Franzman, P. D. Arneson, and J. L. Davis. 1983. An evaluation of visual location data from neck-collared moose. *Journal of Wildlife Management* 47(1): 25–30.

Baker, R. 1984. How to punctuate. *Power of the printed word*. New York: International Paper Company.

Battison, F., and J. Landesman. 1981a. The cost-effectiveness of designing simpler documents. *Simply Stated* 16: 1, 3–4.

———. 1981b. Treating the whole document: A benefits handbook. *Simply Stated* 21: 2–3.

Battison, R., J. Landesman, and F. Pickering. 1982a. Eliminating gender bias in language. *Simply Stated* 28: 1–2.

———. 1982b. Illustrated procedural instructions. *Simply Stated* 26: 1, 3–4.

Bennett, W. 1978. Science goes glossy. *The Sciences* 19(7): 10–15, 22.

Berndt, J. 1984. Interview with authors. July 2.

Bernstein, T. M. 1971. *Miss Thistlebottom's hobgoblins*. New York: Farrar, Straus and Giroux.

———. 1978. *The careful writer*. New York: Atheneum.

Bertrand, J. 1979. Selective avoidance on health topics: A field test. *Communication Research* 6(3): 271–294.

Bibliography of agriculture. 1986. Phoenix, Ariz.: Oryx Press.

Bieger, G. R. 1982. *Comprehension of spatial and contextual information in pictures and texts*. Technical Report No. 6. Ithaca, N.Y.: Cornell University, College of Agriculture and Life Sciences, Department of Education.

Biography master index. 1982. Detroit, Mich.: Gale Research.

Biological abstracts. 1926– . Philadelphia: Biosciences Information Services.

Biological abstracts/RRM. 1980– . Philadelphia: Biosciences Information Services.

Biosis previews. 1986. Philadelphia: Biosciences Information Services.

Bishop, A. 1984. *Slides: Planning and producing slide programs*. Kodak Publication S-30. Rochester, N.Y.: Eastman Kodak.

Black, H. 1951. *Black's law dictionary*. 4th ed. St. Paul, Minn.: West Publishing.

Boeker, E. L., V. E. Scott, H. G. Reynolds, and B. A. Donaldson. 1972. Seasonal food habits of mule deer in southwestern New Mexico. *Journal of Wildlife Management* 36: 56–63.

Boggis, C. 1980. An audiovisual handbook for the scientist: A link between research and application. Master's thesis, Colorado State University.

Bohle, B. 1976. The home book of American quotations. New York: Dodd Mead.

Bolles, R. N. 1986. *What color is your parachute?* rev. ed. Berkeley, Calif.: Ten-Speed Press.

Bond, C. 1973. *Keys to Oregon freshwater fishes*. Technical Bulletin 58 rev. Corvallis: Oregon State University Agricultural Experiment Station.

Books in print. [Current year]. New York: Bowker.

Borland, S. W. 1982. Societal aspects of precipitation modification. In *Hailstorms of the Central High Plains: Final report of the national hail research experiment,* 1, ed. C. A. Knight and P. Squires. Boulder: Colorado Associated University Press.

Bormuth, J. R. 1966. Readability: A new approach. *Reading Research Quarterly* 1: 79–132.

Bostian, L., P. Tichenor, and J. P. Yarbrough. 1977. *Evaluation research for information programs*. Workshop sponsored by NCR-90 Communication Committee. Des Moines: Iowa State University.

Bowen, M. E., and J. A. Mazzeo, eds. 1979. *Writing about science*. New York: Oxford University Press.

Bowser, J. M., ed. 1984. *The CPC annual*. Bethlehem, Pa.: College Placement Council.

Broad, W., and N. Wade. 1982. *Betrayers of the truth: Fraud and deceit in the halls of science*. New York: Simon & Schuster.

Brockmann, R. J. 1986. *Writing better computer user documentation: From paper to screen*. New York: Wiley.

Bruno, M. H., ed. 1984. *Pocket pal: A graphic arts primer*. 13th ed. New York: International Paper Company.

Brush, J., and D. Brush. 1978. Visual communications—from cave to boardroom. *Public Relations Journal* 34(9): 12–15.

Bureau of Land Management. 1981. *Final environmental impact statement on the Energy Transportation System, Inc., coal slurry pipeline transportation project*. Denver, Colo.: U.S. Department of the Interior.

Burnett, C., R. Powers, and J. Ross. 1973. *Agricultural news writing*. Dubuque, Iowa: Kendall/Hunt.

Buxton, D. R., R. C. Dinauer, D. A. Fuccillo, J. J. Mortvedt, and C. O. Qualset. 1984. *Publications handbook and style manual*. Madison, Wis.: American Society of Agronomy.

CA SEARCH. 1986. Columbus, Ohio: Chemical Abstract Service.

Caernarven-Smith, P. 1983. *Audience and response*. Pembroke, Md.: Firman Publications.

Campbell, D. T., and J. C. Stanley. 1973. *Experiments and quasi-experimental designs for research*. Chicago: Rand McNally.

Campbell, S. 1974. *Flaws and fallacies in statistical thinking*. Englewood Cliffs, N.J.: Prentice-Hall.

Canadiana. 1985. Ottawa: National Library of Canada.

Carlin, M. 1982. Charting map makers is life-long avocation. *Rocky Mountain News,* September 15, 61–62.

Carroll, L. 1896. *Through the looking glass*. Rpt. New York: St. Martin's Press, 1977.

Carter, R. 1962. *Communication, understanding and support for education*. Paris-Stanford Studies in Communication. Stanford, Calif.: Stanford Institute for Communication Research.

CBE Style Manual Committee. 1983. *CBE style manual: A guide for authors, editors, and publishers in the biological sciences*. 5th ed. Bethesda, Md.: Council of Biology Editors.

Center, A., and F. Walsh, eds. 1985. *Public relations managerial case studies and problems*. 3rd ed. Englewood Cliffs, N.J.: Prentice-Hall.

Central Flyway Council. 1964. *Kansas waterfowl identification guide*. Pratt, Kans.: Central Flyway Council.

Chandler, H. E. 1978. *The how to write what book*. Metals Park, Ohio: American Society for Metals.

Chemical abstracts. 1907– . Columbus, Ohio: Chemical Abstract Service, American Chemical Society.

Chlamydia epidemic: The campus bug that wrecks your sex life. 1985. *Harper's Bazaar,* June, 74.

Chlamydia—nation's no. 1 sexually transmitted disease. 1985. *Jet,* December 23, 69.

Christensen, C. M. 1985. *Edible wild mushrooms*. AG-BU-1357. St. Paul, Minn.: Agricultural Extension Service.

CLASS. Computerized literature access search service. 1986. Fort Collins: Colorado State University, Morgan Library.

Correspondence management. 1973. Washington, D.C.: General Services Administration, National Archives and Records Service, Office of Records Management.

Could you have chlamydia? 1985. *McCalls,* June, 64–65.

Cowan, G., and E. Cowan. 1980. *Writing*. Glenview, Ill.: Scott, Foresman.

Cox, B., and C. Roland. 1973. How rhetoric confuses scientific issues. *IEEE Transactions on Professional Communication*. PC-16: 140–142.

Crandell, T. L., and M. D. Glock. 1981. *Technical communication—taking the user into account*. Technical Report No. 4. Ithaca, N.Y.: Cornell University, College of Agriculture and Life Sciences, Department of Education.

Crews, F. 1984. *The Random House handbook*. 4th ed. New York: Random House.

Crews, F., and S. Schor. 1985. *The Borzoi handbook for writers*. New York: Knopf.

Current index to journals in education. 1986. Phoenix, Ariz.: Oryx Press.

Cutlip, S., A. Center, and G. Broom. 1985. *Effective public relations*. 6th ed. Englewood Cliffs. N.J.: Prentice-Hall.

Danbom, D. R. 1984. Lecture on copyright. Colorado State University, March 9.

Danbom, R. 1983. Interview with authors. Sept. 23.

Davis, L. 1986. Chlamydia detection and (maybe) protection. *Science News* 129:231.

Davis, R. M. 1978. How important is technical writing?—A survey of the opinions of successful engineers. *Journal of Technical Writing and Communication* 8: 207–216.

DeBakey, L. 1978. The persuasive proposal. *Journal of Technical Writing and Communication* 6(1): 5–25.

DeGeorge, J., G. Olson, and R. Ray. 1984. *Style and readability in technical writing*. New York: Random House.

Delbecq, A. L., A. H. Van de Ven, and D. H. Gustafson. 1975. *Group techniques in program planning: A guide to nominal and Dephi processes*. Glenview, Ill.: Scott, Foresman.

Denver Water Department. 1978. *Public annual report*. Supplement to *Empire Magazine, Denver Post.*

Detwiler, R. M. 1979. How to respond to media queries. *Public Relations Journal* 35(12): 28–29.

Dillman, D. 1978. *Mail and telephone surveys: the total design method*. New York: Wiley-Interscience.

Doomed dino. 1979. *Time,* July 16, 76.

Doris, L., and B. Miller. 1983. *Complete secretary's*

handbook. 5th ed. Englewood Cliffs, N.J.: Prentice-Hall.

Dressel, S. 1984. How humans construct and communicate technical concepts. *Proceedings of the 31st international technical communication conference,* Seattle, Wash. Washington, D.C.: Society for Technical Communication.

Dunwoody, S. 1978. *Science writers at work.* Research Report No. 7. Bloomington: Indiana University, Bureau of Media Research.

———. 1981. Science writers at work. In *Communicating university research,* ed. P. Alberger and V. L. Carter. Washington, D.C.: Council for the Advancement and Support of Education.

Eastman Kodak. 1980. *Legibility—art work to screen.* Publication S-24. Rochester, N.Y.: Eastman Kodak.

———. 1986a. Simple is better. *Audiovisual Notes From Kodak.* Kodak Publication No. V9-2-2-5. Rochester, N.Y.: Eastman Kodak.

———. 1986b. *The communicator's catalog.* Publication S-4. Rochester, N.Y.: Eastman Kodak.

———. [Current year]. *Index to Kodak information.* Rochester, N.Y.: Eastman Kodak.

Ebert, R. H. 1980. A fierce race called medical education. *New York Times,* July 9, 18.

Eckberg, A. R. 1986. Questions you may be asked. In *Spectrum of opportunities.* Chicago: Roosevelt University, Career Planning and Placement Office.

Eiseley, L. 1959. *The immense journey.* New York: Vintage Books.

Eisenberg, A. 1982. *Effective technical communication.* New York: McGraw-Hill.

Eisenstein, E. L. 1981. *The printing press as an agent of change,* 1. New York: Oxford University Press.

Elbow, P. 1973. *Writing without teachers.* New York: Oxford University Press.

Ellis, K. 1976. *The word book.* Boston: Houghton Mifflin.

The engineering index monthly. 1986. New York: Engineering Information.

Enrick, N. 1980. *Handbook of effective graphic and tabular communication.* New York: Robert E. Krieger.

ERIC. 1986. Washington, D.C.: National Institutes of Education.

Faigley, L., and T. P. Miller. 1982. What we learn from writing on the job. *College English* 44: 557–568.

Falk, J. H. 1976. Energetics of a suburban lawn ecosystem. *Ecology* 57: 141–150.

Fazio, J., and D. Gilbert. 1981. *Public relations and communications for natural resource managers.* Dubuque, Iowa: Kendall-Hunt.

Feber, R. 1980. *What is a survey?* Washington, D.C.: American Statistical Association.

Felker, D. 1980. *A review of research relevant to document design.* Washington, D.C.: Document Design Center, American Institutes for Research.

Felker, D., F. Pickering, V. R. Charrow, V. M. Holland, and J. C. Redish. 1982. *Guidelines for document designers.* Washington, D.C.: Document Design Center, American Institutes for Research.

Fields, A. 1982. Getting started: Pattern notes and perspectives. In *The technology of text,* ed. D. Jonassen. Englewood Cliffs, N.J.: Educational Technology Publications.

Fischer, D. 1970. *Historians' fallacies.* New York: Harper & Row.

Fishlein, P. 1981. Personal communication. May 4.

Flower, L. 1985a. Communication strategy in professional writing: Teaching a rhetorical case. In *Courses, components and exercises in technical communication,* ed. D. W. Stevenson, R. R. Betz, D. L. Carson, D. H. Cunningham, and T. M. Sawyer. Urbana, Ill.: National Council of Teachers of English.

———. 1985b. *Problem solving strategies for writing.* 2nd ed. San Diego: Harcourt Brace Jovanovich.

Flower L., and J. R. Hayes. 1980. The cognition of discovery: Defining a rhetorical problem. *College Composition and Communication* 31: 21–32.

———. 1981. A cognitive process theory of writing. *College Composition and Communication* 32: 365–387.

———. 1984. Images, plans, and prose. *Written Communication* 1: 120–160.

Flower, L., J. R. Hayes, and H. Swarts. 1983. Revising functional documents: The scenario principle. In *New essays in technical and scientific communication: research, theory, practice,* ed. P. V. Anderson, R. J. Brockmann, and C. R. Miller. Farmingdale, N.Y.: Baywood.

Fluegelman, A., and J. J. Hewes. 1983. *Writing in*

the computer age. New York: Anchor Press/ Doubleday.

Forbes, M. 1982. How to write a business letter. *Power of the printed word*. New York: International Paper Company.

Franklin, J. 1981. Translating the curious languages of research. In *Communicating university research,* ed. P. Alberger and V. L. Carter. Washington, D.C.: Council for the Advancement and Support of Education.

Friedman, S., S. Dunwoody, and C. Rogers. 1985. *Scientists and journalists: Reporting science as news*. New York: Free Press.

Fry, E. 1977. Fry's readability graph: Clarification, validity and extension to level 17. *Journal of Reading* 21: 242–252.

Fuentes, J. 1985. Granville sees stock crash. *Rocky Mountain News,* March 29, 86.

Garratt, G. A. 1968. Education faces a challenge. *Journal of Forestry* 66: 551–555.

Gates, J. K. 1979. *Guide to the use of books and libraries*. New York: McGraw-Hill.

Gerstenfeld, A. A., and P. Berger. 1980. Analysis of utilization differences for scientific and technical information. *Management Science* 26(2): 165–179.

Gonzalez, J. 1980. *The complete guide to effective dictation*. Boston: Kent Publishing.

Goswami, D., J. C. Redish, D. Felker, and A. Siegel. 1981. *Writing in the professions*. Washington, D.C.: Document Design Center, American Institutes for Research.

Gottlieb, L. 1981. *HEIDI*. Livermore, Calif.: Lawrence Livermore National Laboratory.

———. 1982. How to wake up audiences. Presentation at meeting of the Pacifica Chapter of the Society for Technical Communication, December 3, at Lawrence Livermore National Laboratory, Livermore, Calif.

———. 1984. The Livermore Lab has created a new breed of technical communicator. *Proceedings of the 31st international technical communication conference,* Seattle, Wash. Washington, D.C.: Society for Technical Communication.

Government of Canada publications. 1985. Quebec: Canadian Government Publishing Centre.

Government printing office style manual. 1973. Washington: U.S. Government Printing Office.

Graphics. 1985. *PC Week* 2(46): S/1–S/68.

Gregory, R. L. 1970. *The intelligent eye*. New York: McGraw-Hill.

Greitzer, J. 1985. Setting type with PCs. *PC Week* 2(49): 37–39.

Grimm, S. 1982. *How to write computer manuals for users*. New York: Van Nostrand Reinhold.

Hagstrom, W. 1965. *The scientific community*. New York: Basic Books.

Haig, T. 1975. Problem solving strategies. Presentation to environmental systems monitoring class. Institute for Environmental Studies, University of Wisconsin, Madison, Wisc.

———. 1983. Letter to authors. December 22.

Handbook for authors of papers in American Chemical Society publications. 1978. Washington, D.C.: American Chemical Society.

Handsfield, H. H., L. L. Jasman, P. L. Roberts, V. W. Hanson, R. L. Kothenbeutel, and W. E. Stamm. 1986. Criteria for selective screening for *Chlamydia trachomatis* infection in women attending family planning clinics. *JAMA: Journal of the American Medical Association* 255(13): 1730–1734.

Haney W., ed. 1985. *Communication and interpersonal relations: Text and cases*. 5th ed. Homewood, Ill.: Irwin.

Harris, J., and R. H. Blake. 1976. *Technical writing for social sciences*. Chicago: Nelson-Hall.

Harty, K. 1980. *Strategies for business and technical writing*. New York: Harcourt Brace Jovanovich.

Harvey, R. 1984. Interview with authors. April 29.

Hayes, J. R., and L. S. Flower. 1983. Uncovering cognitive processes in writing: An introduction to protocol analysis. In *Research on writing,* ed. P. Mosenthal, L. Tamor, and S. A. Walmsley. New York: Longman.

Hendon, J. 1984. Random access: Possible VDT harm still being debated. *Rocky Mountain News,* Dec. 15, 9B.

Hewlett-Packard. 1985. *How to design effective overhead transparencies*. San Diego: Hewlett-Packard.

Hill, M., and W. Cochran. 1977. *Into print*. Los Altos, Calif.: William Kaufman.

Hip women. 1979. *Family Weekly,* January 14, 26.

Hodges, J., and M. Whitten. 1986. *Harbrace college*

handbook. 10th ed. New York: Harcourt Brace Jovanovich.

Huckin, T. 1983. A cognitive approach to readability. In *New essays in technical and scientific communication: Research, theory, practice,* ed. P. V. Anderson, R. J. Brockmann, and C. R. Miller. Farmingdale, N.Y.: Baywood.

Humphrey, S., and T. L. Zinn. 1982. Everglades mink in Florida. *Journal of Wildlife Management* 46: 375–387.

Hutchcroft, T. 1983. *Responding to media cheap shots: Observations on the CAST experience.* Paper No. 16. Ames, Iowa: Council for Agricultural Science and Technology.

IBM. 1982. *Instruction sheet for Selectric typewriters.* Lexington, Ky.: International Business Machines.

If you write or use erasers and drive to work daily or sometimes have headaches . . . then read this . . . maybe. 1982. *Simply Stated* 23: 2–3.

Index to publications of the United States Congress. 1986. Washington, D.C.: Congressional Information Service.

Index medicus. 1960– . Bethesda, Md.: National Library of Medicine, U.S. Department of Health and Human Services.

Ingles, D. 1976. *Is it really so?* Philadelphia: Westminster Press.

Internal Revenue Service. 1975. *Effective writing.* Washington, D.C.: U.S. Government Printing Office.

Irving, J. 1983. How to spell. *Power of the printed word*. New York: International Paper Company.

Jaynes, G. 1984. In Alaska: Where chili is chilly. *Time,* June 25, 9.

Jesperson, J., and J. Fitz-Randolph. 1977. *From sundials to atomic clocks*. National Bureau of Standards Monograph 155. Washington, D.C.: U.S. Government Printing Office.

Joenk, R. L. 1980. Public speaking for engineers and scientists. *IEEE transactions on professional communication* PC-23: 1–60.

Jones, R. R. 1986. Free subscription application. *Research and Development* 86(2): 163.

KayPro. 1983. *KAYPRO user's guide.* Solana Beach, Calif.: KayPro Corporation.

Keene, M., and M. Barnes-Ostrander. 1985. Audience analysis and adaptation. In *Research in technical communication,* ed. M. G. Moran and D. Jornet. Westport, Conn.: Greenwood Press.

Keerdoja, E., P. Vercammen, and M. Lord. 1983. The grand old lady of software. *Newsweek,* May 9, 13.

Kelley, H. 1985. The war against Hessian fly. *Agricultural Research* 33(5): 10–13.

Kenny, M. F., and R. F. Schmitt. 1981. *Images, images, images*. Kodak Publication S-12. Rochester, N.Y.: Eastman Kodak.

Kerr, R. A. 1980. Asteroid theory of extinctions strengthened. *Science* 210: 514–517.

Keyworth, G. 1984. Keyworth comments. *SIPIscope* 13(1): 1–9.

Kieffer, J. S. 1984. Nursing diagnosis can make a critical difference. *NursingLife* 4(3): 18–21.

Kisling, J. 1980. Here's Denver: A word to the wise about clichés. *Denver Post,* March 7, 2.

Klare G. 1963. The measurement of readability. Ames: Iowa State University Press.

———. 1980. *A manual for readable writing*. Glen Burnie, Md.: REM Company.

Kramer, S. N. 1953. *The Sumerians*. Chicago: University of Chicago Press.

Krippendorff, K. 1981. *Content analysis: An introduction to its methodology*. Beverly Hills, Calif.: Sage.

Kuhn, T. S. 1970. *The structure of scientific revolutions*. 2nd ed. Chicago: University of Chicago Press.

Landesman, J. 1981. The "FISAP," before and after. *Simply Stated* 15: 1–4.

Lansford, H. 1971. Communicating with the nonspecialist about basic environmental research. In *Proceedings of the 18th international technical communication conference,* San Francisco. Washington, D.C.: Society for Technical Communication.

———. 1977. *Public information for the national hail research experiment*. Sixth Conference on Inadvertent and Planned Weather Modification, Champaign-Urbana, Ill., p. 2.

Lasswell, H. D. 1948. The structure and function of communication in society. Reprinted in *The process and effects of mass communication,* revised ed., ed. W. Schramm and D. F. Roberts. Urbana: University of Illinois Press, 1972.

Leach, R. M., Jr. 1982. The effect of manganese

deficiency upon the ultrastructure of the eggshell. *Poultry Science* 62: 499–504.

———. 1983a. Understanding eggshell formation. *Poultry Science Update* 1: 1. University Park: Pennsylvania State University, Department of Poultry Science.

———. 1983b. Telephone interview with authors. October 27.

Lear, J. 1970. The trouble with science writing. *Columbia Journalism Review* 9(2): 30–33.

Lechner, G. 1981. Gastric bypass and gastroplasty: Treatment for morbid obesity. *Nursing81* 11(1): 59.

Ledbetter, H. 1982. Interview with authors. July 27.

Leffers, R. 1982. *How to prepare charts and graphs for effective reports*. New York: Barnes and Noble.

Leopold, A. 1972. *Sand county almanac*. New York: Oxford University Press.

Lewis, D., and J. Greene. 1982. *Thinking better*. New York: Rawson, Wade.

Lieberman Research. 1977. *How and why people buy magazines: A national study of the consumer market for magazines*. Port Washington, N.Y.: Publishers Clearing House.

Lubov, A. 1979. *Issuing municipal bonds: A primer for local officials*. Agricultural Information Bulletin No. 429. St. Paul: University of Minnesota, Department of Agricultural and Applied Economics.

Lueck, A. 1984. Interview with authors. July 8.

Lukazewski, J. 1984. *Having effective media interviews*. Minneapolis: Brum Anderson Execu-Comm Group.

Lunsford, R. F. 1985. Confessions of a developing writer. In *Writers on writing*, ed. T. Waldrep. New York: Random House.

Lutz, W. 1985. How I write. In *Writers on writing*, ed. T. Waldrep. New York: Random House.

Macdonald-Ross, M. 1977a. Graphics in text. In *Review of Research in Education* No. 5, ed. L. S. Shulman. Itasca, Ill.: F. E. Peacock.

Macdonald-Ross, M. 1977b. How numbers are shown. *AV Communication Review* 25: 359–409.

MacPherson v. *Buick Motor Co.* 1916. 217 N.Y. 382. 111 NE 1050 (1916).

Marbach, W., G. C. Lubenow, W. J. Cook, F. Gibney, Jr., and K. Willenson. 1982. To each his own computer. *Newsweek,* February 22, 50–56.

Marcus, S. J. 1982. How to court a cat. *Newsweek,* March 22, 13.

Martin, P. 1977. *The word watcher's handbook*. New York: D. McKay.

Martin Marietta. 1976. *Developing effective presentations: 8 steps*. Denver: Martin Marietta.

Matkowski, B. S. 1983. *Steps to effective business graphics*. San Diego: Hewlett Packard.

Matlock, F. 1986. Step softly. *Rodale's New Shelter* 7(3): 80–81.

Matrazzo, D. 1980. *The corporate script writing book*. Philadelphia: Media Concepts Press.

———. 1983. Effective slide presentations. Lecture to the basic technical communication class, Colorado State University. Fort Collins, April 13.

Mayer, K. 1986. Colorado's bike psyche. *Rocky Mountain News,* April 20, 68.

McCain, G., and E. M. Segal. 1982. *The game of science*. Monterey, Calif.: Brooks/Cole.

McCulloch Corporation. 1980. *McCulloch owner's manual: PROMAC 610/650 chain saws*. Los Angeles: McCulloch Corporation.

McKee, B. 1974. Types of outlines used by technical writers. *Journal of English Teaching Techniques* 17: 30–36. (Also in *A guide for writing better technical papers,* ed. C. Harkins and D. L. Plung [New York: Institute of Electrical and Electronics Engineers, 1982.])

McWilliams, P. 1983. *The word processing book*. New York: Ballantine.

Meilach, D. Z. 1985. Overhead transparency sales up dramatically. *Graphics, PC Week* 2(46): S–8.

Merrill, J. C., and R. L. Lowenstein, eds. 1979. *Media, messages, and men*. New York: Longman.

Metzler, K. 1977. *Creative interviewing*. Englewood Cliffs, N.J.: Prentice-Hall.

MicroPro. 1983. *KAYPRO WordStar II reference manual*. San Rafael, Calif.: MicroPro Corporation.

Midwest Plan Service. 1974. *Farmstead planning handbook*. Ames: Iowa State University, Midwest Plan Service.

Miller, C. 1985. Invention in technical and scientific discourse: A perspective survey. In *Research in technical communication,* ed. M. Moran and J. Journet. Westport, Conn.: Greenwood Press.

Miller, D. 1971. *Design for lively learning*. Manual

78. Columbia: University of Missouri, Cooperative Extension Service.

Monthly catalog of United States government publications. 1986. Washington, D.C.: Superintendent of Documents.

Moody, W. E., Jr. 1982. How humans read and understand. Workshop presented at the 29th International Technical Communication Conference, Boston, Mass. (The accompanying manual, *Handbook: Human factors guidelines in information transfer,* is available from International Business Machines, Armonk, N.Y.)

Moran, M. G., and D. Journet, eds. 1985. *Research in technical communication.* Westport, Conn.: Greenwood Press.

Morgan, C. 1980. What's wrong with technical writing today? *Byte* 5: 7–12, 294.

Munn, L. 1982. Making the best of word processing. *Technical Communication* 29(2): 3.

Murphy, H., and H. Hildebrandt. 1984. *Effective business communication.* 4th ed. New York: McGraw-Hill.

Murphy, J. 1985. Computers: Convert to the write stuff. *Time,* December 9, 98–100.

Murray, D. 1985. Getting under the lightning. In *Writers on writing,* ed. T. Waldrep. New York: Random House.

Naisbitt, J. 1983. *Megatrends: Ten new directions transforming our lives.* 6th ed. New York: Warner Books.

———. 1985. Trend notes: Display terminal problems growing. *Rocky Mountain News Sunday Magazine,* May 19, 14.

National agricultural library catalog. 1978. New York: Rowman & Littlefield.

National Science Foundation. n.d. *Grants for scientific and engineering research.* NSF 79–81. Washington, D.C.: U. S. Government Printing Office.

Nau, C. 1978. *The driving skills book.* Shell Answer Book #14. Houston: Shell Oil Company.

Nelson, R. P. 1983. *Publication design.* 3rd ed. Dubuque, Ia.: Wm. C. Brown.

Nelson, W. B. 1984. *Inspector's statement of the incident on Rocky Mountain Airways DHC-7, N9058P, Serial Number 5.* Aurora, Colo.: National Transportation Safety Board.

Newsom, D., and A. Scott. 1984. *This is PR: The realities of public relations.* 3rd ed. Belmont, Calif.: Wadsworth.

Newsome, W. L. 1978. *New guide to popular government publications for libraries and home reference.* Littleton, Colo.: Libraries Unlimited.

Newton, R. 1977. *The crime of Claudius Ptolemy.* Baltimore: Johns Hopkins University Press.

Nikon. 1986. *Nikon F3 high-eyepoint instruction manual.* New York: Nikon.

O'Hayre, J. 1978. *Gobbledygook has gotta go.* Washington, D.C.: U.S. Government Printing Office.

O'Keefe, G. 1985. Telephone interview with authors. December 26.

OCLC. 1983. Columbus: Ohio College Library Center.

Odell, L., D. Goswani, A. Herrington, and D. Quick. 1983. Studying writing in non-academic settings. In *New essays in technical and scientific communication: Research, theory, practice,* ed. P. V. Anderson, R. J. Brockman, and C. Miller. Farmingdale, N.Y.: Baywood.

Office of Records Management. 1973. *Correspondence management: Records management handbook.* Washington, D.C.: U.S. Government Printing Office.

Oppenheim, L., C. Kydd, V. P. Carroll, and G. Carroll. 1981. *A study of the effects of the use of overhead transparencies on business meetings.* Philadelphia: University of Pennsylvania, Wharton Applied Research Center.

Page, B., P. Knightly, E. Potter, and A. Perry. 1976. Behind the thalidomide tragedy. *Atlas World Press Review,* September, 15–20.

Parish, D. W. 1981. *State government reference publications: An annotated bibliography.* Littleton, Colo.: Libraries Unlimited.

Payne, S. L. 1980. *The art of asking questions.* Princeton, N.J.: Princeton University Press.

Pember, D. 1984. *Mass media and the law,* 3rd ed. Dubuque, Ia.: Wm. C. Brown.

Petersen, B. 1979. How to make OHP transparencies. *Photomethods* 22(2): 30–31.

Peterson, I. 1982a. Calculators can do funny things. *Science News* 122: 75.

———. 1982b. Can your computer count? *Science News* 122: 72–75.

Pinelli, T., V. Cordle, M. Glassman, and R. Vondran, Jr. 1984a. Report format preferences of technical managers and non-managers. *Technical Communication* 31(2): 4–8.

———. 1984b. Report-reading patterns of technical managers and non-managers. *Technical Communication* 31(3): 20–24.

Pion, G. M., and M. W. Lipsey. 1981. Public attitudes toward science and technology: What have the surveys told us? *Public Opinion Quarterly* 45: 303–316.

Plagiarism, piracy, and principles. 1980. *Nature* 286: 831–832.

Plimpton, G. 1983. How to make a speech. *Power of the printed word*. New York: International Paper Company.

Plotnik, A. 1982. *The elements of editing*. New York: Macmillan.

Pool, I. de S., W. Schramm, F. Frey, N. Maccoby, and E. Parker, eds. 1973. *Handbook of communication*. Chicago: Rand McNally.

Poulton, E. B. 1952. Darwin, Charles Robert. *Encyclopaedia Britannica*.

Powers, R., L. E. Sarbaugh, H. Culbertson, and T. Flores. 1961. *Comprehension of graphs*. Bulletin 31. Madison: University of Wisconsin, Dept. of Agricultural Journalism.

Price, J. 1985. *How to write computer manuals: a handbook for writing software manuals*. Menlo Park, Calif.: Benjamin-Cummings.

Price, D. de Solla. 1963. *Little science, big science*. New York: Columbia University Press.

Product safety sign and label system. 1985. 3rd ed. Santa Clara, Calif.: FMC Corporation.

Quinn, K. T. 1982. Interview with authors. November 30.

Radner, D., and M. Radner. 1982. *Science and unreason*. Belmont, Calif.: Wadsworth.

Reader's guide to periodical literature. 1986. Bronx, N.Y.: H. W. Wilson.

Report of the Warren commission on the assassination of President Kennedy. 1964. New York: McGraw-Hill.

Research libraries information network. 1981. Stanford: Research Libraries Group.

Resources in education. 1986. Washington, D.C.: Department of Education.

Reynolds, L., and D. Simmonds. 1982. *Presentation of data in science*. Boston: Matinus Nijhoff.

Rice, R. E., and W. J. Paisley, eds. 1981. *Public communication campaigns*. Beverly Hills, Calif.: Sage.

Richardson, J. 1981. What the public needs to know about science and technology: A report to the Club of Vienna. *Journal of Technical Writing and Communication* 11: 303–313.

Rivers, W. 1975. *Finding facts*. Englewood Cliffs, N.J.: Prentice-Hall.

———. 1984. *News in print: Writing and reporting*. New York: Harper & Row.

Rockmore, M. 1984. Use concise reply to recruiting ad. *Denver Post,* December 16, 10H.

Rogers, E. M. 1983. *Diffusion of innovations,* 3rd ed. New York: Free Press.

Romano, R. M. 1982. *Pretesting health communication*. NIH Publication No. 83-1493. Bethesda, Md.: National Cancer Institute.

Ross, K. 1981. Legal and practical considerations for the creation of warning labels and instruction books. *Journal of Products Liability* 4(1): 29–45.

Roundy, N., and D. Mair. 1982. The composing process of technical writers: A preliminary study. *Journal of Advanced Composition* 3(1–2): 89–101.

Ruark, H. C. 1981. Amoco's AV department fills up the corporate communications tank. *Technical Photography* 13(2): 24–45.

Rubinstein, M. F., and K. P. Pfeiffer. 1980. *Concepts in problem solving*. Englewood Cliffs, N.J.: Prentice-Hall.

Runyon, R. P. 1981. *How numbers lie*. Lexington, Mass.: Lewis.

Russell, C. 1981. How I cover science. In *Communicating university research,* ed. P. Alberger and V. L. Carter. Washington, D.C.: Council for the Advancement and Support of Education.

Ryan, M. 1974. Problem areas in science news writing. *Journal of Technical Writing and Communication* 4: 225–235.

———. 1979. Attitudes of scientists and journalists toward media coverage of science news. *Journalism Quarterly* 56: 18–26, 30.

Sagan, C. 1985. *Cosmos*. New York: Ballantine.

Salisbury, D. F. 1984. Why the IBM PC spawned a mob of look-alikes. *Christian Science Monitor,* February 9, 26.

Santoli, A. 1982. Hog good are our military officers? *Parade,* November 28, 4–7.

Sarton, G. 1948. *The life of science. Essays in the history of civilization.* New York: Henry Schuman.

Scherer, C. 1978. Interview with the authors. November 18.

———. 1982. Producing effective slide sets—a communication workshop, Department of Technical Journalism, Colorado State University, Fort Collins, Colo.

Schramm, W., 1972. The nature of communication between humans. In *The process and effects of mass communication,* ed. W. Schramm and D. F. Roberts. Urbana: University of Illinois Press.

Schramm, W., and W. Porter. 1982. *Men, women, messages and media.* New York: Harper & Row.

Science citation index. 1986. Philadelphia: Institute for Scientific Information.

Schulz, C. 1976. *You're not elected, Charlie Brown.* CBS. September 9.

Scientists expect huge crop of mosquitoes in nation. 1983. *Rocky Mountain News,* May 13, 56.

Selltiz, C., M. Jahoda, M. Deutsch, and S. W. Cook. 1959. *Research methods in social relations.* New York: Holt, Rinehart & Winston.

Selzer, J. 1981. The composing process of an engineer. Paper read at 96th Annual Convention of the Modern Language Association.

———. 1983a. The composing process of an engineer. *College Composition and Communication* 34: 178–187.

———. 1983b. What constitutes a "readable" technical style. In *New essays in technical and scientific communication: Research, theory, and practice,* ed. P. V. Anderson, R. J. Brockman, and C. Miller. Farmingdale, New York: Baywood.

Settle, L. 1982. Interview with authors. December 12.

Seymour, J. 1985. The corporate video: VDT precautions pay large return for small investment. *PC Week* 2(15): 26.

Shaner, W., W. Schmehl, and P. Philipp. 1982. *Framing systems research and development: Guidelines for developing countries.* Boulder, Colo.: Westview Press.

Shannon, C. E., and W. Weaver. 1949. *The mathematical theory of communication.* Urbana: University of Illinois Press.

Sheehy, E. P. 1982. *Guide to reference books.* Chicago: American Library Association.

Shell-Shocked. 1983. *Family Weekly,* January 9, 34.

Sherwood, T. K. 1974. The treatment and mistreatment of data. *CHEMTECH* 4: 736–740.

Shurter, R. 1954. *Effective letters in business.* New York: McGraw-Hill.

The silent epidemic of Chlamydia. 1985. *Seventeen,* October, 26.

Sindermann, C. 1982. *Winning the game scientists play.* New York: Plenum Press.

Small Homes Council. 1971a. *Heating the home.* Circular G3.1. Champaign: University of Illinois.

———. 1971b. *Fuels and burners.* Circular G3.5. Champaign: University of Illinois.

Smith, E. J. 1981. *Energy and labor use by rural manufacturing industries.* Rural Development Research Report No. 26. Washington, D.C.: Economics and Statistics Service, United States Department of Agriculture.

Smith, C., and K. Kiefer. 1982. *Manual for writing with computer assistance.* Fort Collins: Colorado State University, Department of English.

Smith, M. J. 1982. *Persuasion and human action: A review and critique of social influence theories.* Belmont, Calif.: Wadsworth.

Smith Kline & French. 1964. *Scientist meet the press: A practical guide to press relations for scientists and research managers.* Philadelphia: Smith Kline & French Laboratories.

Sohn, A. B. 1981. Women in newspaper management: An update. *Newspaper Research Journal* 31: 94–106.

Souther, J. 1962. What management wants in the technical report. *Journal of English Education* 52: 498–503. (Reprinted in *Readings in technical communication writing,* ed. D. C. Leonard and P. J. McGuire [New York: Macmillan, 1983].)

———. 1984. Correspondence checklist. In *College placement annual,* ed. J. M. Bowser. Bethlehem, Pa.: College Placement Council.

Souther, J., and M. White. 1977. *Technical report writing.* 2nd ed. New York: Wiley.

Sparrow, W. K., and D. H. Cunningham, eds. 1978. *The practical craft.* Boston: Houghton Mifflin.

Spretnak, C. M. 1982. Reading and writing for engineering students. *Technical Writing Teacher* 9: 133–136.

Statistical abstract of the United States 1982–83.

103rd ed. Washington, D.C.: U.S. Government Printing Office.

Steen, L. A. 1981. Making the arcane plain. In *Communicating University Research,* ed. P. Alberger and V. L. Carter. Washington, D.C.: Council for the Advancement and Support of Education.

Steenrod, N. E., J. L. Dobb, L. Carlitz, F. A. Ficken, and G. Piranian. 1980. *A manual for authors of mathematical papers.* 7th ed. Providence, R.I.: American Mathematical Society.

Steinegger, D. H., R. C. Shearman, T. P. Riordan, and E. J. Kinbacher. 1983. Mower blade sharpness effects on turf. *Agronomy Journal* 75: 479–480.

Stone, D. E., and M. D. Glock. 1981. How do young adults read directions with and without pictures? *Journal of Educational Psychology* 73: 419–426.

Stone, D. E., C. K. Pine, G. R. Bieger, and M. D. Glock. 1981. *Methodological issues in research on reading text with illustrations.* Technical Report No. 2. Ithaca, N.Y.: Cornell University, College of Agriculture and Live Sciences, Department of Education.

Strong, E. K. 1926. Values of white space in advertising. *Journal of Applied Psychology* 10: 107–116.

Strunk, W., and E. White. 1979. *The elements of style.* 3rd ed. New York: Macmillan.

Suomi, V., and T. Haig. 1974. Lectures and class discussions, environmental systems monitoring class, Institute for Environmental Studies, University of Wisconsin, Madison, Wisc.

Sutton, D. 1983. *Placement manual.* Fort Collins: Colorado State University Career Service Center.

———. 1985. CSU's career services. Presentation to the basic technical communication class, March 1985, Colorado State University, Fort Collins, Colo.

Swinehart, J., and J. McLeod. 1960. News about science: Channels, audiences, and effects. *Public Opinion Quarterly* 24: 583–589.

Taylor, W. L. 1956. Recent developments in the use of the Cloze procedure. *Journalism Quarterly* 33: 42–48.

Thomas, L. 1974. *Lives of a cell.* New York: Bantam.

Time/Life. 1970. *Photographer's handbook.* New York: Time.

Tips on picking and using strawberries. 1980. Champaign: University of Illinois, Cooperative Extension Service.

Tufte, E. 1983. *The visual display of quantitative information.* Cheshire, Conn.: Graphic Press.

Turnbull, A. F., and R. N. Baird. 1980. *The graphics of communication.* 4th ed. New York: Holt, Rinehart & Winston.

Ulrich's international periodicals directory. 1986. 27th ed. New York: R. R. Bowker.

United Press International. 1984. 500 fur seals clubbed as annual kill begins. *Rocky Mountain News,* July 3, 24.

———. 1985. Lives saved by A-bombs 'exaggerated.' *Rocky Mountain News,* August 2, 47.

Upton, B., and J. Upton. 1975. *Photography.* Boston: Little, Brown.

U.S. Department of Agriculture. 1985. *Handbook of agricultural charts.* Washington, D.C.: U.S. Government Printing Office.

U.S. Fish and Wildlife Service. 1980. *Terrestrial habitat evaluation criteria handbook. Ecoregion 2522. Edwards Plateau of Texas.* Project Impact Evaluation. Division of Biological Services. Fort Collins, Colo.: U.S. Department of the Interior.

U.S. Forest Service. 1977. *Biological evaluation recreation site examination.* Santa Fe, N.M.: U.S. Department of Agriculture.

———. 1980. *Making fuel management decisions.* National Fuel Inventory and Appraisal Project. Fort Collins, Colo.: U.S. Department of Agriculture.

Venolia, J. 1982. *Better Letters: A handbook of business and personal correspondence.* Berkeley, Calif.: Ten-Speed Press.

Vinci, V. 1975. Ten reporting pitfalls: How to avoid them. *Chemical Engineering* 82(27): 45.

Vogt, H. E. 1985. Wordless instructions—say it with pictures. *Proceedings of the 31st international technical communication conference,* Seattle, Wash. Washington, D.C.: Society for Technical Communication.

Waldrep, T., ed. 1985. *Writers on writing.* New York: Random House.

Wallis, C., and R. Schapiro. 1983. Fraud in a Harvard lab. *Time*, February 28, 49.

Wallis, W. A., and H. V. Roberts. 1962. *The nature of statistics*. New York: Free Press.

Wassell, C. A. 1982. Interview with authors. July 18.

Watters, I. A. 1986. *Index to U.S. Government periodicals*. Chicago: Infordata International.

Weaver, T., and D. R. Lampe. 1984. Communication skills: Top priority for engineers and scientists. *The MIT report* 12: 1–2.

Webb, J., and J. Salancik. 1966. *The interview, or the only wheel in town*. Journalism Monographs no. 2. Columbia: University of South Carolina College of Journalism.

Webster, B. 1976. Your lawn's full of energy. *Wisconsin State Journal*, May 23, sec. 5:2.

Weinstein, A. S., A. D. Twerski, H. P. Piehler, and W. A. Donaher. 1978. *Products liability and the reasonably safe product: A guide for management, design and marketing*. New York: Wiley.

Weinstein, R. W. 1986. *Photographer's market*. Cincinnati: Writer's Digest.

Weisman, H. M. 1985. *Basic technical writing*. Columbus, Ohio: Charles E. Merrill.

Weiss, E. H. 1985. *How to write a usable user's manual*. Philadelphia: ISI Press.

Wheatley, D. W., and A. W. Unwin. 1972. *The algorithm writer's guide*. London: Longman.

Wildi, E. 1985. Cold weather photography. *Photomethods* 28(11): 33.

Wilhoit, G. C., and M. McCombs. 1976. Reporting surveys and polls. In *Handbook of reporting methods*, ed. M. McCombs, D. L. Shaw, and D. Grey. Boston: Houghton Mifflin.

Williams, F. 1968. *Reasoning with statistics*. New York: Holt, Rinehart & Winston.

Williams, J. 1982. Interview with authors. December 12.

Williams, J. M. 1981. *Style. Ten lessons in clarity and grace*. Glenview, Ill.: Scott, Foresman.

Williams, J. B., D. A. Karp, J. R. Dalphin, and P. S. Gray. 1982. *The research craft*. Boston: Little, Brown.

Wilson, N. 1985. Interview with authors. April 2.

Wimmer, R. D., and J. Dominick. 1983. *Mass media research*. Belmont, Calif.: Wadsworth.

Winbush, B., and G. McDowell. 1980. Testing: How to increase the usability of computer manuals. *Technical Communication* 27(4): 20–22.

Winterbottom v. *Wright*. 1842. 10 Mees & W 109, 152 Eng Reprint 402.

Wolverton, V. 1985. *Running MS DOS*. Bellevue, Wash.: Microsoft Press.

Woodford, F. P. 1972. Experiences in teaching scientific writing in the U.S.A. *Journal of Biological Education* 6: 9–12.

Wordperfect 4.1. Computer software. SSI Corp. IBM DOS Version 4.1, disk.

Wright, P. 1977. Presenting technical information: A survey of results and findings. *Instructional Science* 6(2): 94–134.

Writer's market '86. 1986. Cincinnati, Ohio: Writer's Digest.

Young, M. 1982. Interview with authors. July 27.

Zemke, R., and T. Kramlinger. 1982. *Figuring out things—a trainer's guide to needs and task analysis*. Reading, Mass.: Addison-Wesley.

Zimmerman, D. 1981. Hewlett-Packard demonstrations to highlight meeting. *Technicalities*, March, 4.

Zinsser, W. 1980. *On writing well*. 2nd ed. New York: Harper & Row.

———. 1983. *Writing with a word processor*. New York: Harper & Row.

Index

A

abstract journals, 47–49
  annotated bibliography of, 48
abstractions, 225–226
abstracts
  government, 55
  and indexes, 47–49
  in technical documents, 269–270
  types of, 49
  using, 50
action verbs, 195–196
active voice, in verb usage, 193–195
adjectives
  converting prepositional phrases to, 199
  replacing clauses with, 198
adverbs, converting prepositional phrases to, 199
adverse information, withholding, 379
Air Force Institute of Technology, 4
algorithms
  as flow charts, 164
  list of, 164
  and ordinary language, 163
  specific guidelines for, 160–163
  with specific symbols, 163
  types of, 163
Allport, Gordon, 66
Amarino, Neal, 400
American Chemical Society, *Handbook for Authors*, 211
American Institutes for Research, 103
American Psychological Association, *Publication Manual*, 211
American Society of Agronomy, *Publications Handbook and Style Manual*, 212
American Statistical Association, 73
American Statistics Index (ASI), 55
Apple Computer, 4
application letters, 359
applied science, effective public relations campaign in, 395–398
Applied Science and Technology Index, 48, 50
appositions, converting clauses to, 197
artist, hiring of for illustrations, 150
Associated Press Stylebook, The, 211, 212
associative data, 33
audience
  for descriptions, 231–232
  selection and analysis, 16
audience analysis, 85–106
  using biographical information, 97
  characteristics and levels, 90–100
  communication theory and, 85–87
  content of publications, 100
  describing, 91
  describing potential message, 92
  determining, 87–90
  determining communication design, 92–93
  examining publications, 94–97
  highlights and projects, 104–106
  interviews and surveys, 98–99
  observing professionals, 97

audience analysis, *continued*
   professional communicators use of, 103–104
   professional talks and slide presentations, 299–302
   subscription forms, 98
   study methods, 90
   tailoring communications for, 86
   understanding, 85–87
   using information about, 100–104
automatic writing, 119
author card, 42
author-date citation, style of note taking, 60
authorship, sharing credit for, 380

B

Babbie, Earl, 35
babbling, or jabbering, 119
Baltimore Sun, 388
bar graphs
   general vs. specific, 156
   specific guidelines, 156–157
   vs. tables, 147
Basic Technical Writing, 135
Battison, Robbin, 103–104
Baylor College, 112
Bell Laboratories, 10
   Writer's Workbench, 282
Berger, Paul, 63
Berndt, Judy, 61
Betrayers of the Truth: Fraud and Deceit in the Halls of Science, 375
Bibliography of Agriculture, 48
bibliography/reference list in technical documents, 268
Bingo cards: *See* subscription forms
biographical information, in audience analysis, 97
Biological Abstracts, 48, 49, 50
Black's Law Dictionary, 368
Blake, Reed H., 144
body language, in professional talks, 299
Bond, Carl E., 242
books
   in card catalogs, 42–44
   and publications, 55
Books in Print, 59
Borzoi Handbook for Writers, The, 212
Bostian, Lloyd, 406
brainstorming, 31
   in content selection, 130

Broad, William, 375
Brum & Anderson ExecuCom, 400
Buckley, William F., 141
Buick Motor Company, 381
business letters
   average cost, 337
   highlights and projects, 346–348
   major types, 329–335
   parts of, 335–336
business proposals, 276–278
   revising, 278–279
   sample forms, 277, 278
Byte, 112

C

call numbers, on author, title, and subject cards, 42–43
campus job interviews, 350
Canadian documents, 56
card catalogs, 41–45
   classification systems, 42
   computerized, 41
   finding books in, 42–44
   types of, 41
careers, communication skills important to, 4–5
career services, in job search, 351
Carroll, Lewis, 223
Carson, Rachel, 401
Carter, Richard, 227
cartoons, 170
cases
   inaccurate classification of, 35
   misuse in selection, 36
causal data, 33–34
chalkboards, and models, 303
channels
   in communication theory, 11
   tailoring communications for, 86
   see also audiences
charts
   flow and organization, 160
   highlights and projects, 176–177
   PERT, 164
   time and milestone, 164
Chemical Abstracts, 48
chemistry, notations as codes for, 160
Chicago Manual of Style, The, 211, 212
Christensen, Clyde M., 231, 235
Christian Science Monitor, 224

chronological resume
  with internal headings, 358
  with side headings, 357
circle graphs, specific guidelines for, 158
circular definitions, 225
citations
  author-date system, 270
  numbered, 270
  in technical documents, 270
  typical scientific journal style, 272–273
claim or complaint letters, 329–333
  components of, 330
  responding to, 332
classifications, describing, 241–242
Clarke, Arthur C., 346
clauses
  converting to appositions, 197
  creating prepositional phrases from, 198
  eliminating unnecessary, 197
  replacing with adjectives, 198
clichés, 203
closed-end questions, 77
cloud-seeding, public relations campaign, 395–398
colleagues, letters to, 333
College Composition and Communication, 126
College English, 109
College Placement Council (CPC), 351
Colorado State University, 52, 134
  word processing at, 282
communication
  basic theory of, 10–16
  behavioral concepts of, 14
  content and appearance, 6–7
  effectiveness evaluation, 413–415
  evaluation of, 17–18, 406–419
  forms of, 6
  highlights and projects, 18–20
  informal and formal evaluations of, 407–418
  key ingredients in process of, 10–13
  mathematical model of, 10
  preparation of, 32
  principles and theory of, 3–20
  process evaluation of, 412
  readability evaluation of, 412
  revolution, word processing and, 295
  rhetorical theory and, 9
  skills important to career, 4–5
  sender-receiver model of, 11
  surveys and workshops, 406
  tailoring for audiences and channels, 86

  theory applied to example, 15–16
  see also technical communication
Communication and Organizational Behavior, 233, 235
communication products
  correspondence, 328–348
  format, 258–280
  in job searching, 349–364
  preparing professional, 221–364
  printing process, 288–295
  professional talks, 298–312
  slide presentations, 312–325
  word processing, 281–288
communication theory
  additional concepts, 14
  application and example, 15
  and audience analysis, 85–87
  basic, 10–14
  highlights and projects, 18–20
  see also technical communication theory
communicators, burden on, 7
comparison
  in descriptions, 242–245
  defining by, 227
  inappropriate, 35
competitive pressure, ethics and, 377–378
complaint letters: *See* claim or complaint letters
complex tables, basic components, 154
computer graphics, 287–288
  highlights and projects, 295–296
  illustrations using, 150
  kinds of software for, 288
computers
  in technical communication, 281
  see also word processing
computer searches
  and electronic data bases, 49–52
  as sample of access service, 53–54
concepts, comparison test, 225
Concepts in Problem Solving, 23–24
constructions
  highlights and projects, 205–209
  strategies for simplifying, 196–199
content
  appropriate, 190
  data analysis techniques for, 184
  drawing conclusions from, 185–186
  goals and questions, 183
  highlights and projects, 38–39
  improving, 21–82

content, *continued*
  information sources and evidence for, 183–184
  minimizing errors in, 33–37
  problem solving and, 23–39
  selection, journalistic approach to, 129–130
  selection and organization of, 128–136
  strategies for editing, 182–186
  verifying results, 185
content organization, 131–136
  fundamental patterns, 135
  informal and formal outlines, 134–136
  pattern notes and issue trees, 132–133
  for technical articles, 136
content selection, cubing and brainstorming, 130
contrast, in descriptions, 242–245
controlled circulation magazines, 98
Cook, Stuart, 66
copyediting
  symbols, 182, 204
  with word processor, 283–285
copyright law, 367–373
  fair use limitation, 372
  highlights and projects, 385
  and illustrations, 149
  and licensing and photocopying, 372
  and limitations on owner's rights, 369–373
  notices and infringement of, 369
  and obtaining permissions, 373
  and registration form, 370–371
  and technical communication, 367–368
  terms and registration, 368–369
Cordle, Virginnia E., 260
Corporate Scriptwriting Book, 314
correlational data, 33–34
correspondence, 328–348
  avoiding sexist language, 342–343
  electronic, 346
  highlights and projects, 346–348
  in job search, 359–361
  major categories of, 329–335
  and unknown addressee, 343
  see also letters
Cosmos, 218
Council of Biology *Style Manual,* 211
cover letters, 329
  preparing, 359
  in surveys, 78
Cowan, Elizabeth, 130
Cowan, Gregory, 130
Cox, Barbara, 399

Creative Interviewing, 69
Crews, Frederick, 137, 192
Crime of Claudius Ptolemy, The, 376
criteria
  comparing potential solutions against, 32
  for solutions, 29–30
cubing, in content selection, 130
Current Index to Journals in Education, 51

D

Data
  Analysis of in surveys, 80–81
  concocting, 378
  descriptive, 33–34
Davis, Richard M., 4
definitions
  highlights and projects, 254–256
  importance of, 223–224
  need for, 230
  placing of, 229–230
  precise, 224–229
  shifting, 34
  in technical communication, 223–230
Denver Post, 203
department support, in job search, 351
descriptions
  accuracy of, 233–235
  audience considerations and, 231–232
  common, 237–246
  comparison and contrast, 242–245
  events and classifications, 239–242
  format considerations and, 234–236
  functional, 231, 237–245
  hardware and conditions, 237–239
  highlights and projects, 254–256
  narrative, 237
  observation or inference and, 233–235
  preparing, 231–237
  process, 245–246
  purpose and organization of, 232–233
  spatial relationships and, 239
  of teaching technique, 233
  in technical communication, 230–246
  and words and illustrations, 235
descriptors, in electronic data searches, 51
Design for Lively Learning, 233
De Solla Price, Derek J., 377
Deutsch, Morton, 66
Dewey decimal classification system, 42

dictating, 141
Dillman, Donald, 74
direction maps, 167
dispersion, disregard of, 35
distortions
    in observation, 66–68
    psychological view of, 66
Document Design Center, 110, 120, 163, 190
documents
    appearance of, 210–211
    design model for, 120–126
    electronic and photographic storage of, 57
    format of, 258–280
    key parts of, 261–273
    see also technical documents
dot matrix printout, vs. redrawn illustration, 151
drafting manuscripts, 118–143
    committing ideas to paper, 140–141
    content selection and organization, 128–136
    highlights and projects, 142
    paragraphs, 136–140
    planning stage, 128–136
drafts, 83–178
    allowing time between, 181
    appearance of, 210–211
    and audience analysis, 85–106
    content and organization of, 121
    language and graphics for, 126
    tightening and editing of, 191
    and writing processes of professionals, 107–117
Drawings: *See* line art
Duzet, Harriet, 111

E

Ebert, Robert H., 377
Ecology, 402
Edible Wild Mushrooms, 231–232, 235–236
editing, 189–209
    checklist for, 183–191
    chronology and documentation, 192
    content, 182–186, 190
    highlights and projects, 186–187, 205–209
    instructions and completeness, 192
    overall organization, 189–192
    paragraphs, 192–193
    prepositional phrases, 198–199
    sentences, 193–205
    and simplifying constructions, 196–199
    and stacked modifiers, 199–200
    and strategies for speaking plainly, 200–205
    and unnecessary clauses, 197–198
    see also copyediting
editorial process, in printing, 290
editorial style, 212
    highlights and projects, 219–220
    resolving issues, 212
Educational Resources Information Center (ERIC), 51
Effective Letters in Business, 330
Eiseley, Loren, 218
Elbow, Peter, 119–120
electric typewriter, 140, 245
electronic data bases
    computer searches and, 49–52
    national systems of, 52
electronic mail, 346
electronic publishing, 295
employers
    analyzing, 355
    job announcements and candidate evaluations, 352–354
encoding/decoding, in communication theory, 12
Engineering Index Monthly, 48
Engineers of Distinction, 4
English teachers, 113–114
envelopes, addressing, 342
equations: *See* notations
errors
    in generalization, 37
    technical, 35–36
ethical problems, 377–384
    concocting data, 378
    "cooking" results, 378–379
    highlights and projects, 385
    sharing credit for authorship, 380
    withholding adverse information, 379
ethics
    and competitive pressure, 377–378
    defined, 367
etymology, 229; *see also* word history
evaluation techniques
    audience interviews, 410
    feedback, 407
    field experiments, 413–414
    gathering information, 407–411
    highlights and projects, 418–419
    informal and formal, 407–418
    inspection tests, 413
    instructions to reviewers, 408–409

evaluation techniques, *continued*
  interpreting information, 411
  peer reviews and, 410
  personal observation and review, 407
  and precautions, 417–418
  questionnaires, 413
  readability and effectiveness tests, 413–415
  suggested, 411
  workshops in, 416–417
events, describing, 239–241
explication, defining by, 226–227

F

Faigley, Lester, 109
fair use, limitation on copyright owner's rights, 372
Falk, John H., 402
false extrapolation, overgeneralization and, 37
familiarity, as observational barrier, 68, 69
Feber, Robert, 73
Federal Communications Commission (FCC), 258
feedback techniques
  in communication theory, 13
  for evaluating communications, 407
Felker, Daniel, 120
Fields, Alan, 132
field work, in information gathering, 65–82
Fischer, David, 37
Fitz-Randolph, Jane, 226
flip charts, guidelines for, 303
flow charts
  algorithms as, 164
  for problem identification, 162
  specific guidelines for, 160
Flower, Linda, 110, 126, 133
FMC Corporation, 383
footnotes, placing definitions in, 230
Forbes, Malcolm, 344
Forbes Magazine, 344
formal evaluations, 411–418
  highlights and projects, 418–419
formal outlines, 134–136
format
  anticipating readers' needs, 259–261
  of descriptions, 234–236
  document, 258–280
  in effective communication, 258
  highlights and projects, 279
  of instructions, 249

  key parts of, 261–272
  principles for development, 273–276
  schemata and hierarchy, 261
  school vs. job writing, 259
  selection of, 17
form letters, 345
  highlights and projects, 346–348
  ORM test, 345
forms, communication, 6
frame of reference, in communication theory, 14
Franklin, Jon, 388
free writing, 119–120
  exercises and editing, 119–120
  highlights and projects, 142
Fry readability graph, 414–415
functional descriptions, 231
  types of, 237–245
functional resume, 356

G

Garratt, George, 4
gatekeeper, in communication theory, 14
gender bias, eliminating from language, 189, 190
generalization, errors in, 37
general orientation maps, 166
general references, 46–47
Gerstenfeld, Arthur A., 63
Glassman, Myron, 260
Glossary, placing definitions in, 230
Gobbledygook Has Gotta Go, 201
Goswani, Dixie, 110, 120
Gottlieb, Larry, 299
government documents, 55–56
  guides to, 55
Government Printing Office Style Manual, 211, 212
graphs
  circle and isotype, 158–159
  line and bar, 155–157
Green, James, 33
Guide letters, 345–346
  to students, 345–346
Guide to Reference Books, 47

H

Hagstrom, Warren, 380
Haig, Thomas, 25, 29, 31
Haney, William, 233, 235

Harbrace College Handbook, 212
hardware and conditions
   describing, 237–239
   list format for, 238–239
Harris, John S., 144
Harvard Medical School, 377, 378
Hayes, John, 110, 126
headlines, in transparencies, 304
health communications, criteria for evaluating, 406, 412
hedge words, 202
Hemingway, Ernest, 12
Herrington, Anne, 110
Hewlett-Packard, guidelines for overhead transparencies, 310–311
Historians' Fallacies, 37
Holland, Melissa, 163
Honeywell, 73
Hopper, Grace, 229
How to Prepare Charts and Graphs for Effective Reports, 160
Human Ecology Research Services, 396

I

IBM, 111, 223
   evaluation techniques, 413
ideographs, in transparencies, 304
idiomatic style, 215
Idunits, in transparencies, 304
illustrations, 144–178
   borrowing of, 149
   combining narrative and, 150–152, 250–251
   computer graphics, use of, 150, 288
   defining by, 227
   as descriptions, 235
   functions in technical and scientific communications, 145
   general guidelines for, 152–153
   handling of, 152
   highlights and projects, 176–177
   hiring artist for, 150
   in instructions, 249–253
   obtaining, 149–150
   planning for, 148–149
   preparing, 149
   selecting, 146–147
inaccurate measurements, 35
inappropriate comparisons, 35
Index Medicus, 48

Index to Publications of the United States Congress (CIS), 55
indexes
   abstracts and, 47–49
   government, 55
   using, 50
inference, statements of, 235
informal evaluation, 407–411
   highlights and projects, 418–419
informal outlines, 134
information, in technical and scientific communication, 8, 25
Information gathering, 21–82
   brainstorming techniques, 31
   field work, 65–82
   highlights and projects, 38–39, 418–419
   notes and records, 31
   observing and interviewing, 65–72
   problem solving and, 23–39
   surveys, 72–81
   using the library and reviewing literature, 40–64
information overload, as a communication problem, 88
Ingles, Dwight, 37
inquiries or requests for information, 334
institutional voices, in public relations campaigns, 397
instructions
   and audience considerations, 248
   as critical warnings, 252
   evaluating, 254
   highlights and projects, 254–256
   orientation and format of, 248–249
   planning illustrations for, 149
   planning and producing, 248–254
   preparing, 246–248
   products liability and, 382, 383
   to reviewers, 408–409
   for surveys, 78
   in technical communication, 246–254
   using effective illustrations for, 249–253
   without words, 253
   writing style of, 253
instructors, writing processes of, 112–114
interlibrary loans, 57
interpretation, in communication theory, 12–13
interviewers, selecting and training, 80
interviewing, 69–72
   common barriers to, 72
   highlights and projects, 81–82

interviewing, *continued*
   tape recorders in, 71–72
interviews
   in audience analysis, 98
   guidelines for dealing with, 399–400
   handling of, 361–362
   highlights and projects, 363–364, 404
   and interview offers, 361–362
   intimidation and, 69
   note taking during, 71–72
   phases in process of, 70
   ten steps to productive, 69–71
   typical questions in, 362
Irving, John, 216
Is It Really So?, 37
isotypes, specific guidelines for, 159
issue trees, in content organization, 133
Issuing Municipal Bonds, 227

J

Jahoda, Marie, 66
jargon, 200
Jesperson, James, 226
job searches, 349–364
   accepting offers, 362
   application letters, 352–354
   assessing yourself and analyzing employers, 354–355
   career services and department support, 351
   correspondence for, 359–361
   and employer evaluation of candidates, 352–354
   and employer reviews, 354
   highlights and projects, 363–364
   on and off campus, 349–350
   resumes and letters for, 355–359
   strategies and resources for, 349–352
   techniques of, 354–362
journalistic approach to content selection, 129–130
Journal of Advanced Composition, 109
Journal of Products Liability, 382
Journal of Wildlife Management, The, 239
journals, technical and scientific, 52–55
justification/literature review, in technical documents, 267

K

Keys to Oregon Freshwater Fishes, 242–244

key words, in electronic data searches, 52
Keyworth, George A., 387
Kiefer, Kathleen, 282
Kieffer, Judith S., 241
Kisling, Jack, 203
Klare, George, 412
Kuhn, Thomas, 28

L

Landesman, Joanne, 103–104
Langley Research Center, 260
Lansford, Henry, 396
laser printers, 295
Lasswell, Harold, 389
Leach, Roland, 26
Lechner, George, 238
Ledbetter, Hassel, 109
Leffert, Robert, 160–164
legal aspects of technical communications, 367–377
Legibility—Artwork to Screen, 322
Leopold, Aldo, 218
Lewis, David, 33
Letters, 328–346
   addressing envelopes for, 342
   block format, 337
   claim or complaint, 329–333
   to colleagues, 333
   by electronic mail, 346
   excess verbiage in, 344
   form and guide for, 345–346
   formats for, 336–339
   highlights and projects, 346–348
   and inquiries or requests for information, 334
   inside an organization, 339
   major types, 329–335
   memoranda, 334–335
   optional or specialized parts, 336
   outside an organization, 336–338
   page arrangement and spacing of, 339–341
   parts of, 335–336
   punctuation of, 342
   semi-block format, 338
   simplified form of, 338
   of transmittal, 329, 330
   writing effective, 344
   see also correspondence
leveling, as observational barrier, 66, 68
librarians, 41

asking help from, 58
library
   abstracts and indexes, 47–49
   books and publications, 55
   card catalogs, 41–45
   in careers, 62–63
   college, 40
   computer searches and electronic data bases, 49–52
   electronic and photographic data storage, 57
   general references, 46–47
   highlights and projects, 63–64
   resources, 41–57
   serial publications, 52–55
   serials cards, 45
   special collections, 57
   using, 40–64
Library of Congress, 44
   classification system, 42
licensing, as limitation on copyright owner's rights, 372
line art, 228
   guidelines for, 168–169
line graphs
   complex vs. simple, 155
   specific guidelines, 155–156
list algorithm, 164
List of Subject Headings, 44
literary theft: *See* plagiarism
literature
   evaluating, 62
   general to specific, 61–62
   paraphrasing, 61
   reviewing, 40–64
   technical and scientific, 49
literature search
   developing a plan, 58–59
   executing, 59–62
   highlights and projects, 63–64
   note-taking methods for, 59–61
Lives of a Cell, 218
Lowenstein, Ralph L., 12
Lubov, Andrea, 227
Lunsford, Ronald F., 114
Lukazewski, James E., 400
Lutz, William, 113

M

McCombs, Maxwell, 74
Macdonald-Ross, Michael, 155
McDowell, Glenda, 413
McKee, Blaine, 134
McPhee, John, 112
magazines, 52–55
   controlled circulation of, 98
   see also periodicals
mail surveys, 74
Mair, David, 109
manuscript
   appearance of, 211
   making several trips through, 182
Manville Corporation, 379, 400
maps
   direction, 167
   general orientation, 166
   specific guidelines for, 165–167
Marcus, Steven J., 99
Martin, Phyllis, 203
Matrazzo, Donna, 314
Mayo Foundation, 399
MCI, 346
Measurement of Readability, The, 412
measurements, inaccurate, 35
media
   complaining to, 403
   guidelines for dealing with, 399–400
   helping to understand technology and science, 398–400
   highlights and projects, 404
   "no comment" response to, 400
   science and public, 387–388
   science and technical professionals suspicious of, 390
   and scientists and public understanding, 392–394
   skepticism of technical and science professionals, 391–394
   social role of, 387
medical students, ethics and competitive pressure, 377–378
Megatrends, 281
memoranda, 334–335
   formats for, 336–339
   highlights and projects, 346–348
   page arrangement and spacing, 339–341
   printed forms, 339
Mendel, Gregor, 378
Merrill, John C., 12

messages
  in communication theory, 11
  competition between, 7
Metzler, Ken, 69
milestone charts, 164
Miller, David J., 233
Miller, Thomas P., 109
Modern Language Association (MLA), 108
modifiers, eliminating unnecessary, 202
Monthly Catalog of U.S. Government
    Publications, 55, 56
Morgan, Chris, 112
Munn, Lionel, 283
Murray, Donald, 113

N

Naisbitt, John, 281
narrative
  combining illustrations and, 150–152
  placing definitions in, 229
National Aeronautics and Space Administration
    (NASA), 260
National Agricultural Library Catalog, 48
National Bureau of Standards, 108–109
National Center for Atmospheric Research, 395
National Enquirer, 388
National Geographic, 389
National Hail Research Experiment (NHRE), 395
National Institute of Education, 51, 120
National Institutes of Health (NIH), 378, 406
National Science Foundation, 40, 269, 395
National Science Foundation proposal form,
    274–275
Nature of Statistics, The, 34
Nelson, Kenneth E., 107, 179
New England Journal of Medicine, 378
newsletters, 52
newspapers, 52
Newton, Robert R., 376
New Yorker, The, 112, 179, 401
New York Times, 402
noise
  in communication theory, 12
  mechanical vs. semantic, 12
nominalizations: *See* trapped verbs
notations
  as codes for chemistry and statistics, 161
  drawbacks of, 160

specific guidelines for, 160
note taking
  author-date citation style, 60
  during interviews, 71–72
  in literature search, 59–61
numbers
  disregard of dispersion in, 35
  misuse of, 34–36
  technical errors with, 35–36
Nursing81, 238
NursingLife, 241

O

objectives, in problem statements, 26
observation
  collecting information by, 65–69
  defined, 66–69
  developing good techniques of, 68
  distortions in, 66–68
  highlights and projects, 81–82
  overcoming barriers to, 68–69
  skills test of, 67
  statements of, 233
Odell, Lee, 110
Office of Records Management (ORM), 345
O'Hayre, John, 201
On Writing Well, 112
open-ended questions, 77
organizational charts
  basic diagram, 163
  specific guidelines for, 160
orientation, of instructions, 248–249
outlines
  informal and formal, 134–136
  numerical or decimal, 135
overall organization
  highlights and projects, 205–209
  reviewing and editing, 189–192
overgeneralization, and false extrapolation, 37
overhead projectors
  adding photographs to transparencies, 304
  guidelines for, 303
  producing transparencies for, 308–309
overhead transparencies
  adding photographs to, 304
  guidelines for, 310–311
  producing, 308–309

P

padded constructions, in sentences, 196
Paradis, James, 4
paragraphs
   breaking up, 192–193
   developing, 138–139
   drafting, 136–140
   highlights and projects, 205–209
   length of, 138
   reviewing and editing, 192–193
   typical, 137
   unity and continuity of, 138–139
   writing, 140
passive voice, arguments against, 194
pattern notes, in content organization, 132
PC Week, 287, 295
peer reviews, 218–219
   in evaluating communications, 410
   highlights and projects, 219–220
periodicals, 52–55
   comparison of, 96, 260
   identifying and locating, 45
permission form, for using copyrighted material, 374
personal computers
   and electronic mail, 346
   highlights and projects, 295–296
   word processing and, 284–287
personal observation, as evaluation technique, 407
personal style, 218
persuasion, technical and scientific communication as, 9
Pfeiffer, Kenneth R., 23–24
photocopying, as limitation on copyright owner's rights, 372
photographers, hiring and working with, 322–323
Photographer's Market, 322
photographs
   adding to overhead transparencies, 304
   content and composition of, 171–175
   guidelines for, 170–175
   improving technical quality of, 171
   printing process for, 294
   "rule of thirds" and, 171, 172
phrasing, and solutions, 23–39
picture graphs: *See* isotypes
pie charts: *See* circle graphs
Pinelli, Thomas E., 260

plagiarism, 373–377
   avoiding, 377
   example of, 376
   highlights and projects, 385
   inadvertent and deliberate, 375
plagiarists, 375–377
planning, content and purpose, 128–129
Planning and Producing Slide Programs, 322
Plimpton, George, 300
polls
   evaluating, 74
   see also surveys
possessives, converting prepositional phrases to, 199
Postman, Leo, 66
post writing, 126
popularizing science and technology, 401–402
potential solutions
   comparing against criteria, 32
   developing, 31
precise definitions, 224–229
prepositional phrases
   creating from clauses, 198
   eliminating unnecessary, 199
   revising excessive, 198–199
presentations: *See* slide presentations
prewriting, 121–125
   audience and task, 121
   highlights and projects, 142
   scope and purpose, 121
printing processes, 288–295
   basic steps in, 289–295
   editorial and production phases of, 290
   highlights and projects, 295–296
   photographs and, 294
   proofreading stages of, 293
printing terms, 291
problem
   defining and solving, 16, 23–32
   general model of, 24
   statements, 26–29, 266
Problem solving, 23–39
   model for technical writing, 126–128
   seven-step method, 25–32
   strategy for, 25
process descriptions, preparing, 245–246
production process, in printing, 290
products liability, 380–384
   communication's role in, 382–383
   development of laws for, 381

products liability, *continued*
 highlights and projects, 385
 legal assumptions of, 381
 marketing defects and, 382
 and safety signs and labels, 382–383
professionals
 technical and scientific, 107–110
 writing processes of, 107–117
professional talks, 298–312
 and audience analysis, 299–302
 eight-step process for, 299–312
 highlights and projects, 325–326
 humor in, 302
 outlining of, 303
 planning for, 302–304
 preparation and evaluation of, 309–312
 rehearsal and practice for, 307–308
 visuals and notes for, 307
professional writers
 use of audience analyses, 103–104
 writing processes of, 110–112
professors, writing processes of, 112–114
Program Evaluation Review Technique (PERT), 164
progress reports, 271
 key elements of, 271
 planning illustrations for, 149
proofreading marks, 293
proposals, 263–266
 bibliography/reference list, 268
 highlights and projects, 279
 justification/literature review, 267
 and problem statement, 266
 revised student, 266
 sample, 263
 sample form, 274–275
 student research paper, 264–265
Proxmire, William, 403
public, science, media and, 387–388
public health, and ethical problems, 379
Public Opinion Quarterly, 389
public relations, 394–398
 components and campaigns of, 395–398
 campaign in applied science, 395–398
 department, relying on, 400–401
 highlights and projects, 404
 importance of, 394–398
 NHRE campaign and, 395–398
 theory and practice of, 394–395
 and using institutional voices, 397

Public Relations Journal, The, 400
publications
 books and, 55
 comparing readerships, 96
 examining, 94–97
 main types of, 94
public understanding
 actively improving, 398–401
 highlights and projects, 404
 popularizing science and technology for, 401–402
 scientists, media and, 392–394
 of technical fields, 387–405
Publishers Clearing House, 406
publishing, ethics of, 377
punctuation, 213–214
 for letters, 342
purpose, statement of, 26

Q

qualifiers, eliminating unnecessary, 202
question formats, in questionnaires, 77
questionnaires
 development of, 75–78
 pretesting of, 78–79
 production of, 79
 word selection in, 78
 see also surveys
questions
 guidelines for refining, 29
 open-ended and closed-ended, 77
 order of, 78
 with overlapping answers, 78
 research into, 26
Quick, Doris, 110
Quinn, Karen T., 63

R

Random House College Dictionary, The, 229
random sampling, 79–80
 and field work, 80
Rather, Dan, 400
readability tests, 412–415
 Fry graph, 414–415
 highlights and projects, 418–419
Reader's Guide to Periodical Literature, 47
readership studies, 96–260
reading rooms, libraries and, 41
receiver, in communication theory, 12

recipes: *see* process descriptions
Redish, Janice, 120
reference books, 47
reference library, developing personal, 215
references, general, 46–47
rejections, 362
reports
    business vs. scientific, 276
    key parts of, 261–272
    see also technical reports
research
    hypotheses, 26
    questions, 26
    tampering with results of, 378–379
Research Libraries Information Network (RLIN), 52
Research Methods in the Social Sciences, 66
research paper, proposal for, 264–265
research reports, highlights and projects, 279
resources, library, 41–57
Resources in Education, 51
resumes
    chronological or functional, 356–358
    highlights and projects, 363–364
    key elements of, 356
    preparing, 355–359
reviewers, instructions to, 408–409
reviewing, 189–192
reviews, as evaluation technique, 407
revising
    copyediting symbols for, 182, 204
    at different levels, 182
    editorial (stylebook) style, 212–215
    general guidelines for, 181–182
    highlights and projects, 219–220
    and idiomatic style, 215
    to improve appearance, 210–220
    and peer reviews, 218–219
    and personal style, 218
    and rewriting, 179–220
    and typographic style, 215
revisions, 181–188
    assuming editors role, 182
    data analysis techniques for, 184
    drawing conclusions in, 185–186
    highlights and projects, 186–187
    information sources and evidence, 183–184
    and strategies for editing content, 182–186
    and verifying results, 185
    on word processor, 283–285

rewriting
    general guidelines for, 181–182
    highlights and projects, 186–187
    revising and, 179–220
rhetorical method of communication, 9
Richardson, Jacques, 389
Rivers, William, 66
Roberts, Harry V., 34
Rockmore, Milton, 352
Roland, Charles, 399
Romano, Rose M., 406
Rose, Andrew, 163
Ross, Kenneth, 382
Roundy, Nancy, 109
Rubinstein, Moshe E., 23–24

S

safety hazard signs, 383–384
Sagan, Carl, 218
Salancik, Jerry, 69–72
Salisbury, David, 224
Salomon Brothers, 72
sampling, 79–80
    highlights and projects, 81–82
    purposeful, 80
Sand County Almanac, 218
Sarton, George, 65
Scherer, Cliff, 313
school writing assignments, 118
Schor, Sandra, 137, 192
Science, 389
science
    effective public relations campaign in, 395–398
    helping media to understand, 398–400
    and media and public, 387–388
    popularity and distrust of, 389
    popularizing for public understanding of, 401–402
    public support for, 388–390
    social role of, 387
Science Citation Index, 48, 50
Science News, 36
science professionals
    media skepticism of, 391–394
    suspicious of media, 390
science writing, 112
scientific communication
    basic theory of, 10–16
    characteristics of, 6–7

scientific communication, *continued*
  and competition between messages, 7
  definition of, 3–5
  highlights and projects, 18–20
  major functions of, 8–9
  terminology of, 5–6
  see also technical communication
Scientific Community, The, 380
scientific professionals, writing processes of, 107–110
scientific writing, strategies for improving, 179–220
scientists
  and ethics and competitive pressure, 377
  media and public understanding of, 392–394
scripts
  recording and synchronizing, 323
  for slide presentations, 314–317
  split-page, 316
  writing tips for, 317
Scully, John, 4
selected indexes
  annotated bibliography of, 48
  types of, 49
selective attention, as observational barrier, 68
Selltiz, Claire, 66
Selzer, Jack, 107–108
semantic noise, 12
sender, in communication theory, 11
sender-receiver model, in communication theory, 11
sentences
  avoiding padded constructions, 196
  editing guidelines for, 193–205
  highlights and projects, 205–209
  simple, 196–197
serial publications, 52–55
serials cards, 45
Settle, Lyle, 17
sexist language
  avoiding, 190
  in correspondence, 342–343
Shannon, Claude, 10
sharpening, as observational barrier, 66, 68
Sheehy, Eugene P., 47
Shell Oil Company, 233, 234
Sherwood, Thomas K., 33
short words, in plain speaking, 201
Shurter, Robert L., 330
Siegel, Alan, 120

Silent Spring, 401
simple line graphs, specific guidelines for, 155
simple sentences, 196–197
Simply Stated, 103, 190
slide presentations, 312–325
  audience and objectives, 312
  checklist for, 324
  evaluating, 325
  highlights and projects, 325–326
  pace and length, 313
  planning and preparation of, 312–323
  preparation and practice for, 324
  progressive disclosure in, 318
  proposal and treatment of, 313–314
  scripting visuals and narrative for, 315–317
  storyboard approach to, 315
  visual style of, 313
slide projectors, 313
slides
  buying, 322
  duplicating and distributing, 323–324
  obtaining, 319–322
  outlining approach to, 314
  preparing, 314–323
  presenting, 325
  screening and revising, 322–323
Smith, Charles, 282
SmithKline Beckman, 399
Social Research for Consumers, 35
Society for Technical Communication, 111, 134
solutions
  acting on, 32
  phrasing and, 23–39
  selection of, 32
  stating criteria for, 29–30
Souther, James, 260
spatial relationships, describing, 239
speaking plainly
  clichés and wordiness, 203
  highlight and projects, 205–209
  short and specific words, 201–202
  strategies for, 200–205
  unnecessary modifiers and qualifiers, 202
special collections, 57
speeches, 300–301; *see also* professional talks
spelling, 216–217
split-page scripts, 316
Spretnak, Charlene, 4
stacked modifiers, breaking up, 199–200
stage fright, practicing to overcome, 308

INDEX 449

Standard Oil Company, 298
statistics
   notations as codes for, 161
   see also numbers
Steen, Lynn A., 390
storyboard
   for slide presentations, 315
   typical planning card for, 317
strong verbs, 193–196
students
   ethics and competitive pressure, 377
   guide letters to, 345–346
style manuals, 210–211
   highlights and projects, 219–220
   illustration guidelines and, 153
subheads, and headlines, 137
subject card, 44
subscription forms, in audience analysis, 98
successive drafts, 181
summary: *See* abstracts
Sundials to Atomic Clocks, 226
Suomi, Verner, 25
surveys, 72–81
   in audience analysis, 99
   conduct of, 74–80
   cover letters and instructions for, 78
   and data analysis and conclusions, 80–81
   defined, 73
   evaluating, 74
   flow of activities and, 75
   highlights and projects, 81–82
   mail and telephone, 74–75
   objectives and types, 74–76
   pretesting of, 78–79
   reasons for and costs of, 73
   sampling methods for, 79–80
   of technical professionals, 109–110
   writing reports for, 81
symbols
   defining by, 227
   see also notations

T

tables
   and bar graphs, 147
   basic components of, 154
   highlights and projects, 176–177
   and narrative, 145
   preparing, 150
   specific guidelines for, 153–155
Tandem Computers, 17
tape recorders, in interviewing, 71–72
teachers, and technical writing, 112
teaching techniques, descriptions of, 233
technical articles, basic organization of, 136
technical communication, 413
   characteristics of, 6–7
   and competition between messages, 7
   cost-effective production of, 17
   definition of, 3–5
   evaluation of, 406–419
   highlights and projects, 18–20
   informal and formal evaluations of, 407–418
   legal and ethical aspects of, 367–386
   major functions of, 8–9
   major steps in process of, 16–18
   and products liability, 382–383
   terminology of, 5–6
   theory of, 1–20
technical documents
   abstract or summary of, 269–270
   bibliography/reference lists, 268
   citations in, 270
   justification/literature review, 267
   key parts of, 261–272
   method, timetable, and budget for, 267
   and problem statements, 266
   recommendations, 269
   results, discussions, and conclusions, 268–269
technical errors, 35–36
technical fields, public understanding of, 387–405
technical manuscripts, instructions to reviewers of, 408–409
technical professionals
   highlights and projects, 115–116
   media skepticism of, 391–394
   surveys of, 109
   suspicions of media, 390
   writing processes of, 107–110
technical reports
   business vs. scientific, 276
   highlights and projects, 279
   key parts of, 261–272
   planning illustrations for, 148
technical writers, surveys of, 109
technical writing
   stacked modifiers in, 199
   strategies for improving, 179–220
   teachers of, 112

technical writing, *continued*
  vs. technical communications, 5–6
technology
  helping media understand, 398–400
  popularizing for public understanding, 401–402
telephone surveys, 75
terminology, in technical and scientific communication, 5–6
thalidomide, 379
Thinking Better, 33
Thomas, Lewis, 218
Through the Looking-Glass, 223
Tichenor, Phillip, 406
Time, 231, 389
time charts, 165
  specific guidelines for, 163–165
Tips on Picking and Using Strawberries, 246–247
title, technical documents, 262
title card, 43
transmittal, letters of, 329, 330
transparencies
  adding photographs to, 304
  body and conclusion for, 306
  and HEIDI technique, 305
  introductory, 305
  producing, 308–311
  see also slides
trapped verbs, 195–196
2001: A Space Odyssey, 346
typographic style, 215

U

Ulrich's International Periodicals Directory, 46
United Press International (UPI), 241
unnecessary clauses, 197–198

V

verbs
  active voice, 193–195
  action of, 195–196
  highlights and projects, 205–209
  strategies for strengthening, 193–196
  tense vs. voice, 195
  weak or trapped, 195–196
video display terminal (VDT), reducing dangers of, 287
visual descriptions, 236
Vondran, Raymond F., 260

W

Wade, Nicholas, 375
Waldrep, Tom, 113
Wallis, W. A., 34
warning signs, interpretation of, 13
Wassell, Carol, 168
Weaver, Warren, 10
Webb, Eugene, 69, 72
Webster, Bayard, 402
Webster's Standard American Style Manual, 211
Weisman, Herman, 135
White, E. B., 112, 202, 401
Wilhoit, G. Cleveland, 74
Williams, Joe, 13
Wilson, Nancy, 283
Winbush, Barbara, 413
word history, defining by, 229
wordiness, eliminating, 203
wordless instructions, 253
word processing, 281–286
  formatting with, 284
  highlights and projects, 295–296
  and initial entries, 283
  and personal computers, 284–287
  revision and copyediting techniques for, 283–285
  student use of, 281–282
  tips on using, 286
  working with operator, 282–283
word processors, 140–141
words
  abstract vs. concrete, 202
  defining with, 226
  as descriptions, 235
  general vs. specific, 201
  vague vs. definite, 202
working outlines, segments of, 131
workshops to evaluate communications, 416–417
Writer's Market, The, 322
Writers on Writing, 113
writing
  preparations for, 128–136
  school vs. job, 259
writing processes, 119–128
  commonalities across fields, 114–115
  document design model, 120–126
  free writing, 119–120
  highlights and projects, 115–116
  post-writing, 126

prewriting, 121–125
problem-solving model, 126–128
by professionals, 107–117
writing style for instructions, 253
Writing with a Word Processor, 112

X

Xerox, 352

Y

Yarbrough, J. Paul, 406
Young, Matt, 108

Z

Zinsser, William, 112, 179, 284

ABOUT THE AUTHORS

Donald E. Zimmerman is an Associate Professor in the Department of Technical Journalism at Colorado State University. He holds a Ph.D. in communication research and theory from the University of Wisconsin-Madison and an M.S. in technical journalism and B.S. in biological sciences from Kansas State University. He heads CSU's technical/specialized communication concentration and coordinates CSU's technical communication service courses. He has published in the *Transactions of the International Technical Communication Conference,* the *Journal of Technical Writing and Communication,* the *Journal of Environmental Education, Journalism Educator, Technical Communication,* and *The Technical Writing Teacher.* He has written chapters for the *Teaching Technical Editing* and *Guidelines for Inventory and Monitoring Wildlife Habitats;* he also edited and contributed to *Farming Systems Research and Development: Guidelines for Developing Countries.* He has served as a technical communication consultant to the U.S. Forest Service, the U.S. Fish and Wildlife Service, the Bureau of Land Management, the Agency for International Development, the Solar Energy Research Institute, and private businesses and industries. In addition, he has worked as a technical writer, editor, producer, and photographer.

David G. Clark is Professor and Chair in the Department of Technical Journalism at Colorado State University. He received a Ph.D. in mass communications from the University of Wisconsin. He has taught at Stanford University and the Universities of Wisconsin, Nebraska, and Cincinnati. His books include *You and Media: Mass Communication and Society* (with W. B. Blankenburg), *Mass Media and the Law* (with E. R. Hutchison, co-editor), and *The American Newspaper* (with Clifford Weigle, co-editor). His chapter, "Writing for Television," appears in *Writing for the Mass Media* (1986), by E. R. Hutchison. He has published widely in journals, including *Journalism Quarterly* and *Journal of Broadcasting.* He was a consultant to the U.S. Surgeon General's Committee on Television and Social Behavior and was a contributor to *Media Content and Control: A Technical Report to the Surgeon General's Advisory Committee.* He has also been a consultant on technical communication for the Solar Energy Research Institute, Kodak Colorado, and the U.S. Soil Conservation Service. He is a former newspaper reporter and television news editor.

SOUTHEASTERN MASSACHUSETTS UNIVERSITY
T11.Z56 1987
The Random House guide to technical and

3 2922 00031 317 8